Advanced Structured Materials

Volume 154

Common engineering materials reach in many applications their limits and new developments are required to fulfil increasing demands on engineering materials. The performance of materials can be increased by combining different materials to achieve better properties than a single constituent or by shaping the material or constituents in a specific structure. The interaction between material and structure may arise on different length scales, such as micro-, meso- or macroscale, and offers possible applications in quite diverse fields.

This book series addresses the fundamental relationship between materials and their structure on the overall properties (e.g. mechanical, thermal, chemical or magnetic etc.) and applications.

The topics of *Advanced Structured Materials* include but are not limited to

- classical fibre-reinforced composites (e.g. glass, carbon or Aramid reinforced plastics)
- metal matrix composites (MMCs)
- micro porous composites
- micro channel materials
- multilayered materials
- cellular materials (e.g., metallic or polymer foams, sponges, hollow sphere structures)
- porous materials
- truss structures
- nanocomposite materials
- biomaterials
- nanoporous metals
- concrete
- coated materials
- smart materials

Advanced Structured Materials is indexed in Google Scholar and Scopus.

More information about this series at http://www.springer.com/series/8611

Alexander Ya. Grigorenko · Wolfgang H. Müller ·
Igor A. Loza

Selected Problems in the Elastodynamics of Piezoceramic Bodies

Springer

Alexander Ya. Grigorenko
S. P. Timoshenko Institute of Mechanics
National Academy of Sciences of Ukraine
Kyiv, Ukraine

Wolfgang H. Müller
Institut für Mechanik
Technische Universität Berlin
Berlin, Germany

Igor A. Loza
Department of Theoretical and Applied Mechanics
National Transport University
Kyiv, Ukraine

ISSN 1869-8433 ISSN 1869-8441 (electronic)
Advanced Structured Materials
ISBN 978-3-030-74198-3 ISBN 978-3-030-74199-0 (eBook)
https://doi.org/10.1007/978-3-030-74199-0

This Springer imprint is published by the registered company Springer Nature Switzerland AG
The registered company address is: Gewerbestrasse 11, 6330 Cham, Switzerland

Preface

In the second half of the twentieth century, a new field of mechanics of coupled fields named as electroelasticity has been generated and began to develop actively. This field represents in modern science the trend devoted to studying problems being at the junction point of two such classic scientific trends as solids mechanics and electrodynamics (electrostatics) of the continuum [2, 3, 8, 10–12].

The electroelasticity is based on studies of French scientist brothers Pier and Jack (Yacob) Curie which were carried out in 1889 and devoted to study the electrical conductivity of quartz plates made of natural crystals. They revealed that in compressing them electrical charges arise: positive on one face of the plate and negative on the other one, both charges are of similar magnitude. This phenomenon was named as piezoeffect from Greek word $\pi\iota\varepsilon\zeta\omega$ (piezo) that designates press or compress. The meaning of the piezoelectrical effect lies in the fact that the plate of a piezoceramic crystal can be used as the transformer of mechanical energy into electrical one and vice-versa, i.e., from electrical to a mechanical one. The basic equations of the theory of piezoelectricity as applied to natural crystals (quartz, tourmaline, Seignette salt) were derived in detail at the beginning of the twentieth centenary [1, 9].

However, the construction of the main models of electroelasticity is related to producing piezoceramic on the base of ferroelectricity in the middle of the twentieth centenary. The first piezoceramic was produced from the titanium–barium powder $BaTiO_3$ [6, 7]. Also, a number of modern piezoceramic and pyroelectric materials such ones as zinc oxide ZnO, cadmium sulfite CdS, bismuth germanium BiGeO, para tellurium TeO, lithium niobium LiNbO, lithium tantalum LiTaO, barium-nitrate niobium BaNaNbO, and gallium-arsenide GaAs should be noted.

The separate place in the application of piezoelectric transformers is occupied by electroelastic composites such ones as structural members composed of piezoelectric and elastic layers .

Due to their property to transform mechanical energy into electrical one and vice-versa, the piezoelectric materials are widely used in practice. They find an application in the following areas: emitters and antennas in hydroacoustics; frequency stabilizers in radio-engineering facilities and timing references; electric filters and delay lines in radio and telephone communications; measuring transducers for accelerations, vibration levels, and acoustic emission in nondestructive inspection; piezotransformers and piezopropulsions; medical tomographs, as well as medical equipment of various designations.

The wide application of piezoceramic elements and facilities is related to the tendency to account for real processes that occur in piezoceramic structural materials, to reveal and study three-dimensional effects typical for thick-walled members. The mathematical difficulties in solving the boundary-value problems of electroelasticity for bodies of arbitrary shape and natural polarization lead to the formation of separate classes of problems depending on the character of boundary conditions and features of the geometry of the piezoelectric members. For this reason, only single publications performed within the framework of the three-dimensional theory and connected with studying dynamic processes in homogeneous and inhomogeneous piezoceramic finite-length cylinders and spheres as well as in piezoceramic inhomogeneous cylindrical waveguides are known from scientific literature devoted to solving the electroelasticity problems.

In this case, along with the necessity to satisfy boundary conditions on bounding surfaces, the conditions of contact conjugation on the interface of two materials should be met. This feature essentially complicates studying with the help of analytical and numerical approaches. In addition, due to dissimilarity in physical–mechanical properties of materials of layers, the stress concentration arises in a structural member during its operation life. This, in turn, leads to the origin of microcracks which decrease the load-bearing capability of the structural member and may cause its fracture. To deal with the problem, the so-named Gradient Piezoceramic Materials (FGPM) have been developed whose physical–mechanical properties continually vary along one or several coordinate directions. The study of dynamic processes in piezoceramic cylindrical and spherical bodies with such dissimilar inhomogeneity as piecewise continuous or continually continuous ones is connected with large computational difficulties. The present monograph contains solutions to the wide class of problems of the spatial electroelasticity theory devoted to wave and vibrational processes in piecewise continuous and continually continuous piezoceramic cylindrical and spherical bodies. How the inhomogeneity, coupled electric field, and kind of preliminary polarization of piezoceramic materials influence the spectral characteristics of the above electroelastic bodies is analyzed. To this end, the discretely continual numerically analytical approach to solve the wide class of problems of the spatial electroelasticity theory for stationary dynamical processes in piecewise-inhomogeneous (laminated) and continually inhomogeneous piezoceramic cylindrical and spherical bodies is proposed. Early, this approach was used efficiently in solving the elasticity problems [4].

In conclusion, it can be noted that the present monograph is devoted to the development of methods of numerical analysis of processes in the relatively new field of mechanics of coupled fields named as electroelasticity which is the important problem of the modern mechanics of solid deformable body [5].

The monograph consists of four chapters. Below we present the summary of the chapters.

Chapter 1: The principal relations of the refined Maxwell electromagnetic theory of piezoceramic bodies demonstrating a piezoeffect are presented. Also, the chapter includes the principal relations of the linear electroelasticity theory composed of equations describing mechanical component of the piezoeffect (the relations being valid for any linear medium, appear from the energy conservation law as well as from the geometric connections needed) and equations of the forced electrostatics of dielectrics describing electrical phenomena in a medium presented in a curvilinear coordinate system. The connection between variables of the equations of two kinds is defined by physical relations presented by the equations of the piezoeffect. If the function of electrical enthalpy is chosen in the form of homogeneous squared shape of independent thermodynamic variables (electrical and mechanical), these equations become linear. The symmetrical tensors of stresses and strains present mechanical variables while the vectors of electrical induction and strength of electric field present electrical variables. The appropriate state equations are derived accounting for the symmetry of properties of the preliminary polarized piezoceramic material. This symmetry corresponds to the symmetry of the crystals of the 6mm class. The closed systems of electroelasticity equations are written as applied to cylindrical and spherical coordinate systems. The boundary conditions for mechanical and electrical components of the coupled field on the surfaces of the piezoceramic body and on the interface of layers in the case of the piecewise-inhomogeneous material are stated. The efficient numerical–analytical discretely continual approach to studying the dynamical behavior of homogeneous and inhomogeneous piezoceramic cylinders and spheres is outlined in the following chapters.

Chapter 2: Various approaches to the solution of linear problems of electroelasticity of the inhomogeneous finite-length cylinders on the basis of discrete–continuous methods and three-dimensional formulations are presented. The advantage of the method consists in the reduction of the partial differential equations of the considered problems to the associated one-dimensional problems (the spline-collocation method) and exact satisfaction of boundary conditions. The approach leads to practically exact solutions of boundary-value problems and eigenvalue problems described by a system of ordinary differential equations with variable coefficients (by the discrete-orthogonalization method).

The axisymmetric and nonaxisymmetric problems of natural vibrations of hollow cylinders and cylinders with piezoelectric properties based on 3D elasticity and electroelasticity are considered. The properties of the material vary along a radial coordinate. We consider two types of inhomogeneous materials: when the properties of the material are piecewise constant (layered structures with metal and dielectric layers) and when they vary continuously (functionally gradient and functionally gradient piezoelectric materials—FGM and FGPM). The external surface of the cylinder is free of tractions and either insulated or short-circuited by electrodes. After separation of variables and representation of the components of the mechanical displacement vector and electric potential in the form of standing circumferential waves, the initially three-dimensional problem is reduced to a two-dimensional partial differential equation problem. By using the method of spline collocations with respect to a longitudinal coordinate, this two-dimensional problem is reduced to a one-dimensional eigenvalue problem (described by ordinary differential equations). This problem is solved by the stable discrete-orthogonalization technique in combination with a step-by-step search method with respect to the radial coordinate. A nontraditional approach to solving problems of the above class is proposed. Different variants of polarized piezoceramic materials are considered. The effect of variation in mechanical and electrical parameters through thickness and the influence of boundary conditions on natural frequencies and vibration modes of the finite-length cylinders with inhomogeneous elastic and piezoelectric properties are analyzed. Significant attention is paid to the validation of the reliability of the results obtained by numerical calculations.

Chapter 3: The propagation of axisymmetric and nonaxisymmetric electroelastic waves in hollow inhomogeneous piezoceramic cylinders based on 3D electroelasticity is considered. The elastic and electric properties of the material vary in a radial direction. Two variants of materials are considered: piecewise constant properties of the material (layered structures with metal and dielectric layers) and continuously varying properties (functionally gradient piezoelectric materials—FGPM). Free and forced motions are investigated.

In the case of free motion, the surfaces of the cylinder are not loaded and free from electrodes and insulation or short-circuited by electrodes. Two variants of boundary conditions are considered in the case of forced motions: electric excitation—when an electrostatic potential with alternating sign is applied to the external cylindrical surface; and mechanical excitation—when a pressure with alternating sign is applied to the external cylindrical surface. An efficient numerical–analytical method to solving this problem is proposed. Components of the elasticity tensor, of the mechanical and electric displacement vector, the electrostatic potential, and of the components of the stress tensor are presented in the form of standing circumferential waves and by running waves in an axial direction. The three-dimensional system of resolving equation is reduced to a boundary-value problem described by a system of inhomogeneous ordinary differential equations. In the case of free motion, this system represents a differential eigenvalue problem. The discrete-orthogonalization method and a step-by-step search approach method are used to solve the problem. In the case of forced motions, a similar procedure is followed and the problem is solved by discrete-orthogonalization methods. Different variants of polarized piezoceramic materials are considered. The influence of the mechanical and electric parameters of the material on the kinematic (mechanical displacement and electrostatic potential) and dynamic (mechanical stress and electric displacement) characteristics are analyzed. As before significant attention is paid to the validation of the reliability of the results obtained by numerical calculations.

Chapter 4: The axisymmetric and nonaxisymmetric problems of natural and forced vibrations of a hollow sphere made of functionally gradient piezoelectric material theory are considered based on 3D electroelasticity. The properties of the material vary along the radial coordinate. The external surface of the sphere is free of tractions and either insulated or short-circuited by electrodes for the analysis of the natural vibrations of the system. Two cases of forced vibrations are investigated: an electric excitation—when an electrostatic potential with an alternating sign is applied to the external surface of the spheres; and mechanical excitation—when a pressure with an alternating sign is applied to its external surface. Separation of variables and series of the components of the mechanical and electric displacements were used. The electric potential of the stress tensor is expressed in terms of spherical functions. As a result, the initially three-dimensional problem which is described by partial differential equations with variable coefficients is reduced to a boundary-value problem expressed by ordinary differential equations. A boundary-value eigenvalue problem for the case of natural vibrations is arrived at in the process. It is solved by discrete-orthogonalization methods combined with a step-by-step search method. The influence of the geometric, mechanical, and electric parameters on the frequency spectrum in the case of nonaxisymmetric natural vibrations of an inhomogeneous piezoceramic thick-walled sphere was analyzed. An inhomogeneous boundary-value problem is obtained for the case of forced vibrations. This problem is solved by a stable discrete-orthogonalization method. The influence of the geometric and electric parameters on the kinematic (mechanical displacement and electrostatic potential) and dynamic (mechanical stress and electric displacement) characteristics was analyzed. Different variants of polarized piezoceramic materials are considered. Once more significant attention is paid to the validation of the reliability of the results obtained by numerical calculations.

Kyiv, Ukraine — Alexander Ya. Grigorenko
Berlin, Germany — Wolfgang H. Müller
Kyiv, Ukraine — Igor A. Loza

References

1. Berlincourt D (1971) Piezoelectric crystals and ceramics. Ultrasonic transducer materials, Mattiat, O.E. (ed.). New York: Plenum Press: 62–124
2. Dökmeci MC (1992) A dynamic analysis of piezoelectric strained elements. New York: Research Development and Standardization Group
3. Jiashi Yang (2005) An introduction to the theory of piezoelectricity. Advances in mechanics and mathematics volume 9 springer nature switzerland AG 2005
4. Grigorenko AYa, Müller WH, Grigorenko YaM, Vlaikov GG (2016) Recent developments in anisotropic heterogeneous shell theory. General theory and applications of classical theory. Volume I, Springer
5. Guz AN (1985) Modern directions in the mechanics of a solid deformable body. Int Appl Mech 21(9):823–828
6. Mason WP (1948) Electrostrictive effect in barium titanate ceramics. Phys Rew 72:1134–1147
7. Mason WP (1956) Piezoelectricity, its history and applications. J Acoust Soc Am 28:1561–1206
8. Tiersten HF (1969) Linear piezoelectric plate vibrations springer science+business Media, LLC
9. Voigt W (1910) Lerhbuch der Krisstallphysics. Leipzig; Berlin, Teubner
10. Wang J, Shi Z (2016) Models for designing radially polarized multilayer piezoelectric/elastic composite cylindrical transducers. J Intelligent Mater Syst and Struct 27(4):500–511
11. Yang J (2005) An introduction to the theory of piezoelecticity. Springer Nature Switzerland AG
12. Zhen-Bang K (2014) Theory of electroelasticity, springer heidelberg new york dordrecht london. Shanghai and Springer-Verlag Berlin Heidelberg

Contents

Chapter 1
Basic Relations of Linear Electroelasticity Theory

Abstract The principal relations of the refined Maxwell electromagnetic theory of piezoceramic bodies demonstrating a piezoeffect are presented. The chapter also includes the principal relations of linear electroelasticity theory composed of equations describing a mechanical component of the piezoeffect (the relations being valid for any linear medium, appearing from the energy conservation law, as well as from the geometric connections needed) and equations of the forced electrostatics of dielectrics describing electrical phenomena in a medium presented in a curvilinear coordinate system. The connection between the variables of the equations of the two kinds is defined by physical relations presented by the equations of the piezoeffect. If the function of electrical enthalpy is chosen as homogeneous square forms of the independent thermodynamic variables (electrical and mechanical), these equations become linear. The symmetrical tensors of stresses and strains present mechanical variables, while the vectors of electrical induction and the strength of the electric field present the electrical variables. The appropriate state equations are derived accounting for the symmetry of properties of the preliminary polarized piezoceramic material. This symmetry corresponds to the symmetry of the crystals of the 6mm class. The closed systems of electroelasticity equations are written and applied to cylindrical and spherical coordinate systems. The boundary conditions for mechanical and electrical components of the coupled field on the surfaces of the piezoceramic body and on the interface of layers in the case of the piecewise-inhomogeneous material are stated. The efficient numerical–analytical discretely continual approach to studying the dynamical behavior of homogeneous and inhomogeneous piezoceramic cylinders and spheres is outlined in the following chapters.

Keywords Basic relations of electroelasticity · Piezoelectric medium · Preliminary polarization · Curvilinear coordinate system · Analytical–numerical approach

A. Ya. Grigorenko et al., *Selected Problems in the Elastodynamics of Piezoceramic Bodies*, Advanced Structured Materials 154,
https://doi.org/10.1007/978-3-030-74199-0_1

1.1 Equations of Motion

A body bound over a whole contour by a wise-smooth surface represents the solid carrying the field of the stress tensor **T**, which is a differentiable function of the field point and continuous on a boundary surface. Let the vector **u** of the displacement of any point of the body be the time function of the spatial coordinates of this point. Note that ρ is the material density, while $\frac{\partial^2 \mathbf{u}}{\partial t^2}$ and the divergence of the stress tensor $\nabla \cdot \mathbf{T}$ are continuous functions inside the body volume. It is assumed that body forces are absent.

If the vector element of the boundary surface S directed outward from the body volume V bounded by the surface is denoted as ds, the vector of the external elementary force applied to this element of the surface takes the form of the scalar product $d\mathbf{s}\,\mathbf{T}$. Denoting the body element by dV and by using the d'Alembert principle, we arrive at equation [8, 10, 15]:

$$\int\limits_S \mathrm{d}\mathbf{s} \cdot \mathbf{T} - \int\limits_V \rho \frac{\partial^2 \mathbf{u}}{\partial t^2} \mathrm{d}V = 0. \tag{1.1}$$

By using the Gauss–Ostrogradsky formula for the first integral, we obtain

$$\int\limits_V \left(\nabla \cdot \mathbf{T} - \rho \frac{\partial^2 \mathbf{u}}{\partial t^2} \right) \mathrm{d}V = 0. \tag{1.2}$$

Since this equality should be satisfied for any part of the body, it follows that

$$\nabla \cdot \mathbf{T} = \rho \frac{\partial^2 \mathbf{u}}{\partial t^2}. \tag{1.3}$$

1.2 Maxwell's Equations for a Piezoelectric Material

All processes in a continuous medium relating to macroscopic electrodynamics are described by Maxwell's equations [7, 12]:

$$\nabla \cdot \mathbf{E} = \frac{\rho_e}{\varepsilon_0}; \; \nabla \times \mathbf{E} = -\frac{\partial \mathbf{B}}{\partial t}; \; \nabla \cdot \mathbf{B} = 0; \; c^2 \nabla \times B = \frac{\mathbf{j}}{\varepsilon_0} + \frac{\partial \mathbf{E}}{\partial t}; \tag{1.4}$$

where **E** and **B** are the electric- and magnetic strength field vectors, respectively; **j** is the vector of electric current; and ρ_e, ε_0, and c are the constant values of the density of electric charge, dielectric permittivity, and speed of light in vacuum. These equations were formulated for the first time by Maxwell and are a generalization of the experimental data accumulated at that time. The first equation expresses Gauss's law whose physical sense consists in the fact that the flux of the electric strength field vector **E** through a closed surface is proportional to the electric charge ρ_e inside of it. Gauss's law is valid both in dynamic and static fields since it does not contain time derivatives. The second equation represents Faraday's law whose physical meaning is: The integral of the electric strength field vector **E** over a closed contour is proportional to the rate with which the flux of the magnetic strength field vector **B** through this contour varies. From the third equation, it follows that a piezoelectric material does not contain magnetic charges (which is distinct from the electric ones). The fourth equation was supplemented also, beside the known Ampere law $c^2\nabla \times \mathbf{B} = \frac{\mathbf{j}}{\varepsilon_0}$ (for the constant magnetic field), by the so-called displacement current $\frac{\partial \mathbf{E}}{\partial t}$, whose correctness was later on supported experimentally. The physical sense of the fourth equation consists of the fact that the line integral of the magnetic strength field vector B over a closed contour is proportional to the electric current plus the temporal change of the electric field E (i.e., the total flux) entering through the surface encompassed by this contour. The second and fourth equations demonstrate the fact that in the case of a dynamic situation, electric and magnetic fields are coupled. However, the velocity with which the fields vary must be very high and comparable with that of light. In the case of acoustic motions, the velocities are considerably lower while the electric and magnetic fields exist separately. We will consider just the electric field because, in a piezoelectric material, it is coupled with a mechanical field. Moreover, because the piezoelectric material is free of electric charges, we have:

$$\nabla \cdot \mathbf{E} = 0, \nabla \times \mathbf{E} = 0. \tag{1.5}$$

In the case of weak electric fields, the vector of electric induction **D** is proportional to the electric strength field vector **E**, $\mathbf{D} = \varepsilon \mathbf{E}$, where ε is the dielectric permittivity of a piezoelectric material. The second equation in (1.5) indicates that the electric field has a potential. Finally, we have:

$$\nabla \cdot \mathbf{D} = 0\,, \mathbf{E} = -\nabla \Phi, \tag{1.6}$$

where Φ is the electrostatic potential. Thus, the magnetic effects are completely neglected. Note that such a statement is not strictly consequential due to the fact that the third equation in (1.4) is unsatisfied. Setting $\mathbf{B} = \mathbf{0}$, we arrive at the statement that the expression $\frac{\partial \mathbf{D}}{\partial t}$ must be equal to zero. In reality, the time derivative of the electric displacement **D** represents the density of the displacement current through a ceramic body and the integral of its normal component for one electrode is equal to the current which is supplied from a generator (electric case of loading). Then, in the general case, we get:

$$\frac{\partial \mathbf{D}}{\partial t} \neq 0. \tag{1.7}$$

It should be noted that the system (1.6) describes accurately an electric field in piezoceramic material only when applied to problems of static deformation. In the case of problems on vibration of piezoceramic bodies, the accuracy of the solution based on the electrostatics equations depends on the frequency of forced vibrations. This situation is similar to that which holds in the theory of capacitors and hollow resonators. By accurate mathematical calculations, it can be shown that the allowance for the magnetic effects in the first approximation yields the refined value of the electric field strength:

$$\mathbf{E}_1 = \mathbf{E}_0 \left[1 - \left(\frac{\omega r}{2c} \right)^2 \frac{\varepsilon}{\varepsilon_0} \right], \tag{1.8}$$

where ω is the circular vibration frequency and r is is the representative dimension of the piezoceramic body. At the same time, the value of the magnetic induction in the first approximation is:

$$\mathbf{B}_1 = \frac{\omega r}{2c^2} \frac{\varepsilon}{\varepsilon_0} \mathbf{E}_0. \tag{1.9}$$

From (1.8), it follows that the correction for magnetic effects for the electric strength field in the case of acoustic vibrations (10–500 kHz) and for dimensions of piezoceramic specimens $r = 0.1$ m and $\dfrac{\varepsilon}{\varepsilon_0} \approx 10^3$ is equal to

$$\left(\frac{\omega r}{2c} \right)^2 \frac{\varepsilon}{\varepsilon_0} \approx 10^{-7} - 10^{-4} \tag{1.10}$$

in comparison with unit. In this case, such corrections can be neglected. To determine the electric field in piezoceramics in the range of vibrations with acoustic frequencies, the electrostatic Eq. (1.6) can be used.

In the range of vibrations with acoustic frequencies, other relations defining the electric state of piezoelectric bodies are simplified. Hence, the density of the internal electric energy U is calculated by

$$U = \frac{1}{2} \mathbf{E} \cdot \mathbf{D};\ \text{J/m} \tag{1.11}$$

The flux density of the vector of the electric energy **h** from the volume the electric body occupies is defined as

$$\mathbf{h} = \Phi \frac{\partial \mathbf{D}}{\partial t};\ \text{W/m}. \tag{1.12}$$

The law of conservation of electric energy for a piezoelectric body, neglecting losses, is written as

$$\oint\!\!\oint_S \Phi \mathbf{n} \cdot \frac{\partial \mathbf{D}}{\partial t} \mathrm{d}s = \frac{1}{2} \frac{\partial}{\partial t} \iiint_V \mathbf{E} \cdot \mathbf{D} \mathrm{d}V, \tag{1.13}$$

where V is the volume of the piezoceramic body, S is its surface, and $\mathbf{n}$ is the unit vector of the normal to the external surface of the body V.

1.3 Linear State Equations of a Piezoelectric Medium

Due to the presence of the piezoeffect in piezoelectric media, the mechanical fields (**T**, **S**) and electric fields (**E**, **D**) are coupled. This effect is taken into account in appropriate formulas of a mathematical model that describes the medium. How the density of internal pressure varies is defined by [2, 9, 10, 13, 14]:

$$\dot{U} = \dot{U}(\mathbf{S}, \mathbf{D}) = T_{ij}\dot{S}_{ij} + E_i\dot{D}_i. \tag{1.14}$$

Then the increment of the density of internal energy is calculated by

$$\mathrm{d}U = \frac{\partial U}{\partial S_{ij}}\mathrm{d}S_{ij} + \frac{\partial U}{\partial D_{ij}}\mathrm{d}D_{ij} = T_{ij}\mathrm{d}S_{ij} + E_i\mathrm{d}D_i. \tag{1.15}$$

Allowing the fact that the formula (1.15) must include a complete differential for the function U, we define physical relations that describe the medium properties:

$$T_{ij} = \frac{\partial U}{\partial S_{ij}},\ E_i = \frac{\partial U}{\partial D_i}. \tag{1.16}$$

It is clear that when restricting to the consideration of linear constitutive equations, a uniform squared expression for U should be constructed:

$$U(\mathbf{S}, \mathbf{D}) = \frac{1}{2}A^D_{ijkl}S_{ij}S_{kl} - h_{kij}D_kS_{ij} + \frac{1}{2}\beta^S_{ij}D_iD_j. \tag{1.17}$$

With (1.17), the constitutive equations become:

$$T_{ij} = c^D_{ijkl}S_{kl} - h_{kij}D_k,\ \ E_i = -h_{ijk}S_{kl} + \beta^S_{ik}D_k, \tag{1.18}$$

where c^D_{ijkl} are the components of the tensor of elastic moduli defined at constant induction, h_{kij} the components of the tensor of piezoelectric coefficients, and β^S_{ik} the components of the tensor of dielectric permittivities defined at constant strain. All tensors appearing here are symmetrical. Their components satisfy the equalities:

$$A^D_{ijkl} = A^D_{klij} = A^D_{jikl} = A^D_{ijlk},\ h_{ikl} = h_{ilk},\ \beta^S_{ik} = \beta^S_{ki}. \tag{1.19}$$

Choosing other pairs of T_i, S_i, E_i, and D_i as independent variables, we can derive additionally three different forms of constitutive equations. Note that of interest are only coupled fields.The purely elastic (T_i ,S_i) and purely electric (E_i, D_i) cases, which give us additionally three different forms of constitutive relations [9], are neglected. All of them are constructed similarly when one or two summands in the form of $E_i D_i and\ T_{ij} S_{ij}$ are subtracted from the internal energy.

Choosing a thermodynamic function in the form of elastic enthalpy yields [6]:

$$H_1(T_{ij}, S_{ij}) = U - T_{ij} S_{ij}. \tag{1.20}$$

Calculating the time derivative $H_1(T_{ij}, S_{ij})$ and allowing for (1.14), we obtain:

$$\dot{H}_1(T_{ij}, S_{ij}) = -S_{ij}\dot{T}_{ij} + E_i \dot{D}_i. \tag{1.21}$$

Because the elastic enthalpy is a function of mechanical stresses and electric induction, we arrive at the equality:

$$\dot{H}_1 = \frac{\partial H_1}{\partial T_{ij}}\dot{T}_{ij} + \frac{\partial H_1}{\partial D_i}\dot{D}_i. \tag{1.22}$$

Equating the appropriate values, we get:

$$S_{ij} = -\frac{\partial H_1}{\partial T_{ij}}, \ E_i = \frac{\partial H_1}{\partial D_i}. \tag{1.23}$$

By determining the elastic enthalpy as a quadratic form:

$$H_1(\mathbf{T}, \mathbf{D}) = -\frac{1}{2}s^D_{ijkl} T_{ij} T_{kl} - g_{kij} D_k T_{ij} + \frac{1}{2}\beta^T_{ij} D_i D_j, \tag{1.24}$$

we arrive at the following linear constitutive equations:

$$S_{ij} = s^D_{ijkl} T_{kl} + g_{kij} D_k, \ \ E_i = -g_{ikl} T_{kl} + \beta^T_{ik} D_k, \tag{1.25}$$

where s^D_{ijkl} are the components of the tensor of elastic compliances defined at constant induction, g_{kij} are components of the tensor of piezoelectric coefficients, and β^T_{ik} are components of the tensor of dielectric permittivities defined at constant mechanical stress.

Choosing a thermodynamic function in the form of electric enthalpy:

$$H_2(\mathbf{S}, \mathbf{E}) = U - E_i D_i \tag{1.26}$$

and differentiating this equation with respect to time, we obtain:

$$\dot{H}_2 = \dot{U} - \dot{E}_i D_i - E_i \dot{D}_i, \tag{1.27}$$

whence with (1.14), it follows:

$$H_2 = T_{ij}\dot{S}_{ij} - D_i\dot{E}_i. \tag{1.28}$$

Thus, we can conclude that the electric enthalpy is the function of mechanical deformations and strength of electric field H(S, E). Then its complete derivative is calculated by

$$\dot{H}_2 = \frac{\partial H_2}{\partial S_{ij}}\dot{S}_{ij} + \frac{\partial H_2}{\partial E_i}\dot{E}_i. \tag{1.29}$$

Comparing formulas (1.28) and (1.29), we arrive at relations:

$$T_{ij} = \frac{\partial H_2}{\partial S_{ij}}, \quad D_i = -\frac{\partial H_2}{\partial E_i}. \tag{1.30}$$

Then the electric enthalpy is defined as uniform quadratic form:

$$H_2(S, E) = -\frac{1}{2}c^E_{ijkl}S_{ij}S_{kl} - e_{kij}E_kS_{ij} + \frac{1}{2}\varepsilon^S_{ij}E_iE_j, \tag{1.31}$$

whence from (1.30), we get the following constitutive relations for a linear piezoelectric material:

$$T_{ij} = c^E_{ijkl}S_{kl} - e_{kij}E_k, \quad D_i = e_{ikl}S_{kl} + \varepsilon^S_{ik}E_k, \tag{1.32}$$

where c^E_{ijkl} are components of the tensor of elastic moduli, defined at constant strength of electric field, e_{kij} are the components of the tensor of piezoelectric coefficients, and ε^S_{ik} are the components of the tensor of dielectric permittivities defined at constant mechanical deformation.

Finally, the last invariant state function being the enthalpy H is introduced as follows:

$$H(T_{ij}, E_i) = U - T_{ij}S_{ij} - E_iD_i. \tag{1.33}$$

Similarly, we get

$$H = -S_{ij}\dot{T}_{ij} - D_i\dot{E}_i. \tag{1.34}$$

Thus, we can conclude that the enthalpy is the function of mechanical stress and strength of electric field $H(\mathbf{T}, \mathbf{E})$, while its time derivative is:

$$\dot{H} = \frac{\partial H}{\partial T_{ij}}\dot{T}_{ij} + \frac{\partial H}{\partial E_i}\dot{E}_i. \tag{1.35}$$

Comparison of (1.34) and (1.35) yields:

$$S_{ij} = -\frac{\partial H}{\partial T_{ij}}, \quad D_i = -\frac{\partial H}{\partial E_i}. \tag{1.36}$$

Table 1.1 Relations between different constants

Elastic	Piezoelectric	Dielectric
$s^D_{ijkl} = s^E_{ijkl} - d_{mij} g_{mkl}$	$g_{mkl} = \beta^T_{mn} d_{nkl}$	$\beta^T_{mn} \varepsilon^T_{mn} = 1$
$c^E_{ijkl} s^E_{ijkl} = 1$	$e_{mkl} = d_{mij} c^E_{ijkl}$	$\varepsilon^S_{mn} = \varepsilon^T_{mn} - d_{mkl} e_{mkl}$
$c^D_{ijkl} = c^E_{ijkl} + e_{mkl} h_{mij}$	$h_{nkl} = \beta^S_{mn} e_{mkl}$	$\beta^S_{mn} = \beta^T_{mn} + g_{nkl} h_{mkl}$
$c^D_{ijkl} s^D_{ijkl} = 1$	$h_{nkl} = g_{nij} c^D_{ijkl}$	$\beta^S_{mn} \varepsilon^S_{mn} = 1$

Table 1.2 Voigt's matrix form

$ij, kl =$	11	22	33	23,32	13,31	12,21
$\lambda, \mu =$	1	2	3	4	5	6

Determining the enthalpy as the uniform quadratic form:

$$H(\mathbf{T}, \mathbf{E}) = -\frac{1}{2} s^E_{ijkl} T_{ij} T_{kl} - d_{kij} E_k T_{ij} + \frac{1}{2} \varepsilon^T_{ij} E_i E_j, \tag{1.37}$$

and substituting (1.37) into (1.36), we obtain:

$$S_{ij} = s^E_{ijkl} T_{kl} + d_{kij} E_k, \quad D_i = d_{ikl} S_{kl} + \varepsilon^T_{ik} E_k, \tag{1.38}$$

where s^E_{ijkl} are components of the tensor of elastic moduli, defined at constant strength of electric field, d_{kij} are components of the tensor of piezoelectric coefficients, and ε^T_{ik} are the components of the tensor of dielectric permittivities defined at constant mechanical stress (Tables 1.1 and 1.2).

Thus, we have four kinds of equivalent constitutive relations (1.18), (1.25), (1.32), and (1.38) that describe properties of a linear piezoelectric material. Sometimes it is more convenient to use either one or the other relation. The equations presented with respect to some variables can be transformed into the equations presented in a different form with other independent variables. Such transformations make it possible to establish equations relating components of the tensors of elastic and piezoelectric constants in accordance with the data given in [2].

Since the constitutive equations contain symmetrical tensors, it is more convenient to use the matrix form of presentation proposed by Voigt. In the indexes of the mechanical stress **T** and the mechanical deformation **S**, we replace two indexes *ij* either by one index λ or μ by the following scheme [16]:

The transition from one pair of indexes to a single index for deformation and stress tensors is realized as follows:

$$S_\mu = S_{ij}, \text{ if } k = l; \; S_\mu = S_{kl} + S_{ik} = 2S_{kl}, \text{ if } k \neq l.$$

In comprehensive form, we have:

$$S_1 = S_{11},\ \ S_2 = S_{22},\ \ S_3 = S_{33},\ \ S_4 = 2S_{23},\ \ S_5 = 2S_{13},\ \ S_6 = 2S_{12},$$

The components of the stress tensor in matrix presentation are noted as

$$T_\lambda = T_{ij},\ \ \lambda = 1, 2, \ldots, 6,\ \ i, j = 1, 2, 3.$$

In passing from a complete form of presentation of constitutive equations to an abbreviated one, the numerical values of physical constants vary. So by using the scheme presented in Tables 1.1 and 1.2, we have:

$$\begin{aligned}
c_{\lambda\mu}^{D(E)} &= c_{ijkl}^{D(E)},\ \ h_{k\lambda} = \ h_{kij},\ \ ,e_{k\lambda} = e_{kij},\ \ \lambda = 1, 2, \ldots 6,\ \ i, j, k, l = 1, 2, 3.\\
s_{\lambda\mu}^{D(E)} &= s_{ijkl}^{D(E)},\ \text{ if } i = j,\ k = l,\ \lambda, \mu = 1, 2, 3.\\
s_{\lambda\mu}^{D(E)} &= 2s_{ijkl}^{D(E)},\ \text{ if } i \neq j,\ k = l,\ \lambda = 1, 2, 3,\ \ \mu = 4, 5, 6.\\
s_{\lambda\mu}^{D(E)} &= 4s_{ijkl}^{D(E)},\ \text{ if } i \neq j,\ k \neq l,\ \lambda, \mu = 4, 5, 6.\\
g_{k\lambda} &= g_{kij},\ d_{k\lambda} = d_{kij},\ \text{ if } i = j,\ \lambda = 1, 2, 3.\\
g_{k\lambda} &= 2g_{kij},\ d_{k\lambda} = 2d_{kij},\ \text{ if } i \neq j,\ \lambda = 4, 5, 6.
\end{aligned} \tag{1.39}$$

Thus, the general number of physical constants have been reduced as follows: for tensors of elastic moduli $c_{ijkl}^{D(E)}$ and compliances $s_{ijkl}^{D(E)}$ from $3^4 = 81$ to $6^2 = 36$, of piezomoduli h_{kij}, g_{kij}, e_{kij}, d_{kij} from $3^3 = 27$ to $3 \times 6 = 18$; while the number of dielectric permittivities $\beta_{ik}^{S(T)}$, $\varepsilon_{ik}^{S(T)}$ is kept unchanged. The four constitutive equations in matrix form become:

$$T_\lambda = c_{\lambda\mu}^{D} S_\mu - h_{k\lambda} D_k,\ \ E_i = -h_{i\mu} S_\mu + \beta_{ik}^{S} D_k, \tag{1.40}$$

$$S_{ij} = s_{\lambda\mu}^{D} T_\mu + g_{k\lambda} D_k,\ \ E_i = -g_{i\mu} T_\mu + \beta_{ik}^{T} D_k, \tag{1.41}$$

$$T_\lambda = c_{\lambda\mu}^{E} S_\mu - e_{k\lambda} E_k,\ \ D_i = e_{i\mu} S_\mu + \varepsilon_{ik}^{S} E_k, \tag{1.42}$$

$$S_\lambda = s_{\lambda\mu}^{E} T_\mu + d_{k\lambda} E_k,\ \ D_i = d_{i\mu} S_\mu + \varepsilon_{ik}^{T} E_k. \tag{1.43}$$

In the case of a previously polarized piezoceramic, the number of independent physical constants becomes still lesser.

1.4 Constitutive Relationship for Previously Polarized Piezoceramic Material

In the case of a previously polarized ferroelectric (piezoceramic) materials, the constitutive equations (1.40)–(1.43) become considerably simplified in accordance with the symmetry. Concerning elastic and electric materials, previously polarized, ferroelectric materials behave as transversely isotropic bodies whose axis coincide with the direction of the field of previous polarization. The symmetry of such ceramics corresponds to the symmetry of crystals of 6mm classes [8–10]. Assume that the $O\gamma$-axis in the orthogonal curvilinear coordinate system $O\alpha\beta\gamma$ is oriented in the direction of force lines of the electric field of previous polarization. The constitutive equations of a piezoceramic medium can be obtained from (1.42) allowing for the symmetry with the following notations:

$$\begin{aligned}
&c_{22}^E = c_{11}^E,\ c_{23}^E = c_{13}^E,\ c_{55}^E = c_{44}^E,\ c_{66}^E = \frac{1}{2}\left(c_{11}^E - c_{12}^E\right);\\
&c_{14}^E = c_{15}^E = c_{16}^E = c_{24}^E = c_{25}^E = c_{26}^E =\\
&\quad = c_{34}^E = c_{35}^E = c_{36}^E = c_{45}^E = c_{46}^E = c_{56}^E = 0;\\
&e_{32} = e_{31},\ e_{24} = e_{15};\ e_{11} = e_{12} = e_{13} = e_{14} = e_{16} =\\
&\quad = e_{21} = e_{22} = e_{23} = e_{25} = e_{34} = e_{35} = e_{36} = 0;\\
&\varepsilon_{22}^S = \varepsilon_{11}^S;\ \varepsilon_{12}^S = \varepsilon_{13}^S = \varepsilon_{23}^S = 0.
\end{aligned} \tag{1.44}$$

Then:

$$\begin{aligned}
&T_1 = c_{11}^E S_1 + c_{12}^E S_2 + c_{13}^E S_3 - e_{31} E_3,\\
&T_2 = c_{12}^E S_1 + c_{11}^E S_2 + c_{13}^E S_3 - e_{31} E_3,\ T_4 = c_{44}^E S_4 - e_{15} E_2,\\
&T_5 = c_{44}^E S_5 - e_{15} E_1,\ T_6 = \frac{1}{2}\left(c_{11}^E - c_{12}^E\right) S_6,\\
&D_1 = \varepsilon_{11}^S E_1 + e_{15} S_5,\ D_2 = \varepsilon_{11}^S E_2 + e_{15} S_4,\\
&D_3 = \varepsilon_{33}^S E_3 + e_{31}\left(S_1 + S_2\right) + e_{33} S_3.
\end{aligned} \tag{1.45}$$

Thus, the constitutive equations of piezoceramic materials include five independent elastic moduli $\left(c_{11}^E, c_{12}^E, c_{13}^E, c_{33}^E, c_{44}^E\right)$defined at stable (zero) electric field as well as three piezomoduli(e_{31}, e_{15}, e_{33}) and two electric permittivities $\left(\varepsilon_{11}^S, \varepsilon_{33}^S\right)$defined at stable (zero) deformation.

Regarding the constitutive equations (1.45), the following should be noted. First, it is necessary to keep in mind that these equations describe additional coupled electroelastic fields in the previously polarized ceramics which originate due to the action of external mechanical and electric stresses. The fields of initial mechanical stresses and initial field of polarization are accounted for only mediumwise, i.e., in terms of material constants. The above form of constitutive equations assumes that the direction of the vector of previous polarization at each point of the body coincides with the direction of the $O\gamma$-axis of the orthogonal coordinates. If the direction of the previous polarization varies from one point to another inside of the body, the equations must be considered as local piezoeffect ones with the direction of polarization oriented along the $O\gamma$-axis. In this case, the coefficients appearing in (1.45) will be functions of the coordinates of the medium. However, if the external field of previous polarization reveals certain properties of symmetry, the constitutive equations can be presented in correctly chosen curvilinear coordinates with stable coefficients over the whole volume of the body. With regards to Eq. (1.43), we arrive at the linear piezoeffect equations for ceramics in the following form:

$$\begin{aligned}
&S_1 = s_{11}^E T_1 + s_{12}^E T_2 + s_{13}^E T_3 + d_{31} E_3,\ S_2 = s_{12}^E T_1 + s_{11}^E T_2 + s_{13}^E T_3 + d_{31} E_3,\\
&S_3 = s_{13}^E (T_1 + T_2) + s_{33}^E T_3 + d_{33} E_3,\ S_4 = s_{44}^E T_4 + d_{15} E_2,\\
&S_5 = s_{44}^E T_5 + d_{15} E_1,\ S_6 = 2\left(s_{11}^E - s_{12}^E\right) T_6,\ D_1 = \varepsilon_{11}^T E_1 + d_{15} T_5,\\
&D_2 = \varepsilon_{11}^T E_2 + d_{15} T_4,\ D_3 = \varepsilon_{33}^T E_3 + d_{31} (T_1 + T_2) + d_{33} T_3.
\end{aligned} \tag{1.46}$$

The presentation of Eq. (1.46) corresponds to the previous polarization of the ceramics in the direction of the $o\gamma$-axis. Here, the following notations are used: $s_{11}^E, s_{12}^E, s_{13}^E, s_{33}^E, s_{44}^E$ are elastic compliances for a stable (zero) electric field; d_{31}, d_{15}, d_{33} are piezoelectric constants; and $\varepsilon_{11}^T, \varepsilon_{33}^T$ are the dielectric permittivities for stable (zero) mechanical stresses.

It should be noted that Eqs. (1.45) and (1.46) do not regard (in the range of vibrations with acoustic frequencies) magnetic effects that accompany the process of electroelastic deformation as insignificant. Thus, the heat exchange of vibrating piezoelectric members with the environment can be neglected [2, 9, 13, 14]. This makes it possible to consider the deformation as adiabatic. Note that Eqs. (1.45) and (1.46) do not allow for the internal dielectric and mechanical losses.

The qualitative analysis of the above equalities shows that with the electric field being absent the first six relations are transformed into Hooke's dependencies for a transversely isotropic body. Similarly, under the absence of mechanical deformations (stresses), the last three relations coincide with the constitutive equations of transversely isotropic dielectrics. At the same time, the isotropy plane in both cases is perpendicular to the vector of the previous polarization.

1.5 Relations of the Linear Theory of Electroelasticity in a Curvilinear Coordinate System

It is obvious that the equalities (1.3), (1.6), and (1.40)–(1.43) are independent of the choice of coordinate system. Because of this, the equalities being the mathematic formulation of laws that occur in electroelastic bodies must be presented in invariant tensor form. The tensor relations are valid in all coordinate systems. Until the present time, the concept tensor as applied to stresses and deformations has been used formally irrespective of its mathematical sense. Let us present minimal information related to tensor analysis.

As is known, any three linearly independent vectors $\mathbf{e}^1$, $\mathbf{e}^2$, $\mathbf{e}^3$ produce a basis in three-dimensional space. The term "linearly independent vectors" indicates that the mixed product of them is not equal to zero $\mathbf{e}^1\mathbf{e}^2\mathbf{e}^3 \neq 0$. In this case, any vector $\mathbf{r}$ can be presented by the linear combination

$$\mathbf{r} = x_1\mathbf{e}^1 + x_2\mathbf{e}^2 + x_3\mathbf{e}^3, \tag{1.47}$$

where x_i $(i = 1, 2, 3)$ are numbers called covariant coordinates of the vector $\mathbf{r}$ relatively to the basis $\mathbf{e}^1$, $\mathbf{e}^2$, $\mathbf{e}^3$. With Einstein's rule of summation of elements with repeated indexes in the form $\mathbf{r} = x_i\mathbf{e}^i$ being used, formula (1.47) becomes shorter. Let us assume that the summation rule is valid when the indexes are denoted by Roman letters in contrast to the case when Greek letters are used. The numbers x_i $(i = 1, 2, 3)$ are different in another basis. The so-called reciprocal basis is chosen usually as the other one (in the literature also the term associated is used). The basis $\mathbf{e}_1$, $\mathbf{e}_2$, $\mathbf{e}_3$is reciprocal to $\mathbf{e}^1$, $\mathbf{e}^2$, $\mathbf{e}^3$ if

$$\mathbf{e}^i \cdot \mathbf{e}_j = \delta^i_j, \tag{1.48}$$

where $\delta^i_j = \begin{cases} 1, & i = j \\ 0, & i \neq j \end{cases}$ is Kronecker's delta.

In the vector basis $\mathbf{e}_1$, $\mathbf{e}_2$, $\mathbf{e}_3$, the vector $\mathbf{r}$ can be written either as

$$\mathbf{r} = x^1\mathbf{e}_1 + x^2\mathbf{e}_2 + x^3\mathbf{e}_3 \tag{1.49}$$

or in the abbreviated form as $\mathbf{r} = x^i\mathbf{e}_i$.

The numbers x^i $(i = 1, 2, 3)$ are called contravariant coordinates of the vector $\mathbf{r}$ comparatively to the basis $\mathbf{e}_1$, $\mathbf{e}_2$, $\mathbf{e}_3$. Evidently, $x_i \neq x^i$ and the method of construction of the reciprocal basis follows from the equations

$$\mathbf{e}_1 = \frac{\mathbf{e}^2 \times \mathbf{e}^3}{(\mathbf{e}^1 \times \mathbf{e}^2) \cdot \mathbf{e}^3}, \quad \mathbf{e}_2 = \frac{\mathbf{e}^3 \times \mathbf{e}^1}{(\mathbf{e}^1 \times \mathbf{e}^2) \cdot \mathbf{e}^3}, \quad \mathbf{e}_1 = \frac{\mathbf{e}^1 \times \mathbf{e}^2}{(\mathbf{e}^1 \times \mathbf{e}^2) \cdot \mathbf{e}^3}. \tag{1.50}$$

Also we have:

$$\begin{aligned} r \cdot \mathbf{e}^i &= x^j\mathbf{e}_j \cdot \mathbf{e}^i = x^j\delta^i_j = x^1\delta^i_1 + x^2\delta^i_2 + x^3\delta^i_3 = x^i \\ r \cdot \mathbf{e}_i &= x_j\mathbf{e}^j \cdot \mathbf{e}_i = x_j\delta^j_i = x_i. \end{aligned} \tag{1.51}$$

Note that the formulas (1.51) contain the scalar products $\mathbf{e}_j \cdot \mathbf{e}^i$ and $\mathbf{e}^j \cdot \mathbf{e}_i$. Let us consider a metric tensor $\mathbf{g}$:

$$g_{ij} = \mathbf{e}_i \cdot \mathbf{e}_j, \; g^{ij} = \mathbf{e}^i \cdot \mathbf{e}^j. \tag{1.52}$$

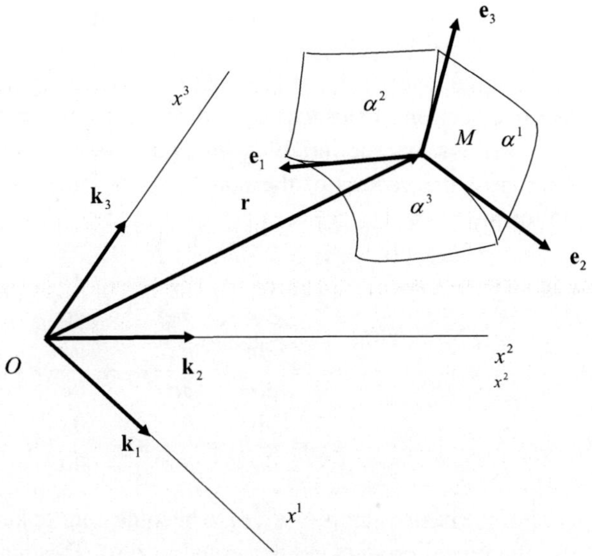

Fig. 1.1 Curvilinear coordinates in an oblique basis

The sense of the metric tensor follows from the expression

$$\mathbf{e}_i = \mathbf{e}_i \cdot 1 = \mathbf{e}_i \cdot (\mathbf{e}_j \cdot \mathbf{e}^j) = (\mathbf{e}_i \cdot \mathbf{e}_j)\, \mathbf{e}^j = g_{ij}\mathbf{e}^j = g_{i1}\mathbf{e}^1 + g_{i2}\mathbf{e}^2 + g_{i3}\mathbf{e}^3. \quad (1.53)$$

From this it follows that components of the metric tensor are the coefficients of expansion of the vectors of the reciprocal basis with respect to the vectors of initial basis. In this case, the symmetric formula is also valid:

$$\mathbf{e}^j = \mathbf{e}^j \cdot (\mathbf{e}^i \cdot \mathbf{e}_i) = (\mathbf{e}^j \cdot \mathbf{e}^i) \cdot \mathbf{e}_i = g^{ji}\mathbf{e}_i = g^{j1}\mathbf{e}_1 + g^{j2}\mathbf{e}_2 + g^{j3}\mathbf{e}_3. \quad (1.54)$$

Let us pass to curvilinear coordinates. Consider the surfaces $\alpha^1, \alpha^2, \alpha^3$, which are the functions of three rectilinear oblique (contravariant) coordinates x^1, x^2, x^3 in the basis $\mathbf{k}_1, \mathbf{k}_2, \mathbf{k}_3$:

$$\alpha^1 = \alpha^1\left(x^1, x^2, x^3\right);\, \alpha^2 = \alpha^2\left(x^1, x^2, x^3\right);\, \alpha^3 = \alpha^3\left(x^1, x^2, x^3\right); \quad (1.55)$$

as it is shown in Fig. 1.1.

Considering the arbitrary point M at a certain time instant, we see that the three surfaces above intersect at this point, while the lines of intersection of the surfaces form coordinate lines. Let us choose a new basis $\mathbf{e}_1, \mathbf{e}_2, \mathbf{e}_3$, whose vectors are directed along tangents to the coordinate lines. As it is known, this is the local reference mark of the curvilinear system. The variables x^1, x^2, x^3 and $\alpha^1, \alpha^2, \alpha^3$ are interrelated by

$$x^1 = x^1\left(\alpha^1, \alpha^2, \alpha^3\right);\, x^2 = x^2\left(\alpha^1, \alpha^2, \alpha^3\right);\, x^3 = x^3\left(\alpha^1, \alpha^2, \alpha^3\right). \quad (1.56)$$

The position vector of the point M can be presented by

$$\mathbf{r} = x^1\mathbf{k}_1 + x^2\mathbf{k}_2 + x^3\mathbf{k}_3. \tag{1.57}$$

Assume that the end of this vector moves along the coordinate line α^1 formed by the intersection of surfaces α^2 and α^3. It is obvious that in this case the coordinate α^1 is the only unique variable, while the two others are constants ($\alpha^2 = \alpha^3$ =const.). Similarly, the vectors of the new basis for other coordinate lines can be found as follows:

$$\begin{aligned}
\mathbf{e}_1 &= \frac{\partial r}{\partial \alpha^1} = \frac{\partial x^1}{\partial \alpha^1}\mathbf{k}_1 + \frac{\partial x^2}{\partial \alpha^1}\mathbf{k}_2 + \frac{\partial x^3}{\partial \alpha^1}\mathbf{k}_3,\\
\mathbf{e}_2 &= \frac{\partial r}{\partial \alpha^2} = \frac{\partial x^1}{\partial \alpha^2}\mathbf{k}_1 + \frac{\partial x^2}{\partial \alpha^2}\mathbf{k}_2 + \frac{\partial x^3}{\partial \alpha^2}\mathbf{k}_3,\\
\mathbf{e}_3 &= \frac{\partial r}{\partial \alpha^3} = \frac{\partial x^1}{\partial \alpha^3}\mathbf{k}_1 + \frac{\partial x^2}{\partial \alpha^3}\mathbf{k}_2 + \frac{\partial x^3}{\partial \alpha^3}\mathbf{k}_3.
\end{aligned} \tag{1.58}$$

For the basis vectors $\mathbf{e}_1$,$\mathbf{e}_2$, $\mathbf{e}_3$ to become noncoplanar, it is necessary and sufficient for the mixed product to be unequal to zero. This leads to the Jacobian

$$J\left(\frac{\partial x^i}{\partial \alpha^j}\right) = X^i_j;\;\; \det X^i_j = \begin{vmatrix} \frac{\partial x^1}{\partial \alpha^1} & \frac{\partial x^2}{\partial \alpha^1} & \frac{\partial x^3}{\partial \alpha^1} \\ \frac{\partial x^1}{\partial \alpha^2} & \frac{\partial x^2}{\partial \alpha^2} & \frac{\partial x^3}{\partial \alpha^2} \\ \frac{\partial x^1}{\partial \alpha^3} & \frac{\partial x^2}{\partial \alpha^3} & \frac{\partial x^3}{\partial \alpha^3} \end{vmatrix} \neq 0 \tag{1.59}$$

It should be noted that the condition (1.59) at the same time is a condition for intersection of all three surfaces $\alpha^1, \alpha^2, \alpha^3$ at one and only one point.

Thus, any vector $\mathbf{a}$ can be presented at each spatial point in the local basis $\mathbf{e}_1$, $\mathbf{e}_2$, $\mathbf{e}_3$by its contravariant coordinates:

$$\mathbf{a} = a^i\frac{\partial \mathbf{r}}{\partial \alpha^i} = a^i\mathbf{e}_i = a^1\mathbf{e}_1 + a^2\mathbf{e}_2 + a^3\mathbf{e}_3. \tag{1.60}$$

Let us construct at the same point a reciprocal local reference mark with vectors of the basis $\mathbf{e}^1$, $\mathbf{e}^2$, $\mathbf{e}^3$ which are now perpendicular to the coordinate surfaces α^1, α^2, α^3 defined by (1.53). With that in view, we will consider these surfaces as ones of the level of the scalar function (1.53). Then, their gradients can be found as follows:

$$\begin{aligned}
\mathbf{e}^1 &= \nabla\alpha^1 = \frac{\partial \alpha^1}{\partial x^1}\mathbf{k}^1 + \frac{\partial \alpha^1}{\partial x^2}\mathbf{k}^2 + \frac{\partial \alpha^1}{\partial x^3}\mathbf{k}^3;\\
\mathbf{e}^2 &= \nabla\alpha^2 = \frac{\partial \alpha^2}{\partial x^1}\mathbf{k}^1 + \frac{\partial \alpha^2}{\partial x^2}\mathbf{k}^2 + \frac{\partial \alpha^2}{\partial x^3}\mathbf{k}^3;\\
\mathbf{e}^3 &= \nabla\alpha^3 = \frac{\partial \alpha^3}{\partial x^1}\mathbf{k}^1 + \frac{\partial \alpha^3}{\partial x^2}\mathbf{k}^2 + \frac{\partial \alpha^3}{\partial x^3}\mathbf{k}^3.
\end{aligned} \tag{1.61}$$

Similarly the condition of noncoplanarity of the vectors $\mathbf{e}^1$, $\mathbf{e}^2$, $\mathbf{e}^3$ and of the reciprocal independence of the variables $\alpha^1, \alpha^2, \alpha^3$ leads to the condition of inequality of the Jacobian of these variables to zero, as in the previous case:

$$J\left(\frac{\partial\alpha^i}{\partial x^j}\right) = Y^i_j;\ \det Y^i_j = \begin{vmatrix} \frac{\partial\alpha^1}{\partial x^1} & \frac{\partial\alpha^1}{\partial x^2} & \frac{\partial\alpha^1}{\partial x^3} \\ \frac{\partial\alpha^2}{\partial x^1} & \frac{\partial\alpha^2}{\partial x^2} & \frac{\partial\alpha^2}{\partial x^3} \\ \frac{\partial\alpha^3}{\partial x^1} & \frac{\partial\alpha^3}{\partial x^2} & \frac{\partial\alpha^3}{\partial x^3} \end{vmatrix} \neq 0. \tag{1.62}$$

The matrix $J\left(\frac{\partial\alpha^i}{\partial x^j}\right) = Y^i_j$ is called Jacobian one, while its determinant is known as Jacobian.

Thus, we can conclude that the vectors (1.56) and (1.58)introduced above are reciprocal ones. To prove this, it is sufficient to form scalar products of two pairs of vectors, which do not refer mutually to the same group:

$$\begin{aligned} \mathbf{e}^i \cdot \mathbf{e}_i &= \frac{\partial\alpha^i}{\partial x^1}\frac{\partial x^1}{\partial\alpha^i} + \frac{\partial\alpha^i}{\partial x^2}\frac{\partial x^2}{\partial\alpha^i} + \frac{\partial\alpha^i}{\partial x^3}\frac{\partial x^3}{\partial\alpha^i} = \frac{\partial\alpha^i}{\partial\alpha^i} = 1; \\ \mathbf{e}^i \cdot \mathbf{e}_j &= \frac{\partial\alpha^i}{\partial x^1}\frac{\partial x^1}{\partial\alpha^j} + \frac{\partial\alpha^i}{\partial x^2}\frac{\partial x^2}{\partial\alpha^j} + \frac{\partial\alpha^i}{\partial x^3}\frac{\partial x^3}{\partial\alpha^j} = \frac{\partial\alpha^i}{\partial\alpha^j} = 0. \end{aligned} \tag{1.63}$$

At the same time, this condition is valid:

$$J\left(\frac{\partial x^i}{\partial\alpha^j}\right) J\left(\frac{\partial\alpha^j}{\partial x^i}\right) = 1. \tag{1.64}$$

The components of the matrix tensor in such local coordinate system are results of the scalar multiplication of the coordinate vectors inside of each triple:

$$g_{ij} = \mathbf{e}_i \cdot \mathbf{e}_j = g_{kl}\frac{\partial x^k}{\partial x^i}\frac{\partial x^l}{\partial x^j};\ g^{ij} = \mathbf{e}^i \cdot \mathbf{e}^j = g^{kl}\frac{\partial x^i}{\partial x^k}\frac{\partial x^j}{\partial x^l};\ g^i_j = \mathbf{e}^i \cdot \mathbf{e}_j = \delta^i_j. \tag{1.65}$$

Using a metrical tensor, we can also to decrease or increase indexes of the vector components because from (1.60) and (1.65) it follows:

$$\begin{aligned} &\mathbf{a} \cdot \mathbf{e}_j = a^i \mathbf{e}_i \cdot \mathbf{e}_j = a^i g_{ij},\ \mathbf{a} \cdot \mathbf{e}_j = a_i \mathbf{e}^i \cdot \mathbf{e}_j = a_i \delta^i_j = a_j, \Rightarrow a_j = a^i g_{ij}, \\ &\mathbf{a} \cdot \mathbf{e}^i = a_j \mathbf{e}^j \cdot \mathbf{e}^i = a_j g^{ji},\ \mathbf{a} \cdot \mathbf{e}^i = a^i \mathbf{e}_j \cdot \mathbf{e}^i = a^i \delta^i_j = a^j, \Rightarrow a^j = a_j g^{ji}. \end{aligned} \tag{1.66}$$

Let us consider how the different-order tensors are transformed when passing from one local coordinate system to another one. Introduce a new curvilinear coordinate system related to the above-mentioned one as follows:

$$\alpha^{i'} = \alpha^{i'}\left(\alpha^1, \alpha^2, \alpha^3\right). \tag{1.67}$$

Suppose that such transformation is invertible at each point:

$$\alpha^{i} = \alpha^{i}\left(\alpha^{1'}, \alpha^{2'}, \alpha^{3'}\right). \tag{1.68}$$

For that, it is necessary and sufficient for the matrix Jacobian $A^{i'}_j \equiv \frac{\partial \alpha^{i'}}{\partial \alpha^j}$, $(i, j = 1, 2, 3))$ to be nonzero:

$$A \equiv \det A^{i'}_j \neq 0. \tag{1.69}$$

In this case, we have the matrix $B^i_{j'}$ reversed to the matrix $A^{i'}_j$, so that
In this case, there is a matrix $B^i_{j'}$ inverse to the matrix $A^{i'}_j$, such that

$$A^{i'}_j B^i_{j'} = \delta^{i'}_{j'}; \quad B^i_{j'} A^{i'}_j = \delta^i_j, \tag{1.70}$$

where B is the Jacobian of the matrix of transformation (1.67).The vector reference marks of the new local basis are defined by

$$\mathbf{e}_{i'} \equiv \frac{\partial \mathbf{r}}{\partial \alpha^{i'}} = \frac{\partial \mathbf{r}}{\partial \alpha^i} B^i_{i'}. \tag{1.71}$$

From this, it follows that vectors of the local reference mark are transformed by the laws:

$$\mathbf{e}_{i'} = \mathbf{e}_i B^i_{i'}, \tag{1.72}$$

$$\mathbf{e}_i = \mathbf{e}_{i'} A^{i'}_i. \tag{1.73}$$

An arbitrary vector $\mathbf{a}$ is presented in the new coordinate system

$$\mathbf{a} = a^{i'} \mathbf{e}_{i'}, \tag{1.74}$$

and as

$$\mathbf{a} = a^i \mathbf{e}_i, \tag{1.75}$$

in the above-mentioned one. From the expressions (1.74), (1.75), and (1.72), we get:

$$a^i \mathbf{e}_i = a^{i'} \mathbf{e}_i B^i_{i'}. \tag{1.76}$$

Since the vectors $\mathbf{e}_i$ are linearly independent, we have

$$a^i = a^{i'} B^i_{i'}. \tag{1.77}$$

Multiplying both sides of equation (1.77) and the first equality from (1.70) by $A_i^{j'}$, we arrive at the following equalities:

$$A_i^{j'} a^i = A_i^{j'} B_{i'}^i a^{i'} = \delta_{i'}^{j'} a^{i'} = a^{j'}. \tag{1.78}$$

Thus, passing from one local coordinate system to other by the law (1.67), the contravariant components of an arbitrary vector are transformed by (1.78) using the Jacobian matrix $A_i^{j'}$. Similarly, we can find how the covariant components of the arbitrary vector are transformed:

$$a_{j'} = \mathbf{a} \cdot \mathbf{e}_{j'} = \mathbf{a} \cdot \mathbf{e}_j B_{j'}^j, \tag{1.79}$$

and consequently,

$$a_{j'} = a_j B_{j'}^j. \tag{1.80}$$

It should be noted that vectors of the reference mark of the new local basis are transformed by the same law as the covariant components of the vector **a**. For this reason, the vectors of the reference mark $\mathbf{e}_j$ are called sometimes the covariant basis of the system. Similarly, we can show that the components of the vector $\mathbf{e}^{j'}$ are transformed into the contravariant components of the vector **a**:

$$\mathbf{e}^{i'} = A_i^{i'} \mathbf{e}^i. \tag{1.81}$$

Sometimes the vectors of the reciprocal basis are called the contravariant basis of a system.

Let us give a formal definition of the covariant and contravariant vector components. A system consisting of three numbers u_i, defined at each point of a space and in each coordinate system, which in passing from one system to other is transformed by (1.80), is called covariant components of the vector **u**. The contravariant components of the vector whose coordinate transformation occurs by (1.78) are determined similarly. The values which do not vary in passing from one coordinate system to the other are called invariants. For example, scalar values are invariants. In our case, the electrostatic potential Φ is scalar. At the same time, the mechanical displacement **u**, electric induction **D**, and the strength of electric field **E** are vector values.

Let us calculate partial derivatives of reference marks of the local basis $\dfrac{\partial \mathbf{e}_i}{\partial \alpha^j}$, ($j = 1, 2, 3$)) in the curvilinear coordinates α^j and expand these vectors in the basis vectors:

$$\frac{\partial \mathbf{e}_i}{\partial \alpha^j} = \Gamma_{ij}^k \mathbf{e}_k. \tag{1.82}$$

The expansion coefficients Γ_{ij}^k are called the second-kind Christoffel symbols. Due to (1.58), they are symmetrical with respect to subscripts:

$$\frac{\partial \mathbf{e}_i}{\partial \alpha^j} \equiv \frac{\partial^2 \mathbf{r}}{\partial \alpha^i \partial \alpha^j} = \frac{\partial^2 \mathbf{r}}{\partial \alpha^j \partial \alpha^i} \Rightarrow \Gamma_{ij}^k \mathbf{e}_k = \Gamma_{ji}^k \mathbf{e}_k. \tag{1.83}$$

Besides that there are different notations of the second-kind Christoffel symbol, for example $\Gamma_{ij}^k \equiv \left\{ \begin{matrix} k \\ ij \end{matrix} \right\}$, $\Gamma_{ij,k} \equiv [k, ij]$. From (1.82), it follows that the Christoffel symbols Γ_{ij}^k are nothing but projections of the components of the vector $\frac{\partial \mathbf{e}_i}{\partial \alpha^j}$ on the direction of the vector $\mathbf{e}_k$. Thus, the Christoffel symbols being important in tensor analysis characterize the curvature level of coordinate lines, although they are not components of any tensor. The Christoffel symbols can be expressed in terms of the metric tensor:

$$\frac{\partial g_{ij}}{\partial \alpha^k} = \frac{\partial \mathbf{e}_i}{\partial \alpha^k} \cdot \mathbf{e}_j + \mathbf{e}_i \cdot \frac{\partial \mathbf{e}_j}{\partial \alpha^k} = \Gamma_{ik}^m \mathbf{e}_m \cdot \mathbf{e}_j + \mathbf{e}_i \cdot \Gamma_{jk}^m \mathbf{e}_m = \Gamma_{ik}^m g_{mj} + \Gamma_{jk}^m g_{mi}. \tag{1.84}$$

Interchanging the indexes i and k in the equality (1.84), we have:

$$\Gamma_{ki}^m g_{mj} + \Gamma_{ji}^m g_{mk} = \frac{\partial g_{kj}}{\partial \alpha^i}. \tag{1.85}$$

Proceeding similarly with $j$and k, we get

$$\Gamma_{ji}^m g_{mk} + \Gamma_{jk}^m g_{mi} = \frac{\partial g_{ik}}{\partial \alpha^j}. \tag{1.86}$$

Adding (1.85) to (1.86) and subtracting (1.84), we arrive at an expression

$$2\Gamma_{ij}^m g_{mk} = \frac{\partial g_{kj}}{\partial \alpha^i} + \frac{\partial g_{ik}}{\partial \alpha^j} - \frac{\partial g_{ij}}{\partial \alpha^k}. \tag{1.87}$$

Multiplying both sides of Eq. (1.87) by $\frac{1}{2} g^{kl}$and summing over k with (1.66), we obtain:

$$\Gamma_{ij}^l = \frac{1}{2} g^{kl} \left(\frac{\partial g_{kj}}{\partial \alpha^i} + \frac{\partial g_{ik}}{\partial \alpha^j} - \frac{\partial g_{ij}}{\partial \alpha^k} \right). \tag{1.88}$$

Since the vectors of a reference mark in the curvilinear coordinate system α^1, α^2, α^3 are unlike for different points of a space, then

$$\begin{aligned} \mathbf{a}_k \equiv \frac{\partial \mathbf{a}}{\partial \alpha^k} = \frac{\partial a^i \mathbf{e}_i}{\partial \alpha^k} = \frac{\partial a^i}{\partial \alpha^k} \mathbf{e}_i + a^i \frac{\partial \mathbf{e}_i}{\partial \alpha^k} = \\ = \frac{\partial a^i}{\partial \alpha^k} \mathbf{e}_i + \Gamma_{ik}^m a^i \mathbf{e}_m = \left(\frac{\partial a^m}{\partial \alpha^k} + \Gamma_{ik}^m a^i \right) \mathbf{e}_m. \end{aligned} \tag{1.89}$$

The value appearing here in brackets is called covariant derivative of the contravariant components of the vector $\mathbf{a}$ and denoted as follows:

$$a^m_{,k} \equiv \frac{\partial a^m}{\partial \alpha^k} + \Gamma^m_{ik} a^i. \tag{1.90}$$

Then formula (1.89) becomes:

$$\mathbf{a}_k \equiv \frac{\partial \mathbf{a}}{\partial \alpha^k} = a^m_{,k} \mathbf{e}_m. \tag{1.91}$$

Similarly, we determine the covariant derivative $a_{m,k}$ of the covariant components of arbitrary vector as

$$a_k \equiv \frac{\partial a}{\partial \alpha^k} = a_{m,k} \mathbf{e}^m. \tag{1.92}$$

Multiplying both sides of Eq. (1.92) by $\mathbf{e}_n$, we get:

$$\frac{\partial \mathbf{a}}{\partial \alpha^k} \cdot \mathbf{e}_n = a_{m,k} \mathbf{e}^m \cdot \mathbf{e}_n = a_{m,k} \delta^m_n = a_{n,k}. \tag{1.93}$$

Differentiating (1.66) with respect to the curvilinear coordinates α^k yields:

$$\frac{\partial a_j}{\partial \alpha^k} = \frac{\partial \left(\mathbf{a} \cdot \mathbf{e}_j\right)}{\partial \alpha^k} = \frac{\partial \mathbf{a}}{\partial \alpha^k} \cdot \mathbf{e}_j + \mathbf{a} \cdot \frac{\partial \mathbf{e}_j}{\partial \alpha^k}. \tag{1.94}$$

Considering Eqs. (1.93), (1.82), and (1.66), we obtain from (1.94):

$$\frac{\partial a_j}{\partial \alpha^k} = a_{j,k} + \mathbf{a} \cdot \Gamma^m_{jk} \mathbf{e}_m = a_{j,k} + \Gamma^m_{jk} a_m. \tag{1.95}$$

Hence, we get a final expression for the covariant derivative with respect to the covariant components of an arbitrary vector:

$$a_{j,k} = \frac{\partial a_j}{\partial \alpha^k} - \Gamma^l_{jk} a_l. \tag{1.96}$$

Along with the second-order Christoffel symbols, the first-order Christoffel symbols are introduced in consideration in tensor analysis as follows:

$$\Gamma_{ij,k} \equiv g_{km} \Gamma^m_{ij}. \tag{1.97}$$

Then from formula (1.88) it follows:

$$\Gamma_{ij,k} = \frac{1}{2} \left(\frac{\partial g_{kj}}{\partial \alpha^i} + \frac{\partial g_{ik}}{\partial \alpha^j} - \frac{\partial g_{ij}}{\partial \alpha^k} \right). \tag{1.98}$$

The first-order Christoffel symbols also are symmetric with respect to the subscripts ij, whence it follows:

$$\Gamma^l_{ij} = g^{lk} \Gamma_{ij,k}. \tag{1.99}$$

Thus, we have realized the transformation of the components of scalar values (zero-rank tensor) and vector values (first-rank tensor) in passing from one curvilinear coordinate system to the other. In tensor analysis, the general rule concerning the transformation of components of the tensor of arbitrary rank in passing from one coordinate system to the other is introduced:

$$Q'^{i'_1 i'_2 \ldots i'_n}_{j'_1 j'_2 \ldots j'_m} = A^{i'_1}_{i_1} A^{i'_2}_{i_2} \ldots A^{i'_n}_{i_n} B^{j_1}_{j'_1} B^{j_2}_{j'_2} \ldots B^{j_m}_{j'_m} Q^{i_1 i_2 \ldots i_n}_{j_1 j_2 \ldots j_m}. \tag{1.100}$$

Exactly, this rule defines a tensor: If the components of a certain geometric object are transformed by (1.100), this tensor is called the tensor of $n + m$ rank (sometimes it is called valence). Such a tensor is called a mixed tensor being contravariant with respect to n superscripts and covariant with respect to m subscripts. It is clear how the covariant and contravariant tensors are determined.

As applied to our case, we have:

a zero-rank tensor, the electrostatic potential:

$$\Phi' = \Phi; \tag{1.101}$$

first-rank tensors, the vectors of displacements, strength of electric field, electric induction:

$$u'_{j'} = B^j_{j'} u_i; \tag{1.102}$$

second-rank tensors, the mechanical stresses and deformations, dielectric permittivities:

$$T'_{i'j'} = B^i_{i'} B^j_{j'} T_{ij}; \tag{1.103}$$

a third-rank tensor, the piezoelectric constants:

$$e'_{i'j'k'} = B^i_{i'} B^j_{j'} B^k_{k'} e_{ijk}; \tag{1.104}$$

a fourth-rank tensor (elastic moduli):

$$c'_{i'j'k'm'} = B^i_{i'} B^j_{j'} B^k_{k'} B^m_{m'} c_{ijkm}; \tag{1.105}$$

$i, j, k, m, i', j', k', m' = 1, 2, 3$.

Besides that, the motion Eq. (1.3) include the operation of differentiation of the second-rank tensor (tensor of mechanical stresses). This operation looks as follows:

$$T^i_{j,k} \equiv \frac{\partial T^i_j}{\partial \alpha^k} + T^m_j \Gamma^i_{km} - T^i_m \Gamma^m_{jk}. \tag{1.106}$$

Considering the above-mentioned, Eqs. (1.3), (1.6), and the Cauchy relations become:

$$\frac{1}{\sqrt{g}}\frac{\partial\left(\sqrt{g}T^{ij}\right)}{\partial x^i}+T^{ki}\Gamma^j_{ki}=\rho\frac{\partial^2u^j}{\partial t^2},\tag{1.107}$$

$$\nabla\cdot\mathbf{D}=\frac{1}{\sqrt{g}}\frac{\partial\left(\sqrt{g}D^i\right)}{\partial x^i}=0,\ \ \mathbf{E}=-\mathbf{e}^i\frac{\partial\Phi}{\partial x^i},\tag{1.108}$$

$$2S_{ij}=\frac{\partial u_j}{\partial x^i}-u_k\Gamma^k_{ij}+\frac{\partial u_i}{\partial x^j}-u_m\Gamma^m_{ji},\ (i,j,k,m=1;2;3)\,.\tag{1.109}$$

The most spread in mechanics and physics are orthogonal curvilinear coordinate systems, i.e., such systems in which three sets of surfaces (1.55) intersect orthogonally with each other. In this case, the appropriate vectors of the local basis $\mathbf{e}_i$ are mutually orthogonal. For the fundamental tensor g_{ij}, we have:

$$g_{ij}=0\ (i\neq j)\,.\tag{1.110}$$

Because of this, only three nonzero independent components of the metric tensor remain:

$$\sqrt{g_{\alpha\alpha}}=H_\alpha\ \ (\alpha=1,2,3)\,,\tag{1.111}$$

where H_α are called Lame's parameters. The components of the conjugated metric tensor take the form:

$$g^{\alpha\alpha}=\frac{1}{H_\alpha^2}.\tag{1.112}$$

The matrix determinant g_{ij} is defined by

$$g=\left|g_{ij}\right|=\begin{vmatrix}g_{11}&0&0\\0&g_{22}&0\\0&0&g_{33}\end{vmatrix}=g_{11}g_{22}g_{33}=H_1^2H_2^2H_3^2.\tag{1.113}$$

The first-order Christoffel symbols are determined by (1.98):

$$\begin{aligned}\Gamma_{\beta\beta,\beta}&=\frac{1}{2}\frac{\partial g_{\beta\beta}}{\partial\alpha^\beta}=\frac{1}{2}\frac{\partial\left(H_\beta^2\right)}{\partial\alpha^\beta}=H_\beta\frac{\partial H_\beta}{\partial\alpha^\beta},\\ \Gamma_{\beta\gamma,\beta}&=-\Gamma_{\beta\beta,\gamma}=\frac{1}{2}\frac{\partial g_{\beta\beta}}{\partial\alpha^\gamma}=H_\beta\frac{\partial H_\beta}{\partial\alpha^\gamma},\ (\beta\neq\gamma;\ \ \gamma=1,2,3)\,,\end{aligned}\tag{1.114}$$

while by (1.99), we find the second-order Christoffel symbols:

$$\Gamma^{\beta}_{\beta\beta} = g^{\beta\beta}\Gamma_{\beta\beta,\beta} = \frac{1}{H_\beta^2} H_\beta \frac{\partial H_\beta}{\partial \alpha^\beta} = \frac{1}{H_\beta}\frac{\partial H_\beta}{\partial \alpha^\beta},$$
$$\Gamma^{\gamma}_{\beta\beta} = g^{\gamma\gamma}\Gamma_{\beta\beta,\gamma} = \frac{1}{H_\gamma^2}\left(-H_\beta \frac{\partial H_\beta}{\partial \alpha^\gamma}\right) = -\frac{H_\beta \partial H_\beta}{H_\gamma^2 \partial \alpha^\gamma}, \tag{1.115}$$
$$\Gamma^{\beta}_{\beta\gamma} = g^{\beta\beta}\Gamma_{\beta\gamma,\beta} = \frac{1}{H_\beta^2}\left(H_\beta \frac{\partial H_\beta}{\partial \alpha^\gamma}\right) = \frac{1}{H_\beta}\frac{\partial H_\beta}{\partial \alpha^\gamma}, \ (\beta, \gamma = 1, 2, 3).$$

Other Christoffel symbols are equal to zero.

The vectors of the local basis $\mathbf{e}_i$ and $\mathbf{e}^i$ define directions of counts and scales of linear measurements in these directions. Due to this fact, the coordinates of geometric objects appearing in these expressions are given in natural metric. However, the dimensionalities of different basis vectors specified in such a way may not coincide between themselves. This causes inconveniences in writing physical relations.

Let us introduce a unit basis consisting of unit vectors k_i:

$$\mathbf{k}_\alpha \equiv \frac{\mathbf{e}_\alpha}{|\mathbf{e}_\alpha|}. \tag{1.116}$$

The length of the vector $\mathbf{e}_\alpha$ is determined by

$$|\mathbf{e}_\alpha| = \sqrt{g_{\alpha\alpha}} = \frac{1}{\sqrt{g^{\alpha\alpha}}}. \tag{1.117}$$

For an arbitrary vector $\mathbf{a}$, we have:

$$\mathbf{a} = a_i \mathbf{e}^i = \sum_{\alpha=1}^{3} a^\alpha \, |\mathbf{e}_\alpha| \frac{\mathbf{e}_\alpha}{|\mathbf{e}_\alpha|} = a^i_{(phys)} \mathbf{k}_i \ , \tag{1.118}$$

where $a^i_{(phys)}$ are the physical components of the vector $\mathbf{a}$ in a unit basis $\mathbf{k}_i$:

$$a^\alpha_{(phys)} = a^\alpha \, |\mathbf{e}_\alpha| = a^\alpha \sqrt{g_{\alpha\alpha}}. \tag{1.119}$$

Similarly, we compose the unit basis of vectors of mutual local basis $\mathbf{k}^i$:

$$\mathbf{k}^\alpha \equiv \frac{\mathbf{e}^\alpha}{|\mathbf{e}^\alpha|} = \frac{\mathbf{e}^\alpha}{\sqrt{g^{\alpha\alpha}}} = \mathbf{e}^\alpha \sqrt{g_{\alpha\alpha}}. \tag{1.120}$$

Using the formula of transition from one local basis to other, we obtain:

$$\mathbf{k}_\alpha = \frac{\mathbf{e}_\alpha}{\sqrt{g_{\alpha\alpha}}} = \sum_{\beta=1}^{3} \frac{g_{\alpha\beta}\mathbf{e}^\beta}{\sqrt{g_{\alpha\alpha}}} = \frac{g_{\alpha\alpha'}\mathbf{e}^\alpha}{\sqrt{g_{\alpha\alpha}}} = \sqrt{g_{\alpha\alpha}}\mathbf{e}^\alpha = \frac{\mathbf{e}^\alpha}{\sqrt{g^{\alpha\alpha}}} = \mathbf{k}^\alpha, \tag{1.121}$$

whence it follows that the unit bases $\mathbf{k}_i$ and $\mathbf{k}_i$ coincide.

For the covariant components of the vector a_i, we have:

$$\mathbf{a} = a_i \mathbf{e}^i = \sum_{\alpha=1}^{3} a^\alpha \, |\mathbf{e}_\alpha| \, \frac{\mathbf{e}_\alpha}{|\mathbf{e}_\alpha|} = a_{i(phys)} \mathbf{k}^i, \tag{1.122}$$

where $a_{\alpha(phys)} = a_\alpha \sqrt{g^{\alpha\alpha}}$.

Considering (1.119), (1.121), and (1.122), we get:

$$a_{\alpha(phys)} = a^\alpha_{(phys)}, \tag{1.123}$$

whence it follows that the physical components of the vector are not transformed, as it holds in the case of the components of the vector when passing from one basis to other, and, consequently, they are not components of the vector. For these tensors, the same rules are introduced. Let us show this by the example of a tensor:

$$T^\beta_{\alpha(phys)} = T^\beta_\alpha \sqrt{g^{\alpha\alpha} g_{\beta\beta}} = T^\beta_\alpha \frac{H_\alpha}{H_\beta}. \tag{1.124}$$

Thus, the tensor components introduced for the basis composed of the unit vectors $\mathbf{k}_i$ and $\mathbf{k}^i$ will be called physical components. They are not transformed by tensor rules; however, they have a uniform dimensionality.

For example, Eq. (1.106) in physical coordinates becomes

$$\frac{1}{\sqrt{g}} \sum_i \frac{\partial}{\partial x^i} \left(T^{ij} \sqrt{\frac{g}{g_{ii} g_{jj}}} \right) + \sum_{k,i} \frac{T^{ki} \Gamma^j_{ki}}{\sqrt{g_{kk} g_{ii}}} = \frac{\rho}{\sqrt{g_{jj}}} \frac{\partial^2 u^j}{\partial t^2} \quad (j = 1; 2; 3)\,. \tag{1.125}$$

Note that cylindrical and spherical coordinate systems are the most spread in science and engineering.

Cylindrical coordinates. The Cartesian (x, y, z) and cylindrical (r, θ, z) coordinates are related as follows:

$$x = r\cos\theta,\ y = r\sin\theta,\ z = z \Leftrightarrow x^2 + y^2 = r^2,\ y = x\mathrm{tg}\theta. \tag{1.126}$$

The covariant components of the metric tensor are defined by

$$g_{11} = \cos^2\theta + \sin^2\theta = H_1^2 = 1;\ g_{22} = (r\sin\theta)^2 + (r\cos\theta)^2 = H_2^2 = r^2;$$
$$g_{33} = H_3^2 = 1;\ g_{23} = g_{31} = g_{12} = 0.\ (1 \to r,\ 2 \to \theta,\ 3 \to z)\,. \tag{1.127}$$

The nonzero Christoffel symbols are:

$$\Gamma_{21,2} = -\Gamma_{22,1} = r;\ \Gamma^2_{21} = \frac{1}{r};\ \Gamma^1_{22} = -r. \tag{1.128}$$

In this case, Eq. (1.107) becomes:

$$\begin{aligned}
&\frac{\partial T^{11}}{\partial r}+\frac{\partial T^{21}}{\partial \theta}+\frac{\partial T^{31}}{\partial z}-rT^{22}+\frac{1}{r}T^{11}=\rho\frac{\partial^2 u^1}{\partial t^2},\\
&\frac{\partial T^{12}}{\partial r}+\frac{\partial T^{22}}{\partial \theta}+\frac{\partial T^{32}}{\partial z}+\frac{2}{r}T^{12}=\rho\frac{\partial^2 u^2}{\partial t^2},\\
&\frac{\partial T^{13}}{\partial r}+\frac{\partial T^{23}}{\partial \theta}+\frac{\partial T^{33}}{\partial z}+\frac{1}{r}T^{13}=\rho\frac{\partial^2 u^3}{\partial t^2}.
\end{aligned}\tag{1.129}$$

Equation (1.108) takes the form:

$$\begin{aligned}
&E_1=-\frac{\partial \Phi}{\partial r},\ E_2=-\frac{\partial \Phi}{\partial \theta},\ E_3=-\frac{\partial \Phi}{\partial z},\\
&\frac{\partial D^1}{\partial r}+\frac{\partial D^2}{\partial \theta}+\frac{\partial D^3}{\partial z}+\frac{1}{r}D^1=0,
\end{aligned}\tag{1.130}$$

while Eq. (1.109) becomes:

$$\begin{aligned}
&S_{11}=-\frac{\partial u_1}{\partial r},\ S_{22}=\frac{\partial u_2}{\partial \theta}+u_1 r,\ S_{33}=-\frac{\partial u_3}{\partial z},\ 2S_{23}=\frac{\partial u_2}{\partial z}+\frac{\partial u_3}{\partial \theta},\\
&2S_{13}=\frac{\partial u_3}{\partial r}+\frac{\partial u_1}{\partial z},\ 2S_{12}=\frac{\partial u_2}{\partial r}+\frac{\partial u_1}{\partial \theta}-\frac{2}{r}u_2.
\end{aligned}\tag{1.131}$$

Since the coordinates $\alpha^1=r$ and $\alpha^3=z$ in cylindrical coordinates have the dimension of a length, while the coordinate $\alpha^2=\theta$ has the dimension of a radian, the necessity arises to use physical components. The unit basis becomes: $\mathbf{k}^1=\mathbf{e}^1$, $\mathbf{k}^2=r\mathbf{e}^2, \mathbf{k}^3=\mathbf{e}^3$. To designate the physical components, we will use the same symbols as in the case of tensor components but with the indexes employed instead of digits:

$$\begin{aligned}
&T_{rr}=T^{11},\ T_{\theta\theta}=T^{22}r^2,\ T_{zz}=T^{33},\ T_{\theta z}=T^{23}r,\ T_{r\theta}=T^{12}r,\ T_{rz}=T^{13},\\
&S_{rr}=S_{11},\ S_{\theta\theta}=\frac{1}{r^2}S_{22},\ S_{zz}=S_{33},\ S_{\theta z}=\frac{1}{r}S_{23},\ S_{r\theta}=\frac{1}{r}S_{12},\ S_{rz}=S_{13},\\
&u_r=u_1,\ u_\theta=\frac{1}{r}u_2,\ u_z=u_3,\ D_r=D^1,\ D_\theta=D^2 r,\ D_z=D^3,\\
&E_r=E_1,\ E_\theta=rE_2,\ E_z=E_3.
\end{aligned}$$

Let us express the equations of motion of a solid electroelastic medium in terms of the physical components:

$$
\begin{aligned}
&\frac{\partial T_{rr}}{\partial r}+\frac{1}{r}\left(\frac{\partial T_{r\theta}}{\partial \theta}+T_{rr}-T_{\theta\theta}\right)+\frac{\partial T_{rz}}{\partial z}=\rho\frac{\partial^2 u_r}{\partial t^2},\\
&\frac{\partial T_{r\theta}}{\partial r}+\frac{1}{r}\left(\frac{\partial T_{\theta\theta}}{\partial \theta}+2T_{r\theta}\right)+\frac{\partial T_{\theta z}}{\partial z}=\rho\frac{\partial^2 u_\theta}{\partial t^2},\\
&\frac{\partial T_{rz}}{\partial r}+\frac{1}{r}\left(\frac{\partial T_{\theta z}}{\partial \theta}+T_{rz}\right)+\frac{\partial T_{zz}}{\partial z}=\rho\frac{\partial^2 u_z}{\partial t^2},
\end{aligned}
\tag{1.132}
$$

the electrostatics equations:

$$
\begin{aligned}
&E_r=-\frac{\partial \Phi}{\partial r},\ E_\theta=-\frac{1}{r}\frac{\partial \Phi}{\partial \theta},\ E_z=-\frac{\partial \Phi}{\partial z},\\
&\frac{\partial D_r}{\partial r}+\frac{1}{r}\left(\frac{\partial D_\theta}{\partial \theta}+D_r\right)+\frac{\partial D_z}{\partial z}=0,
\end{aligned}
\tag{1.133}
$$

and the geometrical relations:

$$
\begin{aligned}
&S_{rr}=\frac{\partial u_r}{\partial r},\ S_{\theta\theta}=\frac{1}{r}\left(\frac{\partial u_\theta}{\partial \theta}+u_r\right),\ S_{zz}=\frac{\partial u_z}{\partial z},\ 2S_{\theta z}=\frac{\partial u_\theta}{\partial z}+\frac{1}{r}\frac{\partial u_z}{\partial \theta},\\
&2S_{rz}=\frac{\partial u_z}{\partial r}+\frac{\partial u_r}{\partial z},\ 2S_{r\theta}=\frac{\partial u_\theta}{\partial r}+\frac{1}{r}\left(\frac{\partial u_r}{\partial \theta}-2u_\theta\right).
\end{aligned}
\tag{1.134}
$$

Applying the matrix form of presentation, we get:
the equations of motion of a solid electroelastic medium:

$$
\begin{aligned}
&\frac{\partial T_1}{\partial r}+\frac{1}{r}\left(\frac{\partial T_6}{\partial \theta}+T_1-T_2\right)+\frac{\partial T_5}{\partial z}=\rho\frac{\partial^2 u_r}{\partial t^2},\\
&\frac{\partial T_6}{\partial r}+\frac{1}{r}\left(\frac{\partial T_2}{\partial \theta}+2T_6\right)+\frac{\partial T_4}{\partial z}=\rho\frac{\partial^2 u_\theta}{\partial t^2},\\
&\frac{\partial T_5}{\partial r}+\frac{1}{r}\left(\frac{\partial T_4}{\partial \theta}+T_5\right)+\frac{\partial T_3}{\partial z}=\rho\frac{\partial^2 u_z}{\partial t^2},
\end{aligned}
\tag{1.135}
$$

the geometric relations:

$$
\begin{aligned}
&S_1=\frac{\partial u_r}{\partial r},\ S_2=\frac{1}{r}\left(\frac{\partial u_\theta}{\partial \theta}+u_r\right),\ S_3=\frac{\partial u_z}{\partial z},\ 2S_4=\frac{\partial u_\theta}{\partial z}+\frac{1}{r}\frac{\partial u_z}{\partial \theta},\\
&2S_5=\frac{\partial u_z}{\partial r}+\frac{\partial u_r}{\partial z},\ 2S_6=\frac{\partial u_\theta}{\partial r}+\frac{1}{r}\left(\frac{\partial u_r}{\partial \theta}-2u_\theta\right).
\end{aligned}
\tag{1.136}
$$

Spherical coordinates. The spherical coordinates (r, ϕ, θ) are related to the Cartesian (x, y, z) ones by

$$x = r\sin\theta\cos\phi,\ y = r\sin\theta\sin\phi,\ z = r\cos\theta;\ \Leftrightarrow$$
$$x^2 + y^2 + z^2 = r^2,\ x^2 + y^2 = z^2\mathrm{tg}^2\theta,\ y = x\mathrm{tg}\phi. \tag{1.137}$$

The covariant components of the metric tensor are expressed as

$$g_{11} = H_1^2 = 1;\ \ g_{22} = H_2^2 = r^2;\ g_{33} = H_3^2 = r^2\sin^2\theta;$$
$$g_{23} = g_{31} = g_{12} = 0.\ \ (1 \to r,\ 2 \to \theta,\ \ 3 \to \phi)\,. \tag{1.138}$$

The nonzero Christoffel symbols are:

$$\Gamma_{21,2} = -\Gamma_{22,1} = r;\ \Gamma_{31,3} = -\Gamma_{33,1} = r\sin^2\theta;$$
$$\Gamma_{32,3} = -\Gamma_{33,2} = \frac{r^2}{2}\sin 2\theta;\ \Gamma^1_{22} = -r;\ \Gamma^1_{33} = -r\sin^2\theta; \tag{1.139}$$
$$\Gamma^2_{33} = -\frac{1}{2}\sin 2\theta;\ \Gamma^2_{21} = \Gamma^3_{31} = \frac{1}{r};\ \Gamma^3_{32} = \mathrm{ctg}\theta.$$

Considering that $\sqrt{g} = H_1H_2H_3$, expression (1.107) becomes:

$$\frac{1}{H_1H_2H_3}\left(\frac{\partial\left(H_1H_2H_3T^{11}\right)}{\partial r} + \frac{\partial\left(H_1H_2H_3T^{21}\right)}{\partial\theta} + \right.$$
$$\left. + \frac{\partial\left(H_1H_2H_3T^{31}\right)}{\partial z}\right) + T^{22}\Gamma^1_{22} + T^{33}\Gamma^1_{33} = -\rho\frac{\partial^2u^1}{\partial t^2},$$
$$\frac{1}{H_1H_2H_3}\left(\frac{\partial\left(H_1H_2H_3T^{12}\right)}{\partial r} + \frac{\partial\left(H_1H_2H_3T^{22}\right)}{\partial\theta} + \right.$$
$$\left. + \frac{\partial\left(H_1H_2H_3T^{32}\right)}{\partial z}\right) + T^{21}\Gamma^2_{21} + T^{33}\Gamma^2_{33} = -\rho\frac{\partial^2u^2}{\partial t^2}, \tag{1.140}$$
$$\frac{1}{H_1H_2H_3}\left(\frac{\partial\left(H_1H_2H_3T^{13}\right)}{\partial r} + \frac{\partial\left(H_1H_2H_3T^{23}\right)}{\partial\theta} + \right.$$
$$\left. + \frac{\partial\left(H_1H_2H_3T^{33}\right)}{\partial z}\right) + T^{31}\Gamma^3_{31} + T^{32}\Gamma^3_{32} = -\rho\frac{\partial^2u^3}{\partial t^2}.$$

With $H_1H_2H_3 = r^2\sin\theta$ and the expression (1.132), we have:

$$\frac{1}{r^2 \sin\theta}\left(\sin\theta\frac{\partial\left(r^2T^{11}\right)}{\partial r}+r^2\frac{\partial\left(T^{21}\sin\theta\right)}{\partial\theta}+\right.$$
$$\left.+r^2\sin\theta\frac{\partial T^{31}}{\partial\phi}\right)-rT^{22}-r\sin^2\theta T^{33}=\rho\frac{\partial^2u^1}{\partial t^2},$$
$$\frac{1}{r^2 \sin\theta}\left(\sin\theta\frac{\partial\left(r^2T^{12}\right)}{\partial r}+r^2\frac{\partial\left(T^{22}\sin\theta\right)}{\partial\theta}+\right.$$
$$\left.+r^2\sin\theta\frac{\partial T^{32}}{\partial\phi}\right)+\frac{1}{r}T^{21}-\frac{T^{33}}{2}\sin 2\theta=\rho\frac{\partial^2u^2}{\partial t^2}, \quad (1.141)$$
$$\frac{1}{r^2 \sin\theta}\left(\sin\theta\frac{\partial\left(r^2T^{13}\right)}{\partial r}+r^2\frac{\partial\left(T^{23}\sin\theta\right)}{\partial\theta}+\right.$$
$$\left.+r^2\sin\theta\frac{\partial T^{33}}{\partial\phi}\right)+\frac{1}{r}T^{31}+T^{32}\text{ctg}\theta=\rho\frac{\partial^2u^3}{\partial t^2}.$$

Differentiation of (1.134) yields:

$$\frac{\partial T^{11}}{\partial r}+\frac{\partial T^{21}}{\partial\theta}+\frac{\partial T^{31}}{\partial\phi}+\frac{2}{r}T^{11}-rT^{22}-rT^{33}\sin^2\theta+T^{21}\text{ctg}\theta=\rho\frac{\partial^2u^1}{\partial t^2},$$
$$\frac{\partial T^{12}}{\partial r}+\frac{\partial T^{22}}{\partial\theta}+\frac{\partial T^{32}}{\partial\phi}-\frac{T^{33}}{2}\sin 2\theta+\frac{3}{r}T^{21}+T^{22}\text{ctg}\theta=\rho\frac{\partial^2u^2}{\partial t^2}, \quad (1.142)$$
$$\frac{\partial T^{13}}{\partial r}+\frac{\partial T^{23}}{\partial\theta}+\frac{\partial T^{33}}{\partial\phi}+\frac{3}{r}T^{31}+2T^{32}\text{ctg}\theta=\rho\frac{\partial^2u^3}{\partial t^2}.$$

The physical components in a spherical coordinate system are expressed as follows:

$$T_{rr}=T^{11},\, T_{\theta\theta}=T^{22}r^2,\, T_{\phi\phi}=T^{33}r^2\sin^2\theta,\, T_{\theta\phi}=T^{23}r^2\sin\theta,\, T_{r\theta}=T^{12}r,$$
$$T_{r\phi}=T^{13}r\sin\theta,\, S_{rr}=S_{11},\, S_{\theta\theta}=\frac{S_{22}}{r^2},\, S_{\phi\phi}=\frac{S_{33}}{r^2\sin^2\theta},\, S_{\theta\phi}=\frac{S_{23}}{r^2\sin\theta},$$
$$S_{r\theta}=\frac{S_{12}}{r},\, S_{r\phi}=\frac{S_{13}}{r\sin\theta},\, u_r=u_1,\, u_\theta=\frac{u_2}{r},\, u_\phi=\frac{u_3}{r\sin\theta},\, D_r=D^1,$$
$$D_\theta=D^2r,\, D_\phi=D^3r\sin\theta,\, E_r=E_1,\, E_\theta=rE_2,\, E_\phi=E_3r\sin\theta.$$

The motion equations being expressed in terms of the physical components become:

$$\begin{gathered}
\frac{\partial T_{rr}}{\partial r}+\frac{1}{r}\frac{\partial T_{r\theta}}{\partial \theta}+\frac{1}{r\sin\theta}\frac{\partial T_{r\phi}}{\partial \phi}+ \\
+\frac{1}{r}\left(2T_{rr}-T_{\theta\theta}-T_{\phi\phi}+T_{r\theta}\mathrm{ctg}\theta\right)=\rho\frac{\partial^2 u_r}{\partial t^2}, \\
\frac{\partial T_{r\theta}}{\partial r}+\frac{1}{r}\frac{\partial T_{\theta\theta}}{\partial \theta}+\frac{1}{r\sin\theta}\frac{\partial T_{\theta\phi}}{\partial \phi}+ \\
+\frac{1}{r}\left(3T_{r\theta}+\left(T_{\theta\theta}-T_{\phi\phi}\right)\mathrm{ctg}\theta\right)=\rho\frac{\partial^2 u_\theta}{\partial t^2}, \\
\frac{\partial T_{r\phi}}{\partial r}+\frac{1}{r}\frac{\partial T_{\theta\phi}}{\partial \theta}+\frac{1}{r\sin\theta}\frac{\partial T_{\phi\phi}}{\partial \phi}+ \\
+\frac{1}{r}\left(3T_{r\phi}+2T_{\theta\phi}\mathrm{ctg}\theta\right)=\rho\frac{\partial^2 u_\phi}{\partial t^2}.
\end{gathered} \tag{1.143}$$

Then the electrostatics relations expressed in terms of the physical components take the form:

$$\begin{gathered}
E_r=-\frac{\partial \Phi}{\partial r},\ E_\theta=-\frac{1}{r}\frac{\partial \Phi}{\partial \theta},\ E_\phi=-\frac{1}{r\sin\theta}\frac{\partial \Phi}{\partial \phi}, \\
\frac{\partial D_r}{\partial r}+\frac{1}{r}\left(\frac{\partial D_\theta}{\partial \theta}+\frac{1}{\sin\theta}\frac{\partial D_\phi}{\partial \phi}+2D_r+D_\theta\mathrm{ctg}\theta\right)=0.
\end{gathered} \tag{1.144}$$

Similarly, the geometric relations become:

$$\begin{aligned}
S_{rr}&=\frac{\partial u_r}{\partial r},\ S_{\theta\theta}=\frac{1}{r}\left(\frac{\partial u_\theta}{\partial \theta}+u_r\right), \\
S_{\phi\phi}&=\frac{1}{r}\left(\frac{1}{\sin\theta}\frac{\partial u_\phi}{\partial \phi}+u_r+u_\theta\mathrm{ctg}\theta\right), \\
2S_{\theta\phi}&=\frac{1}{r}\left(\frac{\partial u_\phi}{\partial \theta}+\frac{1}{\sin\theta}\frac{\partial u_\theta}{\partial \phi}-2u_\phi\mathrm{ctg}\theta\right), \\
2S_{r\phi}&=\frac{\partial u_\phi}{\partial r}+\frac{1}{r}\left(\frac{1}{\sin\theta}\frac{\partial u_r}{\partial \phi}-2u_\phi\right), \\
2S_{r\theta}&=\frac{\partial u_\theta}{\partial r}+\frac{1}{r}\left(\frac{\partial u_r}{\partial \theta}-2u_\theta\right).
\end{aligned} \tag{1.145}$$

Using the matrix form of the Voigt presentation, we arrive at the following system of equations:

the equations of motion:

$$\frac{\partial T_1}{\partial r}+\frac{1}{r}\frac{\partial T_6}{\partial \theta}+\frac{1}{r\sin\theta}\frac{\partial T_5}{\partial \phi}+\frac{1}{r}\left(2T_1-T_2-T_3+T_6\mathrm{ctg}\theta\right)=\rho\frac{\partial^2 u_r}{\partial t^2},$$
$$\frac{\partial T_6}{\partial r}+\frac{1}{r}\frac{\partial T_2}{\partial \theta}+\frac{1}{r\sin\theta}\frac{\partial T_4}{\partial \phi}+\frac{1}{r}\left(3T_6+\left(T_2-T_3\right)\mathrm{ctg}\theta\right)=\rho\frac{\partial^2 u_\theta}{\partial t^2}, \quad (1.146)$$
$$\frac{\partial T_5}{\partial r}+\frac{1}{r}\frac{\partial T_4}{\partial \theta}+\frac{1}{r\sin\theta}\frac{\partial T_3}{\partial \phi}+\frac{1}{r}\left(3T_6+2T_4\mathrm{ctg}\theta\right)=\rho\frac{\partial^2 u_\phi}{\partial t^2},$$

the geometric relations:

$$S_1=\frac{\partial u_r}{\partial r},\ S_2=\frac{1}{r}\left(\frac{\partial u_\theta}{\partial \theta}+u_r\right),$$
$$S_3=\frac{1}{r}\left(\frac{1}{\sin\theta}\frac{\partial u_\phi}{\partial \phi}+u_r+u_\theta\mathrm{ctg}\theta\right),$$
$$2S_4=\frac{1}{r}\left(\frac{\partial u_\phi}{\partial \theta}+\frac{1}{\sin\theta}\frac{\partial u_\theta}{\partial \phi}-2u_\phi\mathrm{ctg}\theta\right), \quad (1.147)$$
$$2S_5=\frac{\partial u_\phi}{\partial r}+\frac{1}{r}\left(\frac{1}{\sin\theta}\frac{\partial u_r}{\partial \phi}-2u_\phi\right),$$
$$2S_6=\frac{\partial u_\theta}{\partial r}+\frac{1}{r}\left(\frac{\partial u_r}{\partial \theta}-2u_\theta\right).$$

These equations should be supplemented with physical relations, which depend on the direction of the previous polarization of piezoceramic. In the case of different kinds of polarization, we will be confronted with different systems of governing equations, boundary conditions as well as dissimilar problems, and use physical relations in the form of (1.45). If the coordinate system is curvilinear and orthogonal $\alpha^1, \alpha^2, \alpha^3$, while the vector of polarization of the material coincides with the α^3-axis, these relations become:

$$T^1=c_{11}S_1+c_{12}S_2+c_{13}S_3-e_{31}E_3;\ T^2=c_{12}S_1+c_{11}S_2+c_{13}S_3-e_{31}E_3;$$
$$T^3=c_{13}S_1+c_{13}S_2+c_{33}S_3-e_{33}E_3;\ T^4=2c_{55}S_4-e_{15}E_2; \quad (1.148)$$
$$T^5=2c_{55}S_5-e_{15}E_1;\ T^6=2\left(c_{11}-c_{12}\right)S_6.$$

In the case of problems on vibrational or wave processes in functionally gradient materials, the physical–mechanical moduli appearing in Eq. (1.149) are not constants but functions of one coordinate:

$$
\begin{aligned}
&T^1 = c_{11}(r)S_1 + c_{12}(r)S_2 + c_{13}(r)S_3 - e_{31}(r)E_3;\\
&T^2 = c_{12}(r)S_1 + c_{11}(r)S_2 + c_{13}(r)S_3 - e_{31}(r)E_3;\\
&T^3 = c_{13}(r)S_1 + c_{13}(r)S_2 + c_{33}(r)S_3 - e_{33}(r)E_3;\\
&T^4 = 2c_{55}(r)S_4 - e_{15}(r)E_2;\ T^5 = 2c_{55}(r)S_5 - e_{15}(r)E_1;\\
&T^6 = 2\left(c_{11}(r) - c_{12}(r)\right)S_6;\ D_1 = \varepsilon_{11}^S(r)E_1 + e_{15}(r)S_5;\\
&D_2 = \varepsilon_{11}^S(r)E_2 + e_{15}(r)S_4,\ D_3 = \varepsilon_{33}^S(r)E_3 + e_{31}(r)\left(S_1 + S_2\right) + e_{33}(r)S_3.
\end{aligned}
\tag{1.149}
$$

The existence of a unique solution is enabled by specifying initial and boundary conditions. Below, we will consider only stable processes. Because of this, we will restrict our consideration to harmonic vibrations and wave processes only. In the case of the usual formulation of boundary electroelasticity problems at any point of the surface bounding the body, it is necessary to specify two conditions for mechanical variables and one for the electrical one. It is assumed that the given loads appearing in the boundary conditions (forces, charges, displacements, potentials) vary harmonically with time.

1.6 Statement of the Principal Boundary-Value Problems of the Electroelasticity Theory

Let us consider boundary conditions separately for mechanical and electric components of a conjugated field on the surfaces of a piezoceramic body and on the interfaces of joint layers. The boundary conditions for the mechanical components of the conjugated field are formulated similarly to the problems of the pure elasticity theory. Equations (1.107)–(1.109) must be satisfied over whole body volume. We will consider the case when the mechanical components of the conjugated field are specified on the surfaces parallel to the coordinate surfaces α, β, γ. For definiteness, assume that the body under consideration is loaded on the surfaces parallel to the surface: $\alpha = \text{const} : \alpha = \alpha_0;\ \alpha = \alpha_N$. If the vector of external displacements of the body particles $v_- = v_-^\alpha e_\alpha + v_-^\beta e_\beta + v_-^\gamma e_\gamma$ is given on the surface $\alpha = \alpha_0$ bounding the body (or on its part), the searched solutions for the displacements following from system (1.107)–(1.109) should obey the condition:

$$
u_\alpha|_{\alpha=\alpha_0} = v_-^\alpha,\quad u_\beta\big|_{\alpha=\alpha_0} = v_-^\beta,\quad u_\gamma\big|_{\alpha=\alpha_0} = v_-^\gamma. \tag{1.150}
$$

If the vector of external displacements of the body particles $v_+ = v_+^\alpha e_\alpha + v_+^\beta e_\beta + v_+^\gamma e_\gamma$ is given on the surface $\alpha = \alpha_N$ (or on its part), the searched solutions for the displacements following from system (1.107)–(1.109) should obey the condition:

$$
u_\alpha|_{\alpha=\alpha_N} = v_+^\alpha,\quad u_\beta\big|_{\alpha=\alpha_N} = v_+^\beta,\quad u_\gamma\big|_{\alpha=\alpha_N} = v_+^\gamma. \tag{1.151}
$$

If the vector of external stresses $F_- = F_-^\alpha e_\alpha + F_-^\beta e_\beta + F_-^\gamma e_\gamma$ is given on the surface $\alpha = \alpha_N$ (or on its part), the searched solutions for the displacements following from system (1.107)–(1.109) should obey the condition:

$$T^{\alpha\alpha}\big|_{\alpha=\alpha_0} = F_-^\alpha, \quad T^{\alpha\beta}\big|_{\alpha=\alpha_0} = F_-^\beta, \quad T^{\alpha\gamma}\big|_{\alpha=\alpha_0} = F_-^\gamma, \tag{1.152}$$

If the vector of external stresses $F_+ = F_+^\alpha e_\alpha + F_+^\beta e_\beta + F_+^\gamma e_\gamma$ is given on the surface $\alpha = \alpha_N$ (or on its part), the searched solutions for the displacements following from system (1.107)–(1.109) should obey to the condition:

$$T^{\alpha\alpha}\big|_{\alpha=\alpha_N} = F_+^\alpha, \quad T^{\alpha\beta}\big|_{\alpha=\alpha_N} = F_+^\beta, \quad T^{\alpha\gamma}\big|_{\alpha=\alpha_N} = F_+^\gamma. \tag{1.153}$$

Alternative boundary conditions can be considered, when on one surface the displacements are given, while the stresses are prescribed on the other. In the case of a piecewise-inhomogeneous structure along the coordinate $\alpha = \alpha_i$, it is necessary to specify contact conditions of adjacent layers. In what follows, we will address the case of tight mechanical contact of adjacent layers without rupture and sliding. At the same time, the continuity conditions for the components of the displacement vector and of appropriate components of the stress tensor should be fulfilled:

$$T_i^{\alpha\alpha} = T_{i+1}^{\alpha\alpha}, T_i^{\alpha\beta} = T_{i+1}^{\alpha\beta}, T_i^{\alpha\gamma} = T_{i+1}^{\alpha\gamma}, u_i^\alpha = u_{i+1}^\alpha, u_i^\beta = u_{i+1}^\beta, u_i^\gamma = u_{i+1}^\gamma. \tag{1.154}$$

The formulation of electrical boundary conditions depends on the kind of loading of a piezoceramic body. To derive the boundary conditions, it would be more convenient to use the integral form of the Maxwell equations, in which their physical sense is more clearly demonstrated. In the case of electrostatic field, they are:

$$\oint_L \mathbf{E} \cdot \mathbf{dl} = 0, \tag{1.155}$$

$$\oint_S \mathbf{D} \cdot \mathbf{dS} = q, \tag{1.156}$$

$$\mathbf{D} = \varepsilon \mathbf{E}, \tag{1.157}$$

where $\mathbf{dl} = \mathbf{n}_0 dl$ and $\mathbf{dS} = \mathbf{n}_0 dS$. At first, we will consider the normal components of the electrostatic fields $\mathbf{E}$ and $\mathbf{D}$. Let us choose on the interface of two media an element ΔS sufficiently small to be considered as flat (Fig. 1.2). Construct on it the elementary cylinder of the height Δh located in two media. In the general case, the interface of two media can contain a charge.

Suppose that the charge does not have a volume and is concentrated in a near-surface layer. Then its value becomes:

Fig. 1.2 Normal component of the vector of electric induction on the media interface

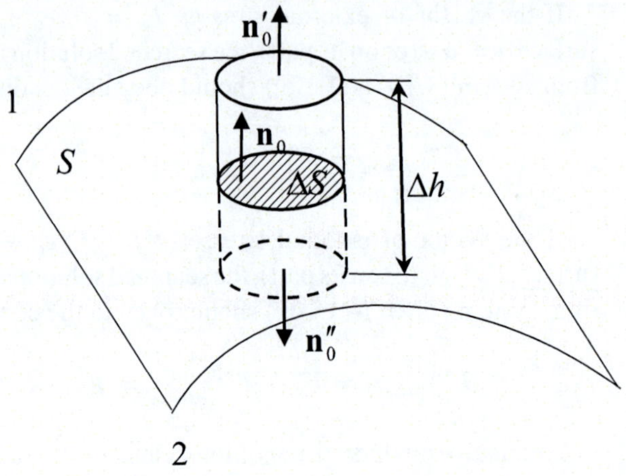

$$\rho_S = \lim_{\Delta S \to 0} \frac{\Delta q}{\Delta S}, \tag{1.158}$$

where $\Delta q$is the charge of the element ΔS.

Let us apply the Gauss theorem (1.156) to the cylindrical volume. Then, due to homogeneity of the electric field, the flux of the vector of electric induction **D** through the upper and lower foundations of the cylinder being considered is defined by the scalar product of this vector and unit vector of the external normal (n_0' or n_0'', respectively) times the area of the cross-section ΔS. Then, in accordance with (1.156) and (1.158), we have:

$$\mathbf{D}_1 \cdot \mathbf{n}_0' \Delta S + \mathbf{D}_2 \cdot \mathbf{n}_0'' \Delta S + \Phi_s^e = \rho_S \Delta S, \tag{1.159}$$

where indices 1 and 2 designate the medium on each side of the boundary, and Φ_s^e is the flux of the vector of electric induction **D** through the lateral surface of the cylinder. Let us decrease unrestrictedly the height of the cylinder Δh so that $\mathbf{n}_0' \to \mathbf{n}_0$, $\mathbf{n}_0'' \to \mathbf{n}_0$, $\Phi_s^e \to 0$, while both foundations of the cylinder will coincide with ΔS. Having divided both sides of the equation by ΔS, we obtain:

$$(\mathbf{D}_1 - \mathbf{D}_2) \cdot \mathbf{n}_0 = \rho_S, \tag{1.160}$$

or

$$D_{1n} - D_{2n} = \rho_S. \tag{1.161}$$

The boundary condition obtained indicates that the component of the vector of electric induction being normal to the interface of two media changes by the value of the surface density of the charge. If the boundary surface is free of charge, the normal component of the vector of the electric induction D_{1n} in passing from medium 1 to 2 is continuous. The normal component of strength of the electric field **E**, as it follows from (1.161) and (1.157), varies proportionally to the dielectric permittivity by

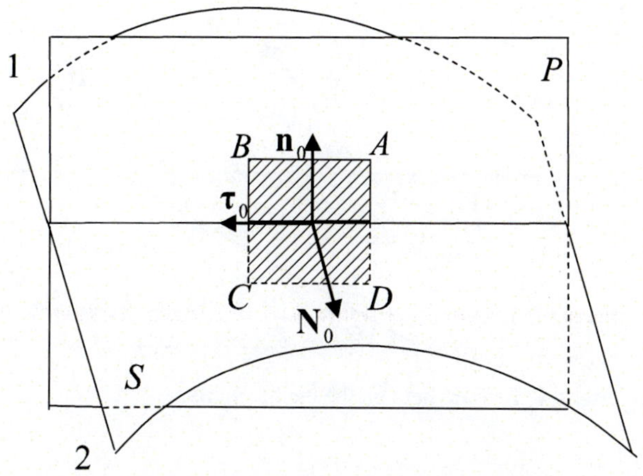

Fig. 1.3 Tangential component of the vector of electric induction on the media interface

$$E_{1n} = \frac{\varepsilon_2}{\varepsilon_1} E_{2n}. \tag{1.162}$$

Consider how the tangential components of the vectors of electrostatic fields **E** and **D** vary. Let us transect the media interface by a plane P (Fig. 1.3), which is perpendicular to some small element of this surface (in the figure it is not shown).

Consider the rectangular contour $L = (ABCD)$ at the plane P which intersects the above elementary element within its boundaries. At that $AB = CD = \Delta l$ and $BC = AD = \Delta h$. The unit vector $\mathbf{n}_0$ is perpendicular to the media interface. The unit vector $\boldsymbol{\tau}_0$ coincides with the line of intersection of the plane P and elementary element on the surface S. The unit vector $\mathbf{N}_0$ is normal to the plane P and is so directed that it formed a right-handed system with the contour of tracking of the contour L. The direction of the unit vector $\boldsymbol{\tau}_0$ is so chosen that it composed three vectors with the unit vectors $\mathbf{n}_0$ and $\mathbf{N}_0$: $\boldsymbol{\tau}_0 = \mathbf{N}_0 \times \mathbf{n}_0$. Applying to the contour L the Faraday law (1.155), we get:

$$\mathbf{E}_1 \cdot \boldsymbol{\tau}_0 \Delta l - \mathbf{E}_2 \cdot \boldsymbol{\tau}_0 \Delta l + C_s^e = 0. \tag{1.163}$$

The first summand appearing in (1.163) is the curl of the vector **E** on the segment AB, the second one on the segment CD, and the third one on the lateral sides BC and DA. If the height Δh decreases unrestrictedly, the sides AB and CD join on the boundary and $C_s^e \to 0$. Dividing by Δl yields:

$$(\mathbf{E}_1 - \mathbf{E}_2) \cdot \boldsymbol{\tau}_0 = 0 \tag{1.164}$$

or

$$E_{1\tau} = E_{2\tau}. \tag{1.165}$$

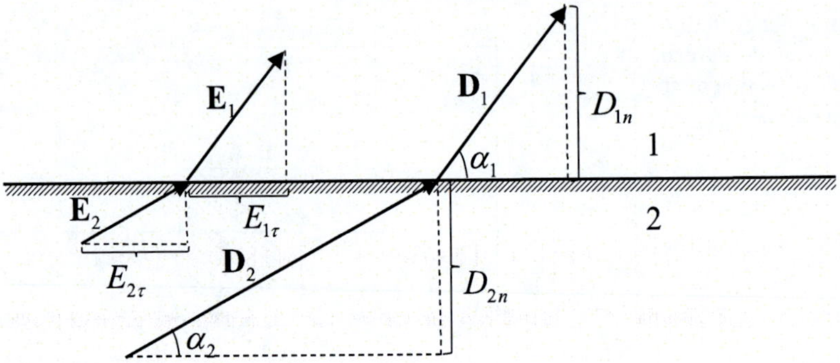

Fig. 1.4 Electric field on the media interface

Thus, the tangential component of the vector **E** is continuous on the media interface (Fig. 1.4).

Provided that $\boldsymbol{\tau}_0 = \mathbf{N}_0 \times \mathbf{n}_0$, Eq. (1.164) becomes:

$$(\mathbf{E}_1 - \mathbf{E}_2) \cdot \mathbf{N}_0 \times \mathbf{n}_0 = \mathbf{n}_0 \times (\mathbf{E}_1 - \mathbf{E}_2) \cdot \mathbf{N}_0 = 0. \tag{1.166}$$

Because equality (1.166) is independent of the direction of the vector $\mathbf{N}_0$, which indicates orientation of the contour L on the media interface, then

$$\mathbf{n}_0 \times (\mathbf{E}_1 - \mathbf{E}_2) = 0. \tag{1.167}$$

From (1.165) and (1.157) it follows:

$$D_{1\tau} = \frac{\varepsilon_1}{\varepsilon_2} D_{2\tau}. \tag{1.168}$$

By comparing formulas (1.161), (1.162), (1.165), and (1.168), we see that the vectors **E** and **D** are refracted on the interface of two media (Fig. 1.4). If the media interface is free of surface charges, we have:

$$\mathrm{tg}\alpha_1 = \frac{\varepsilon_2}{\varepsilon_1} \mathrm{tg}\alpha_2, \tag{1.169}$$

where

$$\mathrm{tg}\alpha_1 = \frac{E_{1n}}{E_{1\tau}} = \frac{D_{1n}}{D_{1\tau}}, \quad \mathrm{tg}\alpha_2 = \frac{E_{2n}}{E_{2\tau}} = \frac{D_{2n}}{D_{2\tau}}.$$

Since the electrostatic field is derived from a potential, $\mathbf{E} = -\nabla\Phi$, and the projection of the gradient on any direction $\mathbf{l}_0$ is equal to the derivative of the function on the same direction, from (1.162), we obtain:

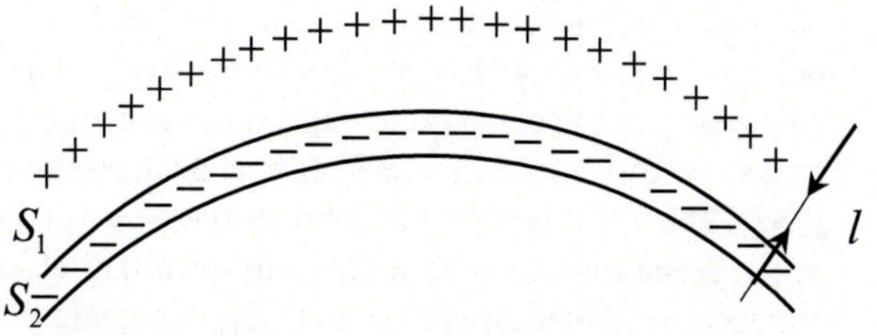

Fig. 1.5 The double charged layer

$$\left(\frac{\partial \Phi}{\partial \tau}\right)_1 = \left(\frac{\partial \Phi}{\partial \tau}\right)_2, \tag{1.170}$$

where the operator $\dfrac{\partial}{\partial \tau}$ means differentiation with respect to any direction in the plane of the tangent to the media interface at the point being considered. Integrating this equality with respect to τ, we get:

$$\Phi_1 = \Phi_2 + C, \tag{1.171}$$

where C is the arbitrary constant, and Φ_1 and Φ_2 are the values of the electrostatic potential in the first and second media, respectively. In many cases, the constant A can be assumed to be equal to zero. Really, the potential Φ produced by body or surface charges is a continuous function. At that, from equality (1.170) it follows:

$$\Phi_1 = \Phi_2. \tag{1.172}$$

If the interface contains a double charged layer, relation (1.171) is violated. This layer can be presented in Fig.1.5 by considering two parallel planes S_1 and S_2 when one of them contains distributed surface charges with density ρ_S, while the other plane contains charges with an opposite sign having the same density.

Bringing together the planes with the value $\rho_S l$ being constant, we arrive in the limit at the double charged layer. The parameter $\rho_S l$ is called layer capacity. In passing through the double charged layer, the potential undergoes a discontinuity, which depends on the layer capacity. In what follows, we will assume that the double charged layers in the domain under consideration are absent.

Moving in the formulas (1.161) to the electrostatic potential Φ, we get the second boundary condition for the potential:

$$\varepsilon_2 \left(\frac{\partial \Phi}{\partial n}\right)_2 - \varepsilon_1 \left(\frac{\partial \Phi}{\partial n}\right)_2 = \rho_S, \tag{1.173}$$

where the operator $\dfrac{\partial}{\partial n}$ means differentiation with respect to the normal to the media interface directed from the second medium to the first one. If one medium is a conductor, the boundary conditions are simplified. Indeed, in analyzing macroscopic properties of the electrostatic field, the conductor can be considered as a closed domain, inside of which charges move freely. The density of the charge flux, i.e., the density of the conduction current in the conductor is proportional to the strength of the electric current, $\mathbf{j} = \sigma\mathbf{E}$. In the electrostatic field, the charges are at rest, i.e., $\mathbf{j} = 0$. Thus, in this case, the conduction $\sigma \neq 0$ and the strength of electrostatic field become equal to zero. Whence it follows that electrostatic field inside of any real conductor is equal to zero. With (1.6), we obtain

$$\nabla\Phi = 0 \;\Rightarrow \Phi = \text{const.} \tag{1.174}$$

Hence in electrostatics, all points of a conductor have the same potential. At the same time, isolated conductors may have different potentials. The boundary conditions on the surface of the conductor of the vectors **E** and **D** of the electrostatic field follow from the formulas (1.161), (1.162), (1.165), and (1.168) under the assumption that the first medium is dielectric, while the second one is a conductor. Then

$$E_{1\tau} = 0, \;\; D_{1\tau} = 0, \tag{1.175}$$

$$E_{1n} = \frac{\rho_S}{\varepsilon_1}, \;\; D_{1n} = \rho_S. \tag{1.176}$$

The conditions (1.175) and (1.176) presented in vector form become:

$$\mathbf{E}_1 = \mathbf{n}_0\frac{\rho_S}{\varepsilon_1}, \; \mathbf{D}_1 = \mathbf{n}_0\rho_S. \tag{1.177}$$

It should be noted that in the case of a variable field, similar conditions are satisfied only on the surface of a perfect conductor while the electrostatic conditions (1.175)–(1.177) are valid for any nonzero specific permittivity of the second medium. Going over to the equations with respect to the electrostatic potential, with (1.175) and (1.176), we have:

$$\Phi|_S = \text{const.}, \tag{1.178}$$

$$\left.\frac{\partial\Phi}{\partial n}\right|_S = -\frac{\rho_S}{\varepsilon_1}. \tag{1.179}$$

In this case, the normal $\mathbf{n}_0$ is considered as external concerning the conducting medium. From the condition (1.178), it follows that the conductor surface is an equipotential surface. Consider some further variants of physically realized boundary conditions for the electric variables. The formulation of boundary conditions depends on the way the electric energy is supplied (picked up) to (by) a piezoceramic body. Both the supply and the picking up of energy from a dynamically deforming body occur by using electrode coatings deposited on the surface of the piezoceramic body. It is supposed that the electrode coatings present are indefinitely thin perfect conductors with a negligibly small mass. The complete statement of the

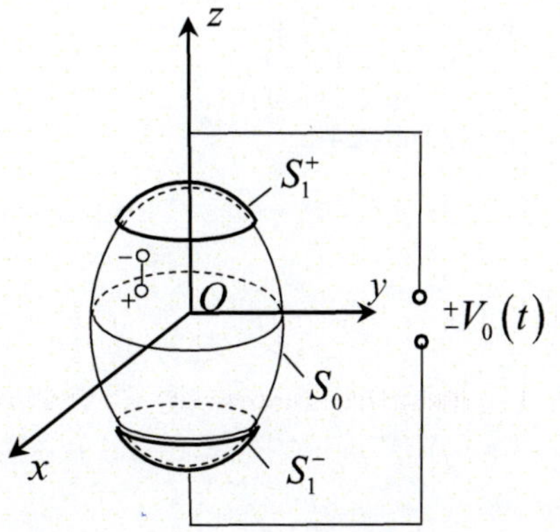

Fig. 1.6 Loading of the piezoceramic body by a given voltage on the electrodes partially covering it

boundary conditions concerning the electric variables requires external electric circuit and energy sources to be considered. At the same time, the supplying elements, such as the electrode coverings, are considered as perfect conductors. The generators of voltage and currents, the model sources of electric energy, are widely employed in electrical engineering. If the piezoceramic body is loaded by the voltage given on the electrodes that partially cover the body, the magnitude of the searched potential is specified on the surfaces $S_1^{\pm}$ (Fig. 1.6). The boundary conditions on these parts of the surface take the form:

$$\Phi|_{S_1^{\pm}} = \pm V_o\,(t)\,. \tag{1.180}$$

Note that in the majority of cases, piezoceramic elements operate inside the environment composed either by air or natural dielectrics (not segnetoelectrics). At the same time, on part of the surface free of electrodes, we have $S_0 = S - S_1^- - S_1^+$. Under such conditions, their dielectric permittivity ε is considerably lesser than the dielectric permittivity of the piezoceramic. In this case, the boundary condition (1.161) may be replaced by

$$D_n|_{S_0} = 0. \tag{1.181}$$

The error that arises in using the approximate boundary condition (1.181) is almost of the same order of smallness as replacement of the exact Maxwell equations by the equations of the forced electrostatics of dielectrics (1.6).

Under loading of a piezoceramic body by a current generator with the given value $I_0\,(t)$, the potential difference on electrodes is an unknown value. In this case, from the condition of the current continuity in a circuit, including the piezoceramic body, the integral condition for the vector of electric induction (1.156) follows:

$$\frac{\mathrm{d}}{\mathrm{d}t}\oint_{S_1^+} \mathbf{D}\cdot \mathrm{d}\mathbf{S} = -\frac{\mathrm{d}q}{\mathrm{d}t} = -I_0(t). \tag{1.182}$$

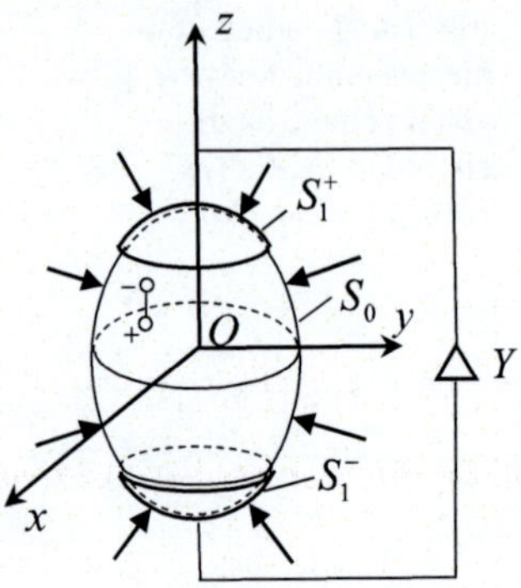

Fig. 1.7 Loading of a piezoceramic body by external mechanical stresses given on its surfaces

The integral condition (1.182) cannot be used directly as a boundary condition. Because of this, at first we will consider the unknown value of the potentials on the electrodes:

$$\Phi|_{S_1^\pm} = \pm V(t). \tag{1.183}$$

Next, the problem is solved in the same way as in the case when the boundary conditions are given by (1.180). Having determined all the components of the conjugate field, the integral condition (1.183) is used to determine the unknown value $V(t)$. Opening the scalar product $\mathbf{D} \cdot \mathrm{d}\mathbf{S}$, using the last three equalities (1.149), Cauchy relations, and solution for the strength of electric field (1.108), it becomes possible to derive an expression for determining the displacement current $I_0(t)$ passing through the ceramic body.

To formulate the boundary conditions on the parts of the surface of the piezoceramic body free of electrodes S_0, we will employ Eq. (1.181).

Let us consider the statement of boundary electroelasticity problems for cases when the piezoelements are employed as the generator of electric energy. Let the deformation of the piezoceramic body be realized mechanically by using the external mechanical stresses $\mathbf{F}_n$ given on its surface, while the electrodes S_1^- and S_1^+ spread on its surface are used to take off electric energy (Fig. 1.7).

The energy is consumed in the external electric circuit and can be presented by losses on the element with a complex conductivity $Y = Y_1 + iY_2$. In this case, both the voltage $V_0(t)$ on the electrodes and the current $I(t)$ in the circuit are unknown. To determine an electric boundary condition for the problem under consideration, we will use Ohm's law for the external circuit with conductivity Y:

$$I = YV. \tag{1.184}$$

With the equality (1.182), we get:

$$\frac{\mathrm{d}}{\mathrm{d}t} \oint\limits_{S_1^+} \mathbf{D} \cdot \mathrm{d}\mathbf{S} = -YV. \tag{1.185}$$

To solve the boundary-value problems, we will specify on the electrodeposited surfaces the unknown voltage $V_0(t)$ (1.183), while the same will be done on the non-electrodeposited surfaces with the normal component of the electric induction (1.181) of the vector boundary condition on parts of the surface, where the mechanical stresses (1.154) are given. If the conjugated problem (1.107)–(1.109) is solved under these boundary conditions, the unknown value of the voltage $V_0(t)$ is determined by (1.183).

From the boundary condition (1.183), two important limiting cases follow. For the electrodes, whose electric energy is not taken off ($Y = 0$ means that the circuit is open), we have:

$$\frac{\mathrm{d}}{\mathrm{d}t} \oint_{S_1^+} \mathbf{D} \cdot \mathrm{d}\mathbf{S} = 0, \tag{1.186}$$

i.e., the total charge on the electrodes during deformation of a piezoceramic body is unchanged. If the conductivity of the external circuit is very high ($Y \to \infty$ means short-circuit), from the condition of passage to the limit and due to the condition of existence of finite limit in the right-hand side in (1.183), it follows that $V(t) = 0$.This condition on the short-circuited electrodes is expressed by

$$\Phi|_{S_1^\pm} = 0, \ \ \Phi|_{S_1^+} - \Phi|_{S_1^-} = 0. \tag{1.187}$$

1.7 Discrete–Continual Analytical–Numerical Approach

Due to the wide application of modern computer facilities—personal computers with large memory and speed—approaches to the calculation of various aspects of the electromechanical behavior of different structures are currently based mainly on numerical methods for solving problems. Universal modern numerical methods appeared to solve various classes of boundary-value problems in the theory of shells. They are called discrete and are based on the reduction of initial partial differential equations to systems of higher-order algebraic equations. There are such methods as: finite differences, variational–difference, and finite element method. The latter method ranks first in application due to its universality and being algorithmic. These features of the finite element method allowed us to create software packages such as LIRA, NX NASTRAN, ANSYS, SIMULA, Abaqus, and PLAXIS that enable solving static and dynamic problems in the theory of electroelasticity and elasticity.

However, along with universal approaches to the solution of problems in electroelasticity theory, so-called discrete–continual approaches are widely used, which makes it possible to reduce the problem to ordinary differential equations on the basis of approximating solutions to other variables by means of analytical tools. To solve the obtained one-dimensional problems, the stable numerical method of discrete orthogonalization was used. Discrete–continual approaches can be considered as an alternative to universal numerical methods, because they are used, as a rule, to investigate electroelasticity objects of a certain class and thus, in this case, can give more effective and accurate results that can be used for testing in the modification of various discrete approaches. To solve the obtained one-dimensional problems, a stable numerical method of discrete orthogonalization [3, 4] is used to solve a wide class of problems of elastic theories and shells for anisotropic inhomogeneous materials.

Currently, for solving problems of computational mathematics, mathematical physics, and mechanics, spline functions are widely used [1]. This is due to the advantages of the spline-approximation technique when compared to others. As basic advantages, the following ones should be mentioned: the stability of splines with respect to local disturbances, i.e., the behavior of the spline near a point does not affect the behavior of the spline as a whole as, for instance, in the case of the polynomial approximation; fast convergence of the spline-interpolation in contrast to the polynomial one; and simplicity and convenience in the realization of algorithms for constructing and calculating splines on personal computers. The use of spline functions in various variational, projective, and other discrete–continuous methods makes it possible to obtain appreciable results in comparison with those obtained with the classical apparatus of polynomials, to simplify essentially their numerical realization, and to obtain the desired solution with a high degree accuracy. In this monograph, this approach was first used to study the dynamic behavior of piezoceramic bodies of cylindrical and spherical shapes with a homogeneous and inhomogeneous structure.

1.7.1 Spline-Collocation Method. Some Spline Functions Information

A function composed of the parts of generalized polynomials over a given basis is called a spline. Polynomial splines, for which the functions $1, x, x^2$ are chosen as basic ones, are the most widely spread. In fact, we will restrict ourselves to polynomial splines.

Let the mesh $\Delta : x_0 < x_1 < x_2 \cdots < x_N$ be given on the segment $[x_0, x_N]$. The function, which is the mth-power polynomial on each interval $[x_i, x_{i+1}]\, i = \overline{0, N-1}$ and $m-1$ times continuously differentiated on the whole segment $[x_0, x_N]$, is called the mth-power $S_m(x)$ polynomial.

The unit Heaviside function

$$\theta(x) = l1 \text{ if } x \geq 0, 0 \text{ if } x < 0. \tag{1.188}$$

is the simplest example of a spline.

As another example we may mention is the power function:

$$x_+^m = x^m \theta(x) = \begin{cases} x^m \text{ if } x \geq 0, \\ 0 \text{ if } x < 0. \end{cases} \tag{1.189}$$

The spline $S_m(x)$ is called an interpolation polynomial spline, which interpolates the function $f(x)$ on the mesh $\bar{\Delta}$, if it is the mth-power polynomial on each interval $[\bar{x}_i, \bar{x}_{i+1}]$ $i = \overline{0, N-1}$ and satisfies the conditions $S_m(\bar{x}_i) = f(\bar{x}_i)$ $(i = \overline{0, N})$. In this case, the nodes of the mesh Δ are called spline nodes and $\bar{\Delta}$ interpolation nodes.

However, in constructing the interpolation spline computational difficulties arise. To avoid them in problem solving B-splines are employed.

Let us expand the mesh $\Delta: \ x_0 < x_1 < x_2 ... < x_N$ with auxiliary points $x_{-m} < \cdots < x_{-1} < x_0$, $x_N < x_{N+1} < \cdots < x_{N+m}$ and consider the mesh $\Delta': \quad x_{-m} < \cdots < x_{-1} < x_0 < \cdots < x_N < x_{N+1} < \cdots < x_{N+m}$, where $x_{-i} = x_0 - i(x_1 - x_0)$, $x_{N+i} = x_N + i(x_N - x_{N-1})$ $i = \overline{1, m}$.

Let us introduce a function $\varphi_m(x, t) = (-1)^{m+1}(m+1)(x-t)_+^m$ and construct for it the separated differences of the (m+1)th order with the values of argument $t = x_i, \ldots, x_{i+m+1}$. Then, we obtain:

$$\tilde{B}_m^i = \varphi_m[x; x_i, \ldots, x_{i+m+1}] \, (i = -m, \ldots, N-1). \tag{1.190}$$

These functions are called basis splines or B-splines of the mth-power.

With the identity

$$(x-t)_+^m = (x-t)^m + (-1)^{m+1}(t-x)_+^m, \tag{1.191}$$

the equality (1.190) becomes:

$$\tilde{B}_m^i(x) = (m+1) \sum_{p=i}^{i+m+1} \frac{(x_p - x)_+^m}{\omega'_{m+1,i}(x_p)} (i = -m, \ldots, N-1), \tag{1.192}$$

where $\omega_{m+1,i}(t) = \prod_{j=1}^{i+m+1} (t - x_j)$.

In calculations, it is convenient to use normalized splines:

$$B_m^i(x) = \frac{x_{i+m+1} - x_i}{m+1} \tilde{B}_m^i(x), \tag{1.193}$$

for which a recurrent formula holds:

$$B_m^i(x) = \frac{x - x_i}{x_{i+m} - x_i} B_{m-1}^i(x) + \frac{x_{i+m+1} - x}{x_{i+m+1} - x_{i+1}} B_{m-1}^{i+1}(x). \tag{1.194}$$

It can be used as value of the B-splines. In this case,

$$B_0^i(x) = \begin{cases} 1 & \text{if} \quad x \in [x_i, x_{i+1}), \\ 0 & \text{if} \quad x \notin [x_i, x_{i+1}). \end{cases} \tag{1.195}$$

The functions $B_m^i(x)$ are defect 1 splines with finite minimal-length carriers. Besides, the system of functions $B_m^i(x)(i = -m, \dots, N-1)$ is linearly independent and forms a basis within the space of splines $S_m(\Delta)$. From this, it follows that each spline $S_m(x) \in S_m(\Delta)$ can be written uniquely as

$$S_m(x) = \sum_{i=-m}^{N-1} b_i B_m^i(x), \tag{1.196}$$

where b_i are some constant coefficients.

The splines $B(x)$ demonstrate the following properties:

(a) $B_m^i(x) = \begin{cases} > 0 & \text{if } x \in [x_i, x_{i+1}), \\ \equiv 0 & \text{if } x \notin [x_i, x_{i+1}); \end{cases}$

(b) $\int_{-\infty}^{\infty} B_m^i(x)dx = \dfrac{x_{i+m+1} - x_i}{m+1}$.

Let us consider the uniform expanded mesh Δ on the interval $[x_0, x_N]$ Δ' : $x_{-m} < \dots < x_{-1} < x_0 < \dots < x_N < x_{N+1} < \dots < x_{N+m}$ $(x_{k+1} - x_k = h = \text{const.})$.

Construct the first three odd-power B-splines, which will be numerated by the mean node of the carriers. Let us mark the odd-power B-splines in terms of $B_m^i(x)$ instead of $B_m^{i-(m+1)/2}(x)$. This indicates that the spline numeration shifts by $(m+1)/2$ units to the right. Then, with (1.194) and (1.195), we get:

– power B-splines:

$$B_1^i(x) = \begin{cases} 0, & -\infty < x < x_{i-1}, \\ t, & x_{i-1} \le x < x_i, \\ 1-t, & x_i \le x < x_{i+1}, \\ 0, & x_{i+1} \le x < \infty; \end{cases} \tag{1.197}$$

– power B-splines:

$$B_3^i(x) = \frac{1}{6} \begin{cases} 0, & -\infty \le x < x_{i-2}, \\ t^3, & x_{i-2} \le x < x_{i-1}, \\ -3t^3 + 3t^2 + 3t + 1, & x_{i-1} \le x < x_i, \\ 3t^3 - 6t^2 + 4, & x_i \le x < x_{i+1}, \\ (1-t)^3, & x_{i+1} \le x < x_{i+2}, \\ 0, & x_{i+2} \le x < \infty; \end{cases} \tag{1.198}$$

– fifth-power B-splines:

Table 1.3 Derivatives of the third-power B-splines

Interval	$(-\infty, x_{i-2})$	$[x_{i-2}, x_{i-1})$	$[x_{i-1}, x_i)$	$[x_i, x_{i+1})$	$[x_{i+1}, x_{i+2})$	$[x_{i+2}, +\infty)$
$\left(B_3^i(x)\right)'$	0	$\dfrac{t^2}{2h}$	$\dfrac{1+2t-3t^2}{2h}$	$\dfrac{3t^2-4t}{2h}$	$-\dfrac{(1-t)^2}{2h}$	0
$\left(B_3^i(x)\right)''$	0	$\dfrac{t}{h^2}$	$\dfrac{1-3t}{h^2}$	$\dfrac{3t-2}{h^2}$	$\dfrac{1-t}{h^2}$	0

$$B_5^i(x) = \frac{1}{120}\begin{cases} 0, & -\infty < x < x_{i-3}, \\ t^5, & x_{i-3} \le x < x_{i-2}, \\ -5t^5 + 5t^4 + 10t^3 + 10t^2 + 5t + 1, & x_{i-2} \le x < x_{i-1}, \\ 10t^5 - 20t^4 - 20t^3 + 20t^2 + 50t + 26, & x_{i-1} \le x < x_i, \\ -10t^5 + 30t^4 - 60t^2 + 66, & x_i \le x < x_{i+1}, \\ 5t^5 - 20t^4 + 20t^3 + 20t^2 - 50t + 26, & x_{i+1} \le x < x_{i+2}, \\ (1-t)^5, & x_{i+2} \le x < x_{i+3}, \\ 0, & x_{i+3} \le x < \infty, \end{cases} \tag{1.199}$$

where $t = \dfrac{x - x_k}{h}$ on the interval $[x_k, x_{k+1}]$, $k = \overline{i - \dfrac{m+1}{2},\; i + \dfrac{m+1}{2} - 1}$.

Tables 1.3 and 1.4 present derivatives of the third- and fifth-power B-splines, respectively, on different intervals of their carriers, while Table 1.5 summarizes the values of the third- and fifth-power B-splines and their derivatives at nodes of the carriers.

To satisfy the boundary conditions, it is convenient to use linear combinations of the B-splines with coefficients chosen in a certain way. Thus, we will search for the solution of a boundary-value problem in the form of the part of $\varphi_n^i(x)$ series by some functions $\varphi_n^i(x)$ being the linear combinations of the nth-power B-splines.

Let us consider, as an example, the use of the spline-collocation method for solving a one-dimensional boundary-value problems. The necessity arises to find the function $y = y(x)$ in the interval $[a; b]$. This function is the solution of the linear differential nth-order equation

$$Ly = \alpha_0(x)y^{(n)} + \alpha_1(x)y^{(n-1)} + \cdots + \alpha_{n-1}(x)y' + \alpha_n(x)y = r(x), \tag{1.200}$$

which satisfies at the ends of the interval boundary conditions

$$\mathbf{A}_1\bar{y}(a) = \bar{c}, \mathbf{A}_2\bar{y}(b) = \bar{d}, \tag{1.201}$$

where $\bar{y} = [y, y', \ldots, y^{(n-1)}]^T$, $\bar{c} = [c_1, c_2, \ldots, c_k]^T$, $\bar{d} = [d_1, d_2, \ldots, d_{n-k}]^T$ are vector-columns, and A_1 and A_2 are given rectangular matrices of $k \times n$ and $(n - k) \times n$ $(k < n)$th order, respectively,

Table 1.4 Derivatives of the fifth-power B-splines

Interval	$\left(B_5^i(x)\right)'$	$\left(B_5^i(x)\right)''$	$\left(B_5^i(x)\right)'''$	$\left(B_5^i(x)\right)^{IV}$
$(-\infty, x_{i-3})$	0	0	0	0
$\left[x_{i-3}, x_{i-2}\right)$	$\frac{t^4}{24h}$	$\frac{t^3}{6h^2}$	$\frac{t^2}{2h^3}$	$\frac{t}{h^4}$
$\left[x_{i-2}, x_{i-1}\right)$	$\frac{-5t^4+4t^3+6t^2+4t+1}{24h}$	$\frac{-5t^3+3t^2+3t+1}{6h^2}$	$\frac{-5t^2+2t+1}{2h^3}$	$\frac{-5t+1}{h^4}$
$\left[x_{i-1}, x_i\right)$	$\frac{5t^4-8t^3-6t^2+4t+5}{12h}$	$\frac{5t^3-6t^2-3t+1}{3h^2}$	$\frac{5t^2-4t-1}{h^3}$	$\frac{10t-4}{h^4}$
$\left[x_i, x_{i+1}\right)$	$\frac{-5t^4+12t^3-12t}{12h}$	$\frac{-5t^3+9t^2-3}{3h^2}$	$\frac{-5t^2+6t}{h^3}$	$\frac{-10t+6}{h^4}$
$\left[x_{i+1}, x_{i+2}\right)$	$\frac{5t^4-16t^3+12t^2+8t-10}{24h}$	$\frac{5t^3-12t^2+6t+2}{6h^2}$	$\frac{5t^2-8t+2}{2h^3}$	$\frac{5t-4}{h^4}$
$\left[x_{i+2}, x_{i+3}\right)$	$-\frac{(1-t)^4}{24h}$	$\frac{(1-t)^3}{6h^2}$	$-\frac{(1-t)^2}{2h^3}$	$\frac{1-t}{h^4}$
$\left[x_{i+3}, \infty\right)$	0	0	0	0

Table 1.5 Values of the B-splines and their derivatives at the carrier nodes

k	$i-3$	$i-2$	$i-1$	i	$i+1$	$i+2$	$i+3$
$B_3^i(x_k)$	0	0	1/6	2/3	1/6	0	0
$B_5^i(x_k)$	0	1/120	13/60	11/20	13/60	1/120	0
$\left(B_3^i(x_k)\right)'$	0	0	$1/2h$	0	$-1/2h$	0	0
$\left(B_5^i(x_k)\right)'$	0	$1/24h$	$5/12h$	0	$-5/12h$	$-1/24h$	0
$\left(B_3^i(x_k)\right)''$	0	0	$1/h^2$	$-2/h^2$	$1/h^2$	0	0
$\left(B_5^i(x_k)\right)''$	0	$1/6h^2$	$1/3h^2$	$-1/h^2$	$1/3h^2$	$1/6h^2$	0
$\left(B_5^i(x_k)\right)'''$	0	$1/2h^3$	$-1/h^3$	0	$1/h^3$	$-1/2h^3$	0
$\left(B_5^i(x_k)\right)^{IV}$	0	$1/h^4$	$-4/h^4$	$6/h^4$	$-4/h^4$	$1/h^4$	0

Let us assume that the problems (1.200) and (1.201) have a unique solution $y = y(x)$, where the coefficients $\alpha_i (i = \overline{0, n})$ appearing in (1.200) and right-hand side $r(x)$ are sufficiently smooth functions.

Let us introduce the mesh $[a; b]{:}\Delta : x_0 < x_1 < x_2 ... < x_N$ and search for the approximate solution of the problems (1.200) – (1.201) in the form of the spline $S(x)$ of the power m ($m > n + 1$, where $m = n + 1$) with nodes on the mesh Δ.

We will require for the spline $S(x)$ that Eq. (1.200) and the boundary conditions (1.201) be satisfied at the points $\xi_k \in [a, b]$, $k = \overline{0, N}$ (conditions of collocation):

$$L[S(\xi_k)] = r(\xi_k);\ \mathbf{A}_1\bar{S}(a) = \bar{c};\ \mathbf{A}_2\bar{S}(b) = \bar{d}, \tag{1.202}$$

where $\bar{S} = \left\{S, S', \ldots, S^{(n-1)}\right\}^T$ is the vector-column.

The relations (1.202) are a system of algebraic equations with respect to the spline parameters. The points ξ_k are called nodes of collocation. Their amount is defined by the dimensionality of the spline space, which is equal to $N + m$. Since $S(x)$ satisfies N boundary conditions, the number of collocation nodes must be equal to $N + 1$. The points at which the coefficients of Eq. (1.200) have features cannot be collocation nodes. In what follows, we will consider that the collocation nodes are ordered: $\xi_0 < \xi_1 < \cdots < \xi_N$.

The specific form of system (1.201) depends on the way how the spline $S(x)$ is presented and collocation nodes are located. The simplest spline-collocation schemes are realized when the spline is constructed on a uniform mesh and collocation and spline nodes coincide.

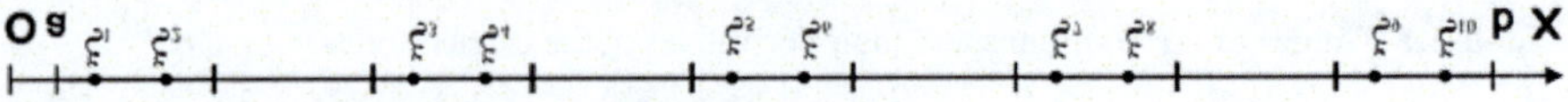

Fig. 1.8 Optimal arrangement of the collocation nodes

The accuracy of the method depends on the value of the error caused by the approximation of the equation at the collocation nodesξ_k:

$$\varepsilon_k = L\left[S\left(\xi_k\right)\right] - r\left(\xi_k\right), k = \overline{0, N} \tag{1.203}$$

and of the approximation error of boundary conditions. The last one can be equal to zero, if the spline $S(x)$ can be constructed such that the boundary conditions (1.201) are satisfied.

The accuracy of the spline-collocation method can be improved by the special choice of the collocation nodes.

Let us consider in detail one special method of arrangement of the collocation points. Let the mesh be formed by nodes Δ with $x_k = a + kh$, $k = \overline{0, N}$, $x_0 = a$, $x_N = b$, $h = \dfrac{b-a}{N}$, $N = 2p + 1$ being an odd number, while the collocation nodes satisfy the conditions:

$$\xi_{2i} \in \left[x_{2i}, x_{2i+1}\right], \xi_{2i+1} \in \left[x_{2i}, x_{2i+1}\right], \tag{1.204}$$

where $\xi_{2i} = x_{2i} + t_1 h$, $\xi_{2i+1} = x_{2i} + t_2 h$ $(i = \overline{0, p})$, and $t_1 = \dfrac{1}{2} - \dfrac{\sqrt{3}}{6}$, $t_2 = \dfrac{1}{2} + \dfrac{\sqrt{3}}{6}$.

The points t_1 and t_2 are the roots of the second-power Legendre polynomial $P_2(t) = 6t^2 - 6t + 1$ on the interval [0, 1]. Thus, each interval $\left[x_{2i}, x_{2i+1}\right]$] contains two collocation nodes, which are absent on the adjacent intervals $\left[x_{2i-1}, x_{2i}\right]$ and $\left[x_{2i+1}, x_{2i+2}\right]$ (see Fig. 1.8).

Such collocation nodes are called optimal. The accuracy can be increased using splines of higher powers. For example, by using fifth-power splines when the functions are four-times differentiable and on each segment $[x_i, x_{i+1}]$ are fifth-power polynomials, we can obtain fourth-order accuracy.

Consider the example with the employment of the spline collocation in solving plane boundary-value problems. Let us find the solution to the equation:

$$L[z] = r(x, y), z = z(x, y), x \in [a_1, b_1], y \in [a_2, b_2], \tag{1.205}$$

where L is the differential operator of two variables. The upper order of the derivative with respect to x is equal to n, while with respect to y to m. The solution of equation (1.205) must satisfy boundary conditions at the contours $x = \text{const.}$ and $y = \text{const.}$:

$$\mathbf{A}_1\bar{z}_1(a_1) = \bar{c}_1, \mathbf{B}_1\bar{z}_1(b_1) = \bar{d}_1, \tag{1.206}$$

$$\mathbf{A}_2\bar{z}_2(a_2) = \bar{c}_2, \mathbf{B}_2\bar{z}_2(b_2) = \bar{d}_2, \tag{1.207}$$

where $\bar{z}_1 = \left\{z, z'_x, \ldots, z^{n-1}_{x\ldots x}\right\}^T, \bar{z}_1 = \left\{z, z'_y, \ldots, z^{m-1}_{y\ldots y}\right\}^T, \bar{c}_1 = \left\{c^1_1, \ldots, c^k_1\right\}^T, \bar{c}_2 = \left\{c^1_2, \ldots, c^{n-k}_2\right\}^T, \bar{d}_1 = \left\{d^1_1, \ldots, d^l_1\right\}^T, \bar{d}_2 = \left\{d^1_2, \ldots, d^{m-l}_2\right\}^T$ are vector-columns, and $\mathbf{A}_1, \mathbf{A}_2, \mathbf{B}_1$ $\mathbf{B}_2$ are rectangular matrices with the dimensionality $k \times n$, $(n-k) \times n$, $l \times m$, $(m-l) \times m$ $(k < n, l < m)$, respectively.

We will assume that the problems (1.205)–(1.207) have a unique solution $z = z(x, y)$, while the coefficients of Eq. (1.205) and the right-hand side $r(x, y)$ are sufficiently smooth functions.

The collocation will be performed with respect to the variable y. Let us introduce a uniform mesh Δ in the interval $[a_2, b_2]$. The mesh is formed by the nodes $y_k = a_2 + kh$ $(k = \overline{0, N}, y_0 = a_2, y_N = b_2; h = \dfrac{b_2 - a_2}{N})$ with $N = 2p + 1$ being an odd number, while the collocation nodes, as in the previous case, satisfy the conditions:

$$\xi_{2i} \in \left[y_{2i}, y_{2i+1}\right], \xi_{2i+1} \in \left[y_{2i}, y_{2i+1}\right], \tag{1.208}$$

where $\xi_{2i} = y_{2i} + t_1 h$, $\xi_{2i+1} = y_{2i} + t_2 h$ $(i = \overline{0, p})$, $t_1 = \dfrac{1}{2} - \dfrac{\sqrt{3}}{6}$ $t_2 = \dfrac{1}{2} + \dfrac{\sqrt{3}}{6}$ are the roots of the second-power Legendre polynomial on the segment [0, 1].

We will search the solution of Eq. (1.205) in the form of the series segment with respect to the linear combinations of the $(m + 1)$th-power B-splines with nodes on the mesh Δ which exactly satisfy

$$z(x, y) = \sum_{j=0}^{N} f_j(x)\psi^j_{m+1}(y). \tag{1.209}$$

We will require that the function (1.209) satisfies exactly Eq. (1.205) at the given points of collocation (1.208). As a result, we obtain a system of ordinary differential equations with respect to the functions f_j $(j = \overline{0, N})$ with appropriate boundary conditions, which allow for the condition (1.206). The same scheme will be employed later on solving plane and spatial boundary-value problems that deal with free electroelastic vibrations of inhomogeneous hollow finite-length cylinders with the spheres polarized in different directions of coordinate axes. The searched functions are approximated in the longitudinal direction by the segments of series with respect to the linear combinations of the B-splines, which by number exceed the order of senior derivatives with respect to the coordinate in whose direction the approximation proceeds. This one-dimensional eigenvalue problem is solved with the stable numerical discrete-orthogonalization method in combination with the incremental search method.

1.7.2 The Discrete-Orthogonalization Method in Combination with the Incremental Search

In contrast to [4], here is a variant of the discrete-orthogonalization method for solving edge increments of eigenvalues.

Consider the system of ordinary linear homogeneous differential equations with variable coefficients:

$$\frac{d\bar{y}}{dx} = \mathbf{A}(x, \lambda)\bar{y}\ (a \le x \le b)\,, \tag{1.210}$$

with homogeneous boundary conditions:

$$\mathbf{B}_1\ (\lambda)\ \bar{y}(a) = \bar{0}, \tag{1.211}$$

$$\mathbf{B}_2\ (\lambda)\ \bar{y}(b) = \bar{0}, \tag{1.212}$$

where $\bar{y} = [y_1, y_2, \ldots, y_n]^T$ is the column-vector of unknown functions; $\mathbf{A}(x, \lambda)$ is the square nth-order matrix; $\mathbf{B}_1\ (\lambda), \mathbf{B}_2\ (\lambda)$ are given rectangular $k \times n$ and $(n-k) \times n(k < n)$th-order matrices; and λ is the system parameter. Values of the parameter λ, at which the nonzero solutions of the system (1.210) satisfying the homogeneous boundary conditions (1.211) and (1.212) exist, are called eigenvalues (numbers) of a differential operator, while the appropriate solutions of the system are called eigenfunctions. The totality of all values λ forms a discrete operator spectrum. The problem consisting in finding the eigenvalues of λ and the appropriate eigenfunctions is called eigenvalue problem for a differential operator.

As is shown in [3, 4], the application of the discrete-orthogonalization method makes it possible to obtain a stable computational process due to orthogonalization of the vector-solutions of the Cauchy problems at the finite number of points of the interval within which the argument varies.

To find the eigenvalues in the problems (1.210)–(1.212), we employ the forward discrete-orthogonalization method in combination with the incremental search method [3, 4].

For determining the eigenvalues, we will use the following algorithm:

Let us give for the parameter λ some initial approximation and a step with which it varies: $\lambda = \lambda_0$ and $h\ (\lambda)$.

Calculate the matrices of the boundary conditions $\mathbf{B}_1\ (\lambda)$ and $\mathbf{B}_2\ (\lambda)$.

Perform the forward stroke of the discrete-orthogonalization method with the parameter λ being given.

To do this, we present the searched solution of problems (1.210)–(1.212) as follows:

$$\bar{y}(x) = \sum_{j=1}^{m} c_j \bar{y}_j(x) \tag{1.213}$$

or

$$\bar{y} = Y^T \bar{C}, \tag{1.214}$$

where $Y^T(x) = \left[\bar{y}_1(x)\ \bar{y}_2(x)\ ...\ \bar{y}_m(x)\right]$, $m = \min(k, n-k)$ (for definiteness let $m = \min(k, n-k)$); $\bar{y}_j$, $j = \overline{1, m}$ is the solution of the Cauchy problem for system (1.210) and the boundary condition (1.211) at $x = a$, $\bar{C} = \{c_1, c_2, ..., c_m\}^T$. Let us divide the segment $a \le x \le b$ into parts by integration points x_s, $(s = \overline{0, N})$ between which we chose orthogonalization points $x_i (i = \overline{0, M})$. The choice of these points is governed by the necessity to enable accuracy to problem solving.

Assume that solutions m of the Cauchy problem at the points x_i, $(i = \overline{0, M})$ are found with certain numerical method (for example, with the Runge–Kutta method). Let us designate the solutions as $\bar{u}_r(x_i)(r = \overline{1, m})$. Next, we orthonormalize the vectors $\bar{u}_1(x_i)$, $\bar{u}_2(x_i)$, ..., $\bar{u}_m(x_i)$ and notate them in terms of $\bar{z}_1(x_i)$, $\bar{z}_2(x_i)$, ..., $\bar{z}_m(x_i)$). The vectors z are expressed in terms of the vectors au using the formulas:

$$\bar{z}_r = \frac{1}{\omega_{rr}} \left(\bar{u}_r - \sum_{j=1}^{r-1} \omega_{rj} \bar{z}_j \right) (r = \overline{1, m}), \tag{1.215}$$

where

$$\omega_{rr} = \sqrt{(\bar{u}_r, \bar{u}_r) - \sum_{j=1}^{r-1} \omega_{rj}^2}, \ \omega_{rj} = (\bar{u}_r, \bar{z}_j)(j < r, r = \overline{1, m}). \tag{1.216}$$

From here it follows:

$$\begin{bmatrix} \bar{u}_1(x_i) \\ \bar{u}_2(x_i) \\ ... \\ \bar{u}_m(x_i) \end{bmatrix} = \mathbf{\Omega}_i \begin{bmatrix} \bar{z}_1(x_i) \\ \bar{z}_2(x_i) \\ ... \\ \bar{z}_m(x_i) \end{bmatrix}, \tag{1.217}$$

or

$$\mathbf{U}_i^T = \mathbf{Z}_i^T \mathbf{\Omega}_i^T, \tag{1.218}$$

where

$$\mathbf{U}^T(x_i) = \mathbf{U}_i^T = \left[\bar{u}_1(x_i)\ \bar{u}_2(x_i)\ ...\ \bar{u}_m(x_i)\right],$$

$$\mathbf{Z}^T(x_i) = \mathbf{Z}_i^T = \left[\bar{z}_1(x_i)\ \bar{z}_2(x_i)\ ...\ \bar{z}_m(x_i)\right],$$

$$\mathbf{\Omega}(x_i) = \mathbf{\Omega}_i = \begin{bmatrix} \omega_{11}(x_i) & 0 & ... & 0 \\ \omega_{21}(x_i) & \omega_{22}(x_i) & ... & 0 \\ ... & ... & ... & ... \\ \omega_{m1}(x_i) & \omega_{m2}(x_i) & ... & \omega_{mm}(x_i) \end{bmatrix}. \tag{1.219}$$

The components of the vectors $\bar{z}_r(x_i)$ are initial values, which are required to find solutions to the Cauchy problems on the segment $[x_i, x_{i+1}]$.

Thus, the solution of system (1.219) on the interval $x_i < x < x_{i+1}$ can be presented as follows:

$$\bar{y}(x) = \sum_{j=1}^{m} c_j^{(i)} \bar{z}_j(x), \tag{1.220}$$

or by

$$\bar{y}(x) = \bar{y} = \mathbf{Z}^T \bar{C}^i. \tag{1.221}$$

The solution of system (1.210), which satisfies the boundary condition (1.211), at each point of orthogonalization x_i $(i = \overline{0, M})$ takes the form:

$$\bar{y}(x_i) = \bar{y}_i = \sum_{j=1}^{m} c_j^{(i)} \bar{z}_j(x_i) \tag{1.222}$$

or

$$\bar{y}(x_i) = \mathbf{Z}^T(x_i)\, \bar{C}^{(i)} \Leftrightarrow \bar{y}_i = \mathbf{Z}_i^T \bar{C}^{(i)}. \tag{1.223}$$

The unknown components $c_1^{(M)}, c_2^{(M)}, \ldots\ldots, c_m^{(M)}$ of the vector $\bar{c}^{(M)}$ at the point $x = x_N = x_M = b$ are calculated provided that the vector $\bar{y}(x_M)$ satisfies the boundary condition (1.212):

$$\mathbf{B}_2 \bar{y}_M = \bar{0} \Rightarrow \mathbf{B}_2 Z_M^T \bar{C}^{(M)} = \bar{0}. \tag{1.224}$$

As a result, we get the homogeneous linear system of algebraic equations with respect to the unknown constants $\bar{C}^M$ on the segment $\left[x_{M-1}, x_M\right]$. It is known that the homogeneous system of equations in the form (1.224) has nontrivial solutions only in the case when the determinant of the system is equal to zero. When solving the numerical problems (1.213) and (1.214) and by discrete changing of the parameter λ, it is impossible to establish accurate equality to zero of the determinant of system (1.214) for the given parameter λ. To determine and localize the search eigenvalue of system (1.210), we will fixate two sequential values λ_k and λ_{k+1}, at which the determinant of system (1.214) has opposite signs. The approximate value of the eigennumber $\tilde{\lambda}$ can be defined with the given accuracy using the methods of bisection, chords, etc.(but with the required forward run by the method of discrete orthogonalization).

If the determinants of system (1.214) for two sequential values λ_k and λ_{k+1} have the same signs, then increasing λ_{k+1} by the value of the step $h(\lambda)$, remembering the previous value λ_k, we will return to the onset of the algorithm outlined until the predesigned amount of the eigenvalues of problems (1.210)–(1.212) becomes defined.

To define the eigenform corresponding to the defined eigenvalue, we will use the forward and backward strokes of the discrete-orthogonalization methods.

The algorithm of search of the eigenform is as follows:

(i) Using the Gauss method and choosing the main element, we define the determinant of system (1.224). At the same time we fixate indexes p and q corresponding by modulus to element of the matrix $\mathbf{B}_2\mathbf{Z}_M^T$.

(ii) Let for the parameter $\bar{C}^{(M)}$ in the column-vector be $c_q^{(M)} = 1$ (instead of 1 it may be any number).

(iii) In the matrix $\mathbf{B}_2\mathbf{Z}_M^T$, the row with index p is eliminated.

(iv) The column with index q is carried out on the right-hand side of equality (1.224) (with the exception of the element notated by the indexes p and q.

(v) The linear inhomogeneous system of the $(m-1)$th-order algebraic equations is solved with the Gauss method. The vector of the constant $\bar{C}^{(M)}$ at the point $x_M = b$ can be found with accuracy to arbitrary constant.

(vi) System (1.223) is solved on the interval $x_i \le x \le x_{i+1}$ by formulas (1.220) and (1.221). Requiring that the solving be continuous at the point x_i, we get: $\bar{y}(x_i - 0) = \bar{y}(x_i + 0) \Rightarrow \mathbf{Y}_i^T\bar{C}^{(i-1)} = \mathbf{Z}_i^T\bar{C}^{(i)}$. With (1.217) and (1.218) being taken into account, we obtain:

$$\mathbf{\Omega}_i^T\bar{C}^{(i-1)} = \bar{C}^{(i)} i = M, M-1, \ldots, 1 \tag{1.225}$$

Since the matrix $\mathbf{\Omega}_i^T$ represents an upper triangle matrix, system (1.225) can be solved having conducted only backward stroke with the Gauss method.

The value of the eigenfunction at some point can be determined by the formulas (1.218) and (1.219). It should be noted that in solving the problems, the solution must be found only at certain points, which are the points of the separation of the results. The number of these points is considerably smaller than the number of those in orthogonalization. At the same time, it is necessary to keep the matrix $\mathbf{Z}_i^T$ at points of the separation of the results, and for the matrices $\mathbf{\Omega}_i^T$ to keep their products instead. By using (1.225), we get a recurrent formula needed to define constants of integration on spaces between points of separation

$$\bar{C}^{(i-1)} = \left[\prod_{j=1}^{p}\mathbf{\Omega}_{i+j}^T\right]^{-1}\bar{C}^{(i+p)}, \tag{1.226}$$

where $p = \overline{1, M-j}$.

This technique makes it possible to define with the predetermined accuracy any number of eigenfunctions of problems (1.210)–(1.212) by constructing a stable computational process. To do this, the choice of the initial approximation for λ and step $h(\lambda)$ must be unobjectionable.

References

1. Alberg JH, Nielson E, Walsh J (1967) Theory of splines and their applications. Academic Press, New York
2. Cady WG (2018) Piezoelectricity. Volume two: an introduction to the theory and applications of electromechanical phenomena in crystals. Dover Publications, New York, USA
3. Grigorenko YM, Grigorenko AY (2013) Static and dynamic problems for anisotropicinhomogeneous shells with variable parameters and their numerical solution (review). Int Appl Mech 49(2):123–193
4. Grigorenko AY, Müller WH, Grigorenko YM, Vlaikov GG (2016) Recent developments in anisotropic heterogeneous shell theory. General theory and applications of classical theory, vol. I. Springer, Berlin
5. Jeffreys H (1931) Cartesian tensors. Cambridge University Press, Cambridge
6. Lagally M (1945) Vorlesungen über Vektor-Rechnung. Verlag, Leipzig, Becker and Erler
7. Landau LD, Lifshitz EM (1960) Electrodynamics of continuous media (Volume 8 of a course of theoretical physics) Pergamon Press
8. Love AEH (1927) A treatise on the mathematical theory of elasticity. University Press
9. Mason WP (1950) Piezoelectric crystals and their application to ultrasonics. New York
10. Madelung E (1943) Die mathematischen hilfsmittel des physikers (Mathematical tools for the physicist) Published by Dover Publications
11. Mc Connell AJ (1957) Application of tensor analysis. Dover Publications Inc, New York
12. Panofsky WKH, Philips M (2012) Classical electricity and magnetism, 2nd edn. Addison-Wasley Publishing Company Inc, Reading, Massachusetts USA, London, England
13. Royer D, Dieulesaint E (2000) Elastic waves in solids I: free and guided propagation. Springer, Berlin
14. Tiersten HF (1968) Linear piezoelectric plate vibrations. Plenum press, New York
15. Timoshenko SP, Goodier JN (1951) Theory of elasticity. McGraw-Hill Book Company
16. Voigt W (1910) Lerhbuch der Krisstallphysics. Berlin, Teubner, Leipzig

Chapter 2
Free Axisymmetric and Nonaxisymmetric Vibrations of Hollow Homogeneous and Inhomogeneous Piezoceramic Cylinders of Finite Length with Different Polarization

Abstract Various approaches to the solution of linear problems of elasticity and electroelasticity of anisotropic inhomogeneous finite-length cylinders based on discrete–continuous methods and three-dimensional formulations are presented. The advantage of these method consists in the reduction of the partial differential equations of the considered problems to the associated one-dimensional problems (the spline-collocation method) and exact satisfaction of the boundary conditions. The approach leads to practically exact solutions of boundary-value problems and eigenvalue problems described by a system of ordinary differential equations with variable coefficients (by the discrete-orthogonalization method). Axisymmetric and nonaxisymmetric problems of natural vibrations of hollow inhomogeneous elastic cylinders and cylinders with piezoelectric properties based on 3D elasticity and electroelasticity are considered. The properties of the material vary along a radial coordinate. We consider two types of inhomogeneous materials: when the properties of the material are piecewise constant (layered structures with metal and dielectric layers) and when they vary continuously (functionally gradient and functionally gradient piezoelectric materials FGM and FGPM). The external surface of the cylinder is free of tractions and either insulated or short-circuited by electrodes. After separation of variables and representation of the components of the mechanical displacement vector and electric potential in the form of standing circumferential waves, the initially three-dimensional problem is reduced to a two-dimensional partial differential equation problem. By using the method of spline-collocations with respect to the longitudinal coordinate, this two-dimensional problem is reduced to a one-dimensional eigenvalue problem (described by ordinary differential equations). This problem is solved by the stable discrete-orthogonalization technique in combination with a step-by-step search method with respect to the radial coordinate. A nontraditional approach to solving problems of the above class is proposed. Different variants of polarized piezoceramic materials are considered. The effect of variation in mechanical and electrical parameters through thickness, and the influence of boundary conditions on natural frequencies and vibration modes of the finite-length cylinders with inhomo-

A. Ya. Grigorenko et al., *Selected Problems in the Elastodynamics of Piezoceramic Bodies*, Advanced Structured Materials 154,
https://doi.org/10.1007/978-3-030-74199-0_2

geneous elastic and piezoelectric properties is analyzed. Significant attention is paid to the validation of the reliability of the results obtained by numerical calculations.

Keywords 3D electroelasticity theory · Piezoceramic cylinder · Finite length · Free vibrations · Analytical–numerical method

2.1 Introduction

Cylindrically-shaped piezoceramic active elements, which are elements of functional electronics, are extensively used in radioelectronics, automatic and computer equipment, and measuring devices with the use of nanotechnologies. The fast progress of this relatively novel scientific and technical trend caused the development of highly efficient piezoelectric sources of high and low voltage, multipurpose piezodrivers with translational, rotational, and combined types of motion, piezoceramic matrices, and storage devices. The high efficiency of transformation of electric energy to mechanical energy and vice versa in combination with extremely small sizes makes it possible to use such materials in nanotechnologies. One important aspect of providing efficient operating conditions of the indicated bodies is the acquisition of information on the characteristics of their free vibrations. The high-strength-assessment requirement, the attempts to take complete account of the real properties of structural materials, and the detection and study of three-dimensional effects that take place on thick-walled elements lead to the necessity to perform calculations based on the theory of electroelasticity (3D model). The solution of dynamic problems for thick-walled elements as spatial problems of the theory of elasticity is connected with substantial difficulties caused by the complexity of the system of initial differential equations in partial derivatives and the necessity to satisfy the boundary conditions on surfaces that confine the body. These difficulties substantially increase provided that the fields (the electric and mechanic fields) are bound and that the piezoceramic materials are anisotropic. It should be noted that, in the scientific literature, only individual works on vibrations of piezoceramic cylinders of finite length, performed within the framework of the three-dimensional theory of elasticity [1, 2, 8–10, 14, 16] are known. Along with homogeneous piezoceramic cylinders, piezoceramic cylinders made of the continuously heterogeneous ceramics in the book based on the analytical–numerical technique proposed by the authors are considered [3, 12, 13, 17]. A solution with a high degree of accuracy was obtained for problems of free axisymmetric and nonaxisymmetric vibrations of homogeneous and inhomogeneous hollow piezoceramic cylinders of finite length with different polarization directions of the piezoceramics, and the effect of inhomogeneity factors and associated electric field on the behavior of dynamic characteristics of the corresponding cylinders was analyzed.

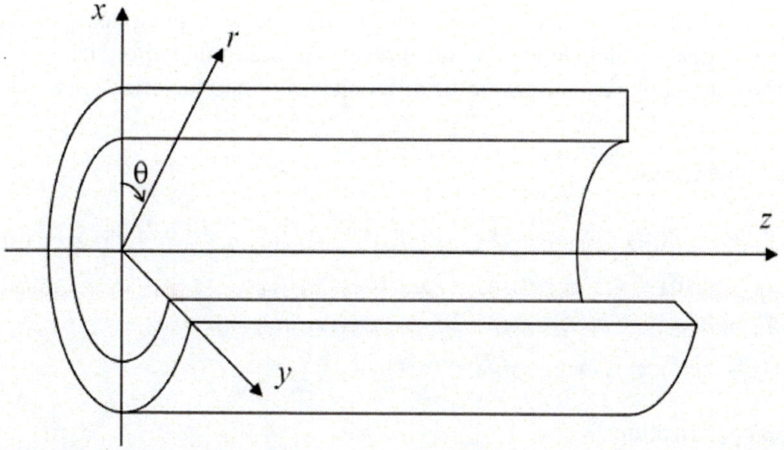

Fig. 2.1 Cylindrical coordinate system

2.2 Basic Relations

The closed system of equations describing the nonaxisymmetric free vibrations of the piezoceramic cylinder of the finite length in cylindrical coordinates system $(r,\ \theta,\ z)$ consists of (1.133), (1.135), (1.136) (Fig. 2.1):

Equations of motion are

$$\begin{aligned}
&\frac{\partial T_1}{\partial r}+\frac{1}{r}\left(\frac{\partial T_6}{\partial \theta}+T_1-T_2\right)+\frac{\partial T_5}{\partial z}=\rho\frac{\partial^2 u_1}{\partial t^2},\\
&\frac{\partial T_6}{\partial r}+\frac{1}{r}\left(\frac{\partial T_2}{\partial \theta}+2T_6\right)+\frac{\partial T_4}{\partial z}=\rho\frac{\partial^2 u_2}{\partial t^2},\\
&\frac{\partial T_5}{\partial r}+\frac{1}{r}\left(\frac{\partial T_4}{\partial \theta}+T_5\right)+\frac{\partial T_3}{\partial z}=\rho\frac{\partial^2 u_3}{\partial t^2};
\end{aligned}\tag{2.1}$$

Equations of electrostatics are

$$E_1=-\frac{\partial \Phi}{\partial r},\ E_2=-\frac{1}{r}\frac{\partial \Phi}{\partial \theta},\ E_3=-\frac{\partial \Phi}{\partial z},\ \frac{\partial D_1}{\partial r}+\frac{1}{r}\left(\frac{\partial D_2}{\partial \theta}+D_1\right)+\frac{\partial D_3}{\partial z}=0;\tag{2.2}$$

Kinematic relations are

$$\begin{aligned}
&S_1=-\frac{\partial u_1}{\partial r},\ S_2=\frac{\partial u_2}{\partial \theta}+u_1 r,\ S_3=-\frac{\partial u_3}{\partial z},\\
&2S_4=\frac{\partial u_2}{\partial z}+\frac{\partial u_3}{\partial \theta},\ 2S_5=\frac{\partial u_3}{\partial r}+\frac{\partial u_1}{\partial z},\ 2S_6=\frac{\partial u_2}{\partial r}+\frac{\partial u_1}{\partial \theta}-\frac{2}{r}u_2.
\end{aligned}\tag{2.3}$$

The presented relations should be supplemented with polarized piezoceramics state equations, which depend on the direction of preliminary polarization.

The constitutive equations for the different preliminary polarization are described as

Axial polarization:

$$\begin{aligned}
&T_1 = c_{11}S_1 + c_{12}S_2 + c_{13}S_3 - e_{13}E_3,\, T_2 = c_{12}S_1 + c_{11}S_2 + c_{13}S_3 - e_{31}E_3,\\
&T_3 = c_{13}S_1 + c_{13}S_2 + c_{33}S_3 - e_{33}E_3,\, T_6^i = 2c_{66}^iS_6^i,\, T_5 = 2c_{55}S_5 - e_{15}E_1,\\
&T_4 = 2c_{55}S_4 - e_{15}E_2,\, D_1 = 2e_{15}S_5 + \varepsilon_{11}E_1,\, D_2 = 2e_{15}S_4 + \varepsilon_{11}E_2,\\
&D_3 = e_{13}S_1 + e_{13}S_2 + e_{33}S_3 + \varepsilon_{33}E_3.
\end{aligned} \tag{2.4}$$

Radial polarization:

$$\begin{aligned}
&T_1 = c_{33}S_1 + c_{13}S_2 + c_{13}S_3 - e_{33}E_1,\, T_2 = c_{13}S_1 + c_{11}S_2 + c_{12}S_3 - e_{31}E_1,\\
&T_3 = c_{13}S_1 + c_{12}S_2 + c_{11}S_3 - e_{31}E_1,\, T_6 = 2c_{55}S_6^i - e_{15}E_2,\\
&T_5 = 2c_{55}S_5 - e_{51}E_3;\, T_4 = 2c_{66}S_4,\, D_1 = e_{33}S_1 + e_{31}S_2 + e_{31}S_3 + \varepsilon_{33}E_1,\\
&D_2 = 2e_{15}S_4 + \varepsilon_{11}E_2,\, D_3 = 2e_{51}S_5 + \varepsilon_{33}E_3;
\end{aligned} \tag{2.5}$$

Circumferential polarization:

$$\begin{aligned}
&T_1 = c_{11}S_1 + c_{13}S_2 + c_{12}S_3 - e_{13}E_2,\, T_2 = c_{13}S_1 + c_{33}S_2 + c_{13}S_3 - e_{33}E_2,\\
&T_3 = c_{12}S_1 + c_{13}S_2 + c_{11}S_3 - e_{13}E_2,\, T_6 = 2c_{55}S_6 - e_{15}E_1,\\
&T_5 = 2c_{66}S_5;\, T_4 = 2c_{55}S_4 - e_{15}E_3,\, D_1 = 2e_{15}S_6 + \varepsilon_{11}E_1,\\
&D_2 = e_{13}S_1 + e_{33}S_2 + e_{13}S_3 + \varepsilon_{33}E_2,\, D_3 = 2e_{15}S_4 + \varepsilon_{11}E_3.
\end{aligned} \tag{2.6}$$

All physical–mechanical moduli will be functions of the radial coordinate in the considered case of a continuously inhomogeneous piezoceramic material of the cylinder. The lateral surfaces of the cylinder (at $r = R_0 \pm h$) are free of external forces ($T_1 = T_6 = T_5 = 0$) and covered with short-circuited electrodes ($\Phi = 0$).

The resolving system of equations will depend on the direction of preliminary polarization of ceramics. Let us consider problems for different directions of the preliminary polarization of the piezoceramics. We will perform the same transformations of the basic relations (2.1)–(2.6). Let us present the resolving systems of the partial differential equations in the general case with variable coefficients.

2.3 Axisymmetric Problem

2.3.1 *Basic Relations and Resolving Systems*

When studying an axisymmetric problem ($u_2 = 0,\ \partial u_1/\partial\theta = \partial u_3/\partial\theta = 0$) about the longitudinal harmonic vibrations of the piezoceramic cylinder ($u_i\,(r,\theta,z,t) = u_i\,(r,\theta,z)\,e^{i\omega t}$, $\Phi\,(r,\theta,z,t) = \Phi_i\,(r,\theta,z)\,e^{i\omega t}$, $(i = 1, 2)$), the basic relationships (2.1)–(2.6) take the following form:

Equations of motion are

$$\frac{\partial T_1}{\partial r} + \frac{1}{r}(T_1 - T_2) + \frac{\partial T_5}{\partial z} + \rho\omega^2\tilde{u}_1 = 0,\ \frac{\partial T_5}{\partial r} + \frac{1}{r}T_5 + \frac{\partial T_3}{\partial z} + \rho\omega^2\tilde{u}_3 = 0; \tag{2.7}$$

Equations of electrostatics are

$$\frac{\partial D_1}{\partial r} + \frac{1}{r}D_1 + \frac{\partial D_3}{\partial z} = 0,\ E_r = -\frac{\partial\tilde{\Phi}}{\partial r},\ E_z = -\frac{\partial\tilde{\Phi}}{\partial z}; \tag{2.8}$$

Kinematic relations are

$$S_1 = \frac{\partial\tilde{u}_1}{\partial r},\ S_2 = \frac{1}{r}\tilde{u}_1,\ S_3 = \frac{\partial\tilde{u}_3}{\partial z},\ S_5 = \frac{\partial\tilde{u}_3}{\partial r} + \frac{\partial\tilde{u}_1}{\partial z}. \tag{2.9}$$

The constitutive equations for the different directions of the preliminary polarization of the piezoceramics assume the form
Axial polarization:

$$\begin{aligned}
&T_1 = c_{11}S_1 + c_{12}S_2 + c_{13}S_3 - e_{13}E_3,\ T_2 = c_{12}S_1 + c_{11}S_2 + c_{13}S_3 - e_{13}E_3,\\
&T_3 = c_{13}S_1 + c_{13}S_2 + c_{33}S_3 - e_{33}E_3,\ T_5 = 2c_{55}S_5 - e_{15}E_1,\\
&D_1 = 2e_{15}S_5 + S_{11}E_1,\ D_3 = e_{13}S_1 + e_{13}S_2 + e_{33}S_3 + S_{33}E_3;
\end{aligned} \tag{2.10}$$

Radial polarization:

$$\begin{aligned}
&T_1 = c_{33}S_1 + c_{13}S_2 + c_{13}S_3 - e_{33}E_1,\ T_2 = c_{13}S_1 + c_{11}S_2 + c_{12}S_3 - e_{31}E_1,\\
&T_3 = c_{13}S_1 + c_{12}S_2 + c_{11}S_3 - e_{31}E_1,\ T_5 = 2c_{55}S_5 - e_{51}E_3,\\
&D_1 = e_{33}S_1 + e_{31}S_2 + e_{31}S_3 + \varepsilon_{33}E_1,\ D_3 = 2e_{51}S_5 + \varepsilon_{33}E_3.
\end{aligned} \tag{2.11}$$

We set the following boundary conditions on the surfaces of the cylinder ($r = R_0 \pm h$):

the lateral surfaces $r = R_0 + h = R_+$ and $r = R_0 - h = R_-$are free from external forces, i.e.,: $T_1 = T_5 = 0$, and covered with short-circuited thin electrode, i.e.,$\Phi = 0$.

The different boundary conditions at the ends of the cylinder ($z = \pm L/2$) are:

(i) both end faces are simply supported ($u_1 = 0;\ T_3 = 0$) and are free from electrodes $D_3 = 0$;

(ii) both end faces are rigidly fixed ($\tilde{u}_r = 0,\ \tilde{u}_z = 0$) and covered with short-circuited electrodes ($\Phi = 0$).

The resolving system depends on preliminary polarization of piezoceramics. First, consider the case of the axial polarization of piezoceramics. Need to the basic relations resolved allow relative to $\partial^2\tilde{\Phi}/\partial r^2$, $\partial^2\tilde{u}_1/\partial r^2$, $\partial^2\tilde{u}_3/\partial r^2$.

After the same transformation (2.7)–(2.10), we get the resolving system of the partial differential equations in the general case with variable coefficients.

Axial polarization:

$$\begin{aligned}
\frac{\partial^2\tilde{u}_1}{\partial r^2} &= \frac{1}{\tilde{c}_{11}}\left(\frac{\tilde{c}_{11}}{r^2} - \Omega^2\right)\tilde{u}_1 - \frac{\tilde{c}_{55}}{\tilde{c}_{11}}\frac{\partial^2\tilde{u}_1}{\partial z^2} - \frac{1}{r}\frac{\partial\tilde{u}_1}{\partial r} - \frac{\tilde{c}_{55}}{\tilde{c}_{11}}\frac{\partial^2\tilde{u}_3}{\partial z^2} - \\
&\quad - \frac{\tilde{c}_{13}+\tilde{c}_{55}}{\tilde{c}_{11}}\frac{\partial^2\tilde{u}_3}{\partial r\partial z} - \frac{\tilde{e}_{13}+\tilde{e}_{15}}{\tilde{c}_{11}}\frac{\partial^2\tilde{\Phi}}{\partial r\partial z}, \\
\frac{\partial^2\tilde{u}_3}{\partial r^2} &= -\left(1+\frac{\delta_1}{\delta}\right)\frac{1}{r}\frac{\partial\tilde{u}_1}{\partial z} - \left(1+\frac{\delta_1}{\delta}\right)\frac{\partial^2\tilde{u}_1}{\partial r\partial z} - \frac{\tilde{\varepsilon}_{11}\Omega^2}{\delta}\tilde{u}_3 - \\
&\quad - \frac{\delta_2}{\delta}\frac{\partial^2\tilde{u}_3}{\partial z^2} - \frac{1}{r}\frac{\partial\tilde{u}_3}{\partial r} - \frac{\delta_3}{\delta}\frac{\partial^2\tilde{\Phi}}{\partial z^2}, \\
\frac{\partial^2\tilde{\Phi}}{\partial r^2} &= -\frac{\delta_4}{\delta}\frac{\partial^2\tilde{u}_1}{\partial r\partial z} - \frac{\delta_4}{\delta}\frac{1}{r}\frac{\partial\tilde{u}_1}{\partial z} - \frac{e_{15}}{\delta}\Omega^2\tilde{u}_3 + \frac{\delta_5}{\delta}\frac{\partial^2\tilde{u}_3}{\partial z^2} - \\
&\quad - \frac{1}{r}\frac{\partial\tilde{\Phi}}{\partial r} - \frac{\tilde{c}_{55}\tilde{\varepsilon}_{33}}{\delta}\frac{\partial^2\tilde{\Phi}}{\partial z^2};
\end{aligned} \tag{2.12}$$

where

$$\begin{aligned}
&\delta = c_{55}\varepsilon_{11} + e_{15}^2,\ \delta_1 = c_{13}\varepsilon_{11} + e_{13}e_{15}, \\
&\delta_2 = c_{33}\varepsilon_{11} + e_{15}e_{33},\ \delta_3 = e_{15}\varepsilon_{33} - e_{33}\varepsilon_{11}, \\
&\delta_4 = c_{13}e_{15} - c_{55}e_{13},\ \delta_5 = c_{55}e_{33} - c_{33}e_{15},
\end{aligned} \tag{2.13}$$

And dimensionless values:

$$\varepsilon = \frac{h}{R_0},\ \Omega = \omega h\sqrt{\frac{\rho}{\lambda}},\ \tilde{c}_{ij} = \frac{c_{ij}}{\lambda},\ \tilde{e}_{ij} = \frac{e_{ij}}{\sqrt{\varepsilon_0\lambda}},\ \tilde{\varepsilon}_{ij} = \frac{\varepsilon_{ij}}{\varepsilon_0}. \tag{2.14}$$

Here, R_0—is the radius of the middle surface of the cylinder; h— is the half-thickness of the cylinder; ω—circular frequency; $\lambda = 10^{10}$ Pa, ε_0—is the vacuum permittivity.

In the case of the radial polarization of piezoceramics, after set same transformation the basic relation (2.7)–(2.9), (2.11) have the form:

radial polarization:

$$\frac{\partial^2\tilde{u}_r}{\partial r^2}=\left(\frac{\tilde{c}_{11}}{r^2}-\Omega^2\right)\frac{\tilde{\varepsilon}_{33}}{\delta}\tilde{u}_1-\frac{\delta_1}{\delta}\frac{\partial^2\tilde{u}_1}{\partial z^2}-\left(1+\frac{\tilde{e}_{13}\tilde{e}_{33}}{\delta}\right)\frac{1}{r}\frac{\partial\tilde{u}_1}{\partial r}-$$
$$-\frac{\delta_2-\tilde{c}_{12}\tilde{\varepsilon}_{33}}{\delta}\frac{1}{r}\frac{\partial\tilde{u}_3}{\partial z}-\frac{\delta_1+\delta_2}{\delta}\frac{1}{r}\frac{\partial^2\tilde{u}_3}{\partial r\partial z}-\frac{\delta_3}{\delta}\frac{\partial^2\tilde{\Phi}}{\partial z^2}+\frac{\tilde{e}_{13}\tilde{\varepsilon}_{33}}{\delta}\frac{1}{r}\frac{\partial\tilde{\Phi}}{\partial r},$$
$$\frac{\partial^2\tilde{u}_3}{\partial r^2}=-\left(1+\frac{\tilde{c}_{12}}{\tilde{c}_{55}}\right)\frac{1}{r}\frac{\partial\tilde{u}_1}{\partial z}-\left(1+\frac{\tilde{c}_{13}}{\tilde{c}_{55}}\right)\frac{\partial^2\tilde{u}_1}{\partial r\,\partial z}-\frac{\Omega^2}{\tilde{c}_{55}}u_3-\frac{\tilde{c}_{11}}{\tilde{c}_{55}}\frac{\partial^2\tilde{u}_3}{\partial z^2}-$$
$$-\frac{1}{r}\frac{\partial\tilde{u}_3}{\partial r}-\frac{\tilde{e}_{15}}{\tilde{c}_{55}}\frac{1}{r}\frac{\partial\tilde{\Phi}}{\partial z}-\frac{\tilde{e}_{13}+\tilde{e}_{15}}{\tilde{c}_{55}}\frac{\partial^2\tilde{\Phi}}{\partial r\,\partial z},\tag{2.15}$$
$$\frac{\partial^2\tilde{\Phi}}{\partial r^2}=\left(\frac{\tilde{c}_{11}}{r^2}-\Omega^2\right)\frac{\tilde{e}_{33}}{\delta}\tilde{u}_1-\frac{\delta_4}{\delta}\frac{\partial^2\tilde{u}_1}{\partial z^2}+\frac{\tilde{c}_{33}\tilde{e}_{13}}{\delta}\frac{1}{r}\frac{\partial\tilde{u}_1}{\partial r}-\frac{\delta_5-\tilde{c}_{12}\tilde{e}_{33}}{\delta}\frac{1}{r}\frac{\partial\tilde{u}_3}{\partial z}-$$
$$-\frac{\delta_4+\delta_5}{\delta}\frac{\partial^2\tilde{u}_3}{\partial r\partial z}-\frac{\delta_6}{\delta}\frac{\partial^2\Phi}{\partial z^2}-\left(1-\frac{\tilde{e}_{13}\tilde{e}_{33}}{\delta}\right)\frac{1}{r}\frac{\partial\tilde{\Phi}}{\partial r};$$

Here,

$$\delta=\tilde{c}_{33}\tilde{\varepsilon}_{33}+\tilde{e}_{33}^2,\ \delta_1=\tilde{c}_{55}\tilde{\varepsilon}_{33}+\tilde{e}_{15}\tilde{e}_{33},\ \delta_2=\tilde{c}_{13}\tilde{\varepsilon}_{33}+\tilde{e}_{13}\tilde{e}_{33},$$
$$\delta_3=\tilde{e}_{33}\tilde{\varepsilon}_{11}-\tilde{e}_{15}\varepsilon_{33},\ \delta_4=\tilde{c}_{55}\tilde{e}_{33}-\tilde{c}_{33}\tilde{e}_{15},\tag{2.16}$$
$$\delta_5=\tilde{c}_{13}\tilde{e}_{33}-\tilde{c}_{33}\tilde{e}_{13},\ \delta_6=\tilde{c}_{33}\tilde{\varepsilon}_{11}+\tilde{e}_{15}\tilde{e}_{33}.$$

2.3.2 *Solution of the Problem*

To reduce the differential equations in the partial derivatives (2.12) and (2.15) to systems of the ordinary differential equations, we will use the spline-collocation method [5, 7]. We will now search for functions $\tilde{u}_1(r,z)$, $\tilde{u}_3(r,z)$, $\Phi(r,z)$ of the form:

for the case of axial polarization of a piezoceramic:

$$\tilde{u}_1(r,z)=\sum_{i=0}^{N}u_i(x)\,\psi_{1i}(z),\ \tilde{u}_3(r,z)=\sum_{i=0}^{N}w_i(x)\,\psi_{2i}(z),$$
$$\Phi(r,z)=\sum_{i=0}^{N}\Phi_i(x)\,\psi_{2i}(z);\tag{2.17}$$

For the case of radial polarization of a piezoceramic:

$$\tilde{u}_1(r,z)=\sum_{i=0}^{N}u_i(x)\,\psi_{1i}(z);\ \tilde{u}_3(r,z)=\sum_{i=0}^{N}w_i(x)\,\psi_{2i}(z),$$
$$\Phi(r,z)=\sum_{i=0}^{N}\Phi_i(x)\,\psi_{1i}(z).\tag{2.18}$$

where $x = (r - R_0)/h, u_i(x), v_i(x), w_i(x)$, the unknown functions with respect to the variable x, $\psi_{ji}(z)(j = 1, 2; i = 0, 1, \ldots, N)$, are linear combinations of B-splines on the uniform mesh $\Delta : -L/2 = z_0 < z_1 < \cdots < z_n = L/2$. By considering the boundary conditions on the ends of the cylinder (at $z = -L/2$ and at $z = L/2$), we can see that the system involves derivatives of the solution vector that are not higher than second order. Therefore, we can restrict ourselves to an approximation by third-order spline-functions:

$$B_3^i = \frac{1}{6}\begin{cases} 0; & -\infty < z < z_{i-2} \\ \tilde{z}^3; & \tilde{z}_{i-2} \le \tilde{z} < \tilde{z}_{i-1} \\ -3\tilde{z}^2 + 3\tilde{z}^2 + 3\tilde{z} + 1; & z_{i-1} \le z < z_i \\ 3\tilde{z}^3 - 6\tilde{z}^2 + 4; & z_i \le z < z_{i+1} \\ (1-\tilde{z})^3; & z_{i+1} \le z < z_{i+2} \\ 0; & \tilde{z}_{i+2} \le z < \infty, \end{cases} \tag{2.19}$$

where $\tilde{z} = \dfrac{(z - z_k)}{h_z}$ on the interval $[z_k; z_{k+1}]$,. $k = \overline{i-2, i+1}; i = \overline{-1, N+1}; h_z = z_{k+1} - z_k = \text{const}$.

At the same time, the functions ψ_{ji} will be written in this way:

if the corresponding vector component of the solution is equal to zero, then:

$$\psi_{j0}(z) = -4B_3^{-1}(z) + B_3^0(z),\ \psi_{j1}(z) = -4B_3^{-1}(z) - \frac{1}{2}B_3^0(z) + B_3^1(z),$$
$$\psi_{ji}(z) = B_3^i(z),\ \left(i = \overline{2, N-2}\right). \tag{2.20}$$

If the derivative of the corresponding vector component of the solution with respect to z is equal to zero, then:

$$\psi_{j0}(z) = B_3^0(z),\ \psi_{j1}(z) = B_3^{-1}(z) - \frac{1}{2}B_3^0(z) + B_3^1(z),$$
$$\psi_{ji}(z) = B_3^i(z),\ \left(i = \overline{2, N-2}\right). \tag{2.21}$$

Analogous formulas hold for $\psi_{jN-1}(z)$ and $\psi_{jN}(z)$.

Let us substitute the solution (2.17) in Eq. (2.18) and demand that they be satisfied at the set of collocation points $\xi_k \in [-L/2,\ L/2]$, $k = 0, \ldots, N$. We consider the case where the number of nodes of the mesh is even, i.e., $N = 2n + 1$ $(n \ge 3)$ (is taken into account z_0), at the collocation points we choose in the following way $\xi_{2i} \in [z_{2i}, z_{2i+1}]$, $\xi_{2i=1} \in [z_{2i}, z_{2i+1}]$, $\left(i = \overline{1, n}\right)$. Then, on the segment $[z_{2i}, z_{2i+1}]$, we have two collocation nodes and no collocation node on the neighboring $[z_{2i+1}, z_{2i+2}]$ segments. In each of the segments $[z_{2i+1}, z_{2i+2}]$, we choose collocation points in the following way: $\xi_{2i} = z_{2i} + w_1 h_z$, $\xi_{2i+1} = z_{2i} + w_2 h_z$, $i = 1, \ldots, n$, where w_1 and w_2 are the roots of the Legendre polynomial of the second order on the segment $[0, 1]$, which are equal to $w_1 = 1/2 - \sqrt{3}/6$, $w_2 = 1/2 + \sqrt{3}/6$. This choice of the collocation points is optimal and substantially increases the degree of accuracy of the approximation.

If we introduce the notation:

$$\begin{aligned}
\Psi_j &= \left[\psi_{ji}\left(\xi_k\right)\right],\ k,i=\overline{0,N},\ j=1,2,\\
u &= \left[u_0,u_1,\ldots,u_N\right]^{\mathrm{T}},\ \ v=\left[v_0,v_1,\ldots,v_N\right]^{\mathrm{T}},\\
w &= \left[w_0,w_1,\ldots,w_N\right]^{\mathrm{T}},\ (k,l)\in\left\{(k,l)\,|\,k,l=\overline{1,6}\right\},\\
a_{kl}^T &= \left\{a_{kl}\left(x,\xi_0,\Omega^2\right),a_{kl}\left(x,\xi_1,\Omega^2\right),\ldots,a_{kl}\left(x,\xi_N,\Omega^2\right)\right\},
\end{aligned}$$

then the system (2.12) is transformed in the following system linear differential equations for the functions u, $\tilde{u}$, w, $\tilde{w}$, Φ, $\tilde{\Phi}$:

$$\begin{aligned}
&\frac{du}{dx}=\tilde{u},\ \frac{dw}{dx}=\tilde{w},\ \frac{d\Phi}{dx}=\tilde{\Phi},\\
&\frac{d\tilde{u}}{dx}=\Psi_1^{-1}\left[\left(a_{11}\Psi_1+a_{12}\Psi_1''\right)u+a_{13}\Psi_1\tilde{u}+\right.\\
&\qquad\left.+\,a_{14}\Psi_2''w+a_{15}\Psi_2'\tilde{w}+a_{16}\Psi_2'\tilde{\Phi}\right],\\
&\frac{d\tilde{w}}{dx}=\Psi_2^{-1}\left[a_{21}\Psi_1'u+a_{22}\Psi_1'\tilde{u}+\left(a_{23}\Psi_2+a_{24}\Psi_2''\right)w+\right.\\
&\qquad\left.+\,a_{25}\Psi_2\tilde{w}+a_{26}\Psi_2''\Phi\right],\\
&\frac{d\tilde{\Phi}}{dx}=\Psi_2^{-1}\left[a_{31}\Psi_1'u+a_{32}\Psi_1'\tilde{u}+\left(a_{33}\Psi_2+a_{34}\Psi_2''\right)w+\right.\\
&\qquad\left.+\,a_{35}\Psi_2''\Phi+a_{36}\Psi_2\tilde{\Phi}\right],
\end{aligned}\tag{2.22}$$

where

$$\begin{aligned}
&a_{11}=\frac{1}{x^2}-\frac{\Omega^2}{\tilde{A}_{11}},\ a_{12}=-\frac{A_{55}}{A_{11}},\ a_{13}=-\frac{1}{x},\ a_{14}=-\frac{A_{55}}{A_{11}},\ a_{15}=-\frac{A_{13}+A_{55}}{A_{11}},\\
&a_{16}=-\frac{e_{13}+e_{15}}{c_{11}},\ a_{21}=-\left(1+\frac{\delta_1}{\delta}\right)\frac{1}{x},\ a_{22}=-\left(1+\frac{\delta_1}{\delta}\right),\\
&a_{23}=-\frac{\varepsilon_{11}\Omega^2}{\delta},\ a_{24}=-\frac{\delta_2}{\delta},\ a_{25}=-\frac{1}{x},\ a_{26}=-\frac{\delta_3}{\delta},\ a_{31}=-\frac{\delta_4}{\delta}\frac{1}{x},\\
&a_{32}=-\frac{\delta_4}{\delta},\ a_{33}=-\frac{e_{15}\Omega^2}{\delta},\ a_{34}=-\frac{\delta_5}{\delta},\ a_{35}=-\frac{c_{55}\varepsilon_{33}}{\delta},\ a_{36}=-\frac{1}{x}.
\end{aligned}$$

Then, the system (2.15) will take the form of relatively unknown functions u, $\tilde{u}$, w, $\tilde{w}$, Φ , $\tilde{\Phi}$:

$$
\begin{aligned}
&\frac{\mathrm{d}u}{\mathrm{d}x}=\tilde{u},\ \frac{\mathrm{d}w}{\mathrm{d}x}=\tilde{w},\ \frac{\mathrm{d}\Phi}{\mathrm{d}x}=\tilde{\Phi},\\
&\frac{\mathrm{d}\tilde{u}}{\mathrm{d}x}=\Psi_1^{-1}\left[\left(a_{11}\Psi_1+a_{12}\Psi_1''\right)u+a_{13}\Psi_1\tilde{u}+a_{14}\Psi_2'w+\right.\\
&\qquad\left.+\ a_{15}\Psi_2'\tilde{w}+a_{16}\Psi_1''\Phi+a_{17}\Psi_1\tilde{\Phi}\right],\\
&\frac{\mathrm{d}\tilde{w}}{\mathrm{d}x}=\Psi_2^{-1}\left[a_{21}\Psi_1'u+a_{22}\Psi_{12}'\tilde{u}+\left(a_{23}\Psi_2+a_{24}\Psi_2''\right)w+\right.\\
&\qquad\left.+\ a_{25}\Psi_2\tilde{w}+a_{26}\Psi_1'\Phi+a_{27}\Psi_{11}'\tilde{\Phi}\right],\\
&\frac{\mathrm{d}\tilde{\Phi}}{\mathrm{d}x}=\Psi_1^{-1}\left[\left(a_{31}\Psi_1+a_{32}\Psi_1''\right)u+a_{33}\Psi_1'\tilde{u}+\right.\\
&\qquad\left.+\ a_{34}\Psi_2'w+a_{35}\Psi_2'\tilde{w}+a_{36}\Psi_1''\Phi+a_{37}\Psi_1\tilde{\Phi}\right],
\end{aligned}
\tag{2.23}
$$

where

$$
\begin{aligned}
&a_{11}=\left(\frac{\tilde{c}_{11}}{r^2}-\Omega^2\right)\frac{\tilde{\varepsilon}_{33}}{\delta},\ a_{12}=-\frac{\delta_1}{\delta},\ a_{13}=-\left(1+\frac{\tilde{e}_{13}\tilde{e}_{33}}{\delta}\right)\frac{1}{x},\ a_{14}=-\frac{\delta_2-\tilde{c}_{12}\tilde{\varepsilon}_{33}}{\delta}\frac{1}{x},\\
&a_{15}=-\frac{\delta_1+\delta_2}{\delta},\ a_{16}=-\frac{\delta_3}{\delta},\ a_{17}=\frac{\tilde{e}_{13}\tilde{\varepsilon}_{33}}{\delta}\frac{1}{x},\ a_{21}=-\left(1+\frac{c_{12}}{c_{55}}\right)\frac{1}{x},\\
&a_{22}=-\left(1+\frac{c_{13}}{c_{55}}\right),\ a_{23}=-\frac{\Omega^2}{c_{55}},\ a_{24}=-\frac{c_{11}}{c_{55}},\ a_{25}=-\frac{1}{x},\ a_{26}=-\frac{e_{15}}{c_{55}}\frac{1}{x},\\
&a_{27}=-\frac{e_{13}+e_{15}}{c_{55}},\ a_{31}=\left(\frac{c_{11}}{x^2}-\Omega^2\right)\frac{e_{33}}{\delta},\ a_{32}=-\frac{\delta_4}{\delta},\ a_{33}=\frac{c_{33}e_{13}}{\delta}\frac{1}{x},\\
&a_{34}=-\frac{\delta_5-\tilde{c}_{12}\tilde{e}_{33}}{\delta}\frac{1}{x},\ a_{35}=-\frac{\delta_4+\delta_5}{\delta},\ a_{36}=-\frac{\delta_6}{\delta},\ a_{37}=-\left(1-\frac{e_{13}e_{33}}{\Delta}\right)\frac{1}{x}.
\end{aligned}
$$

The systems (2.22) and (2.23) can be represented as

$$
\frac{\mathrm{d}R}{\mathrm{d}x}=A\left(x,\Omega\right)R,
\tag{2.24}
$$

where

$$
\begin{aligned}
R=\{&u_0\ ,u_1,\ldots,u_N,\tilde{u}_0,\tilde{u}_1,\ldots,\tilde{u}_N,v_0,v_1,\ldots,v_N,\tilde{v}_0,\tilde{v}_1,\ldots,\tilde{v}_N,\\
&w_0,w_1,\ldots,w_N,\tilde{w}_0,\tilde{w}_1,\ldots,\tilde{w}_N\}^T.
\end{aligned}
$$

If we apply

$$
\begin{aligned}
&\frac{1}{\chi_{12}}=\frac{\Psi_1}{\Psi_2},\ \frac{1}{\chi_{21}}=\frac{\Phi_2}{\Phi_1},\ \frac{1}{\chi_{11}'}=\frac{\Phi_1'}{\Phi_1},\ \frac{1}{\chi_{12}'}=\frac{\Phi_1'}{\Phi_2},\ \frac{1}{\chi_{21}'}=\frac{\Phi_2'}{\Phi_1},\\
&\frac{1}{\chi_{22}'}=\frac{\Phi_2'}{\Phi_2},\ \frac{1}{\chi_{11}''}=\frac{\Phi_1''}{\Phi_1},\ \frac{1}{\chi_{12}''}=\frac{\Phi_1''}{\Phi_2},\ \frac{1}{\chi_{21}''}=\frac{\Phi_2''}{\Phi_1},\ \frac{1}{\chi_{22}''}=\frac{\Phi_2''}{\Phi_2},
\end{aligned}
$$

Then the matrix A of the system (2.22) has the form:

$$A = \left\| \begin{matrix} 0 & 1 & 0 & 0 & 0 & 0 \\ \frac{1}{x^2} - \frac{\Omega^2}{\tilde{A}_{11}} - \frac{A_{55}}{A_{11}\chi_{11}''} & -\frac{1}{x} & -\frac{c_{55}}{c_{11}\chi_{21}''} & -\frac{A_{13}+A_{55}}{A_{11}\chi_{21}'} & 0 & -\frac{e_{13}+e_{15}}{c_{11}\chi_{21}'} \\ 0 & 0 & 0 & 1 & 0 & 0 \\ -\left(1+\frac{\delta_1}{\delta}\right)\frac{1}{x\chi_{12}'} & -\left(1+\frac{\delta_1}{\delta}\right)\frac{1}{\chi_{12}'} & -\frac{\varepsilon_{11}\Omega^2}{\delta} - \frac{\delta_2}{\delta\chi_{22}''} & -\frac{1}{x} & -\frac{\delta_3}{\delta\chi_{22}''} & 0 \\ 0 & 0 & 0 & 0 & 0 & 1 \\ -\frac{\delta_4}{\delta}\frac{1}{x\chi_{12}'} & -\frac{\delta_4}{\delta\chi_{12}'} & -\frac{e_{15}\Omega^2}{\delta} - \frac{\delta_5}{\delta\chi_{22}''} & 0 & -\frac{c_{55}\varepsilon_{33}}{\delta\chi_{22}''} & -\frac{1}{x} \end{matrix} \right\|$$

And the matrix A of the system (2.23) reads:

$$A = \left\| \begin{matrix} 0 & 1 & 0 & 0 & 0 & 0 \\ \left(\frac{\tilde{c}_{11}}{x^2} - \Omega^2\right)\frac{\tilde{\varepsilon}_{33}}{\delta} - \frac{\delta_1}{\delta\chi_{11}''} & -\left(1+\frac{\tilde{e}_{13}\tilde{e}_{33}}{\delta}\right)\frac{1}{x} & -\frac{\delta_2 - \tilde{c}_{12}\tilde{\varepsilon}_{33}}{\delta x} & -\frac{\delta_1+\delta_2}{\delta\chi_{21}'} & -\frac{\delta_3}{\delta\chi_{11}''} & \frac{\tilde{e}_{13}\tilde{\varepsilon}_{33}}{\delta}\frac{1}{x} \\ 0 & 0 & 0 & 1 & 0 & 0 \\ -\left(1+\frac{c_{12}}{c_{55}}\right)\frac{1}{x\chi_{12}'} & -\left(1+\frac{c_{13}}{c_{55}}\right)\frac{1}{\chi_{12}'} & -\frac{\Omega^2}{c_{55}} - \frac{c_{11}}{c_{55}\chi_{22}''} & -\frac{1}{x} & -\frac{e_{15}}{c_{55}x\chi_{12}'} & -\frac{e_{13}+e_{15}}{c_{55}\chi_{12}'} \\ 0 & 0 & 0 & 0 & 0 & 1 \\ \left(\frac{c_{11}}{x^2} - \Omega^2\right)\frac{e_{33}}{\delta} - \frac{\delta_4}{\delta\chi_{11}''} & \frac{c_{33}e_{13}}{\delta x\chi_{11}'} & -\frac{\delta_5 - \tilde{c}_{12}\tilde{e}_{33}}{\delta x\chi_{21}'} & -\frac{\delta_4+\delta_5}{\delta\chi_{21}'} & -\frac{\delta_6}{\delta\chi_{11}''} & -\left(1-\frac{e_{13}e_{33}}{\Delta}\right)\frac{1}{x} \end{matrix} \right\|.$$

The boundary conditions are represented by

$$B_1 R(-1) = 0, \;\; B_2 R(1) = 0. \tag{2.25}$$

The matrices B_1 and B_2 for the problem (2.22) are given by

$$B_1 = \begin{pmatrix} -c_{12}\Psi_1 & c_{11}\Psi_1 & c_{13}\Psi_2' & 0 & e_{13}\Psi_2' & 0 \\ c_{55}\Psi_1' & 0 & 0 & c_{55}\Psi_2 & 0 & e_{15}\Psi_2 \\ 0 & 0 & 0 & 0 & \Psi_2 & 0 \end{pmatrix},$$
$$B_2 = \begin{pmatrix} c_{12}\Psi_1 & c_{11}\Psi_1 & c_{13}\Psi_2' & 0 & e_{13}\Psi_2' & 0 \\ c_{55}\Psi_1' & 0 & 0 & c_{55}\Psi_2 & 0 & e_{15}\Psi_2 \\ 0 & 0 & 0 & 0 & \Psi_1 & 0 \end{pmatrix}. \tag{2.26}$$

And the matrices B_1 and B_2 for the problem (2.23) are given by

$$B_1 = \begin{pmatrix} -c_{13}\Psi_1 & c_{33}\Psi_1 & c_{13}\Psi_2' & 0 & 0 & e_{33}\Psi_2 \\ c_{55}\Psi_1' & 0 & 0 & c_{55}\Psi_2 & e_{15}\Psi_2' & 0 \\ 0 & 0 & 0 & 0 & \Psi_2 & 0 \end{pmatrix},$$
$$B_2 = \begin{pmatrix} c_{13}\Psi_1 & c_{33}\Psi_1 & c_{13}\Psi_2' & 0 & 0 & e_{33}\Psi_2 \\ c_{55}\Psi_1' & 0 & 0 & c_{55}\Psi_2 & e_{15}\Psi_2' & 0 \\ 0 & 0 & 0 & 0 & \Psi_2 & 0 \end{pmatrix}. \tag{2.27}$$

Table 2.1 Comparative analysis of cylinder free vibrations frequencies based on numerical and analytical approaches

No. of frequency	Numerical approach, N =24	Numerical approach, N =30	Analytical approach
1	0.6586	0.6429	0.6429
2	0.8300	0.8275	0.8276
3	0.9983	0.9798	0.9798
4	1.1491	1.1518	1.1517
5	1.5733	1.5815	1.5815
6	1.8115	1.7641	1.7639

The solution of the above boundary problems for eigenvalues for systems of ordinary differential equations is obtained by the stable numerical method of discrete orthogonalization in combination with the step-by-step search method.

2.3.3 *Numerical Analysis of Axisymmetric Free Vibrations Frequencies of the Piezoceramic Cylinder with Finite Length in the Case of Axial Polarization*

To assess the accuracy of the proposed method, we compare in Table 2.1 the dimensionless vibration frequencies of a homogeneous piezoceramic cylinder simply supported at the edges, obtained by using this approach for a different number of collocation points, with the analytical solution [17]. The cylinder is made of the piezoceramic PZT-4. The geometric parameters of the cylinder have the following values: length of cylinder $L = 10$, inner radius $R_{-} = 3$, outer radius $R_{+} = 5$ $(\varepsilon = 0, 25)$.

The comparative analysis of the first six eigenvalues of the calculated based on the spline-collocation method and the analytical method are presented in Table 2.1.

The results presented in Table 2.1 indicate convergence and sufficient accuracy of the spline-collocation method used.

Figure 2.2 shows the dependence of the first five free vibrations frequencies of the piezoceramic cylinder on relative length. The geometric parameters of the cylinder have the following values: length of cylinder $L = 10$, inner radius $R_{-} = 3$, outer radius $R_{+} = 5$. Both end faces of the cylinders are rigidly fixed. We show the values of the natural frequencies with regard to the piezoeffect by solid lines and the values of the natural frequencies without regard to the piezoeffect $(e_{ij} = 0)$ by dashed lines. The natural frequencies decrease as the relative length of the cylinder increases. The electric field does practically not affect the first frequency of free vibrations of the cylinder. However, at higher frequencies, the presence of an electric field has a more significant effect on the value of the natural frequencies of the cylinder. In Fig. 2.2, it can be seen that the influence of the piezoeffect leads to a rigidifying of

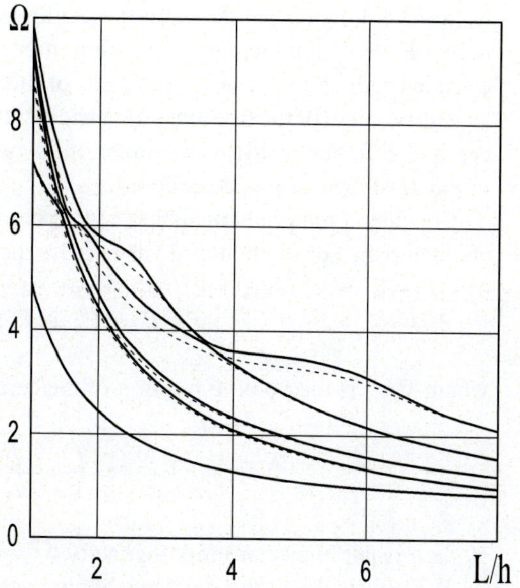

Fig. 2.2 Dependence of the first five free vibrations frequencies on relative cylinder length

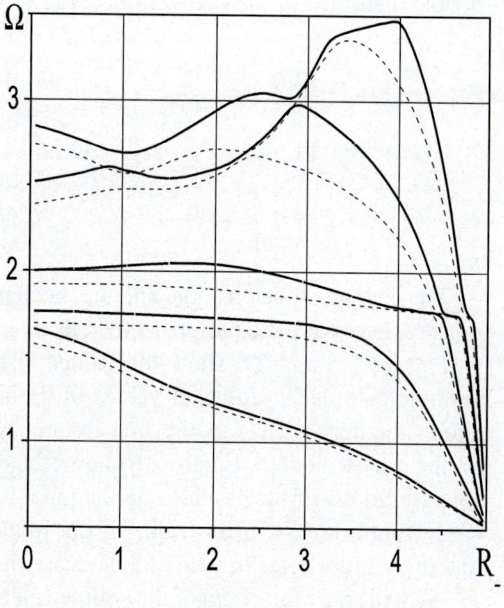

Fig. 2.3 Dependence of the first five free vibrations frequencies on the inner radius of the cylinder

the material, i.e., an increase in the value of the natural frequencies. In this case, in the determination of the first natural frequency, the influence of the piezoeffect can be neglected up to a relative length $L/h = 5$. For the second frequency, a noticeable influence of the piezoeffect is observed for fairly long cylinders ($L/h < 8$). For higher frequencies, this influence is pronounced for the long cylinders (Fig. 2.2).

Figure 2.3 shows the dependence of the first five free vibration frequencies on the inner radius of the piezoceramic cylinder. The geometric parameters of the cylinder have the following values: length of cylinder $L = 10$, inner radius $R_- = 3$, outer radius $R_+ = 5$. An increase of the first three frequencies of natural vibrations of the cylinder with an increase in the radius of the cylinder is observed.

The problem of the free axisymmetric electroelastic vibrations of the hollow piezoceramic cylinders made of the functionally graded piezoceramic material polarized in the axial direction is considered.

Consider a material consisting of two components: steel and piezoceramics. The characteristics of the material vary across the thickness as follows:

$$P(r) = (P_m - P_p)V(r) + P_p, \tag{2.28}$$

where $V(z)$ is the volume fraction of the ceramics, which is described as

$$V(r) = \left(\frac{r - R_0}{2h} + \frac{1}{2}\right)^n, \tag{2.29}$$

Here, n is parameter of inhomogeneity.

The physical and the mechanical characteristics of the material with index 0 on the outer surface of the cylinder have the values:

$$c_{11}^0 = 13.9 \cdot 10^{10}\frac{\mathrm{N}}{\mathrm{m}^2}, c_{12}^0 = 7.43 \cdot 10^{10}\frac{\mathrm{N}}{\mathrm{m}^2}, c_{13}^0 = 7.78 \cdot 10^{10}\frac{\mathrm{N}}{\mathrm{m}^2},$$
$$c_{33}^0 = 11.5 \cdot 10^{10}\frac{\mathrm{N}}{\mathrm{m}^2}, c_{55}^0 = 2.56 \cdot 10^{10}\frac{\mathrm{N}}{\mathrm{m}^2}, e_{13}^0 = -5.2\frac{\mathrm{C}}{\mathrm{m}^2},$$
$$e_{15}^0 = 12.7\frac{\mathrm{C}}{\mathrm{m}^2}, e_{33}^0 = 15,1\frac{\mathrm{C}}{\mathrm{m}^2}, \varepsilon_{11}^0/\varepsilon_0 = 730, \ \varepsilon_{33}^0/\varepsilon_0 = 635.$$

The change of the physical and the mechanical characteristics along the thickness of the sphere for the example of the elastic module c_{11} is shown in Fig. 2.4.

Figures 2.5 and 2.6 show the values of the vibration frequencies of the piezoceramic cylinder at different values of the inhomogeneity parameter n. Figure 2.5 shows the dependence of the first frequency of the free vibrations of the cylinder on the relative length. Figure 2.6 shows the dependence of the first frequency of the free vibrations of the cylinder on the inner radius, outer radius is $R_+ = 5$. Note the significant increase of the value of the frequency for the short cylinders (Fig. 2.5) and high frequencies in case of a decrease in the thickness of the cylinder (Fig. 2.6).

We will study the effect of the coupled electric field and the inhomogeneity parameter on the values of the natural vibrations frequencies of the cylinder. Let us consider the free vibration frequencies of the inhomogeneous and the homogeneous the cylinder with averaged characteristics calculated by the theory of effective moduli, when the moduli change by an exponential law:

$$\mu = \frac{\mu^0}{R_+ - R_-}\int_{R_-}^{R_+} e^{nr}\,dr = \frac{\mu^0}{2nh}e^{nR_-}\left(e^{2nh} - 1\right). \tag{2.30}$$

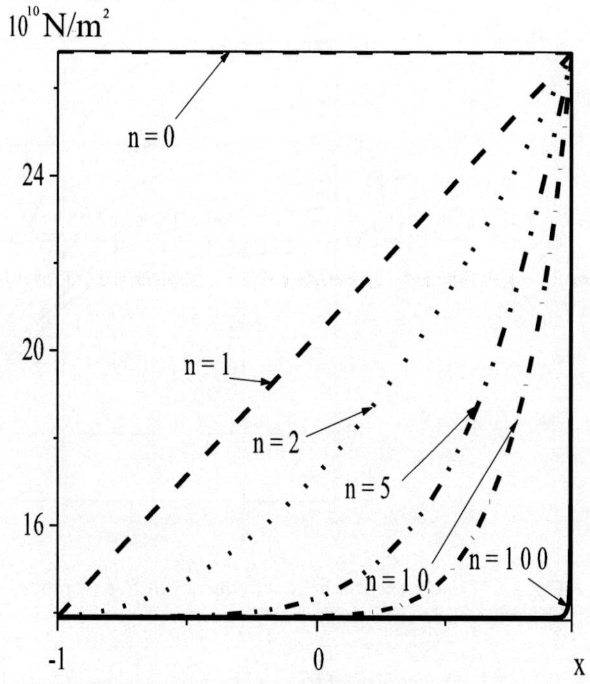

Fig. 2.4 Elastic modulus c_{11} vs. normalized thickness of the cylinder

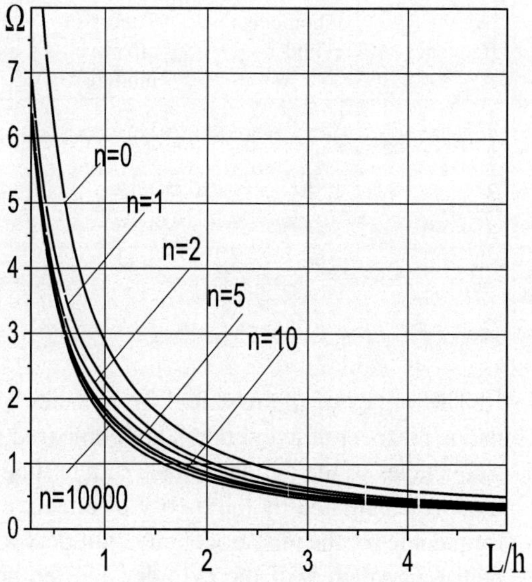

Fig. 2.5 Dependence of the first frequency of free vibrations of the cylinder on relative length for different values of heterogeneity parameter

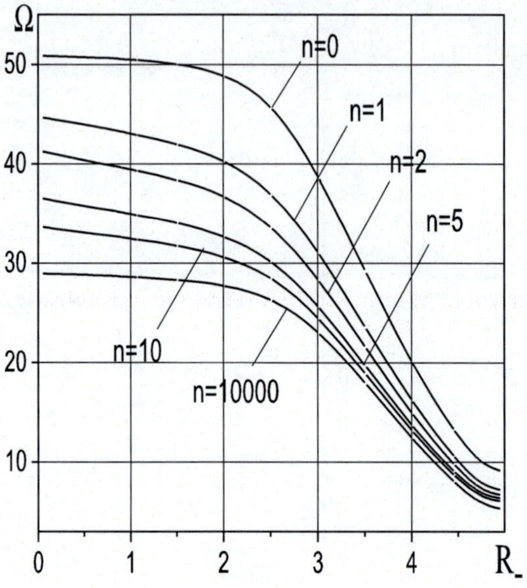

Fig. 2.6 Dependence of the first frequency of free vibrations of the cylinder on the inner radius for different values of heterogeneity parameter

Table 2.2 Comparison of free vibration frequencies of homogeneous and inhomogeneous cylinder of the finite length

No. of frequency	Inhomogeneous cylinder	Theory of effective moduli	Relative error %	Disregarding the piezoeffect	Relative error %
1	0.74	0.73	1.4	0.74	0.0
2	0.90	0.91	1.1	0.88	2.2
3	1.30	1.12	13.8	1.23	5.4
4	1.71	1.75	2.3	1.64	4.1
5	1.99	1.88	5.5	1.88	5.5

The values of the first five frequencies of the free vibrations of the inhomogeneous piezoceramic cylinders are compared with approximate values of the natural frequencies of the cylinder made of a homogeneous material with averaged characteristics calculated by the theory of effective moduli and with values of the natural frequencies of the inhomogeneity cylinders without piezoeffect (Table 2.2). The geometric parameters of the cylinder and the heterogeneity parameter of the cylinder material have values $L = 5$, $R_+ = 5$, $n = 1, 5$.

Based on the data of Table 2.2 note that the influence of the inhomogeneous structure and the coupled electric field on the values of the free vibration frequencies of the cylinder is practically the same. The exception is the effect of a non-uniform cylinder structure on the third natural frequency. In this case, the difference in the frequency value for the homogeneous and inhomogeneous cylinder is 13,8%, which is explained by the large value of the heterogeneity parameter ($n = 1.5$). The greatest influence of the electric field is observed for the value of the fifth natural frequency. In this case, the difference in the frequency value for and inhomogeneous cylinders with the coupled electric field and the disregarding piezoeffect is 5.5%.

2.3.4 Numerical Analysis of Axisymmetric Free Vibration Frequencies of the Piezoceramic Cylinder with Finite Length in the Case of Radial Polarization

To assess the accuracy of the proposed method, we compared (Table 2.3) the dimensionless vibration frequencies of a homogeneous piezoceramic cylinder simply supported at the edges. They are obtained using this approach for different numbers of collocation points, with the analytical solution [13]. The cylinder is made of the piezoceramic PZT-4. The geometric parameters of the cylinder have the following values: length of cylinder $L = 10$, inner radius $R_- = 3$, outer radius $R_+ = 5$ ($\varepsilon = 0, 25$). The comparative analysis of the first six eigenvalues of the calculated based on the spline-collocation method and the analytical method is presented in Table 2.3.

The results presented in Table 2.3 indicate convergence and sufficient accuracy of the spline-collocation method that was used.

Figure 2.7 shows the dependence of the first five free vibration frequencies of the piezoceramic cylinder on relative length. The geometric parameters of the cylinder have the following values: length of cylinder $L = 10$, inner radius $R_- = 3$, outer radius $R_+ = 5$. Both end faces of the cylinders are rigidly fixed. We show the values of the natural frequencies with piezoeffect by solid lines and the values of the natural frequencies without piezoeffect ($e_{ij} = 0$) by dashed lines. The natural frequencies decrease as the relative length of the cylinder increases. The electric field practically does not affect the first frequency of free vibrations of the cylinder. However, at higher frequencies, the presence of an electric field has a more significant effect on the value of the natural frequencies of the cylinder

Table 2.3 Comparative analysis of cylinder free vibration frequencies based on numerical and analytical approaches

No. of frequency	Spline-collocation method, $N = 24$	Spline-collocation method, $N = 30$	Analytical solution
1	0.2586	0.2487	0.2487
2	0.3811	0.3722	0.3721
3	0.3321	0.3303	0.3301
4	0.4958	0.4943	0.4943
5	0.6851	0.6842	0.6841
6	0.7083	0.7078	0.7075

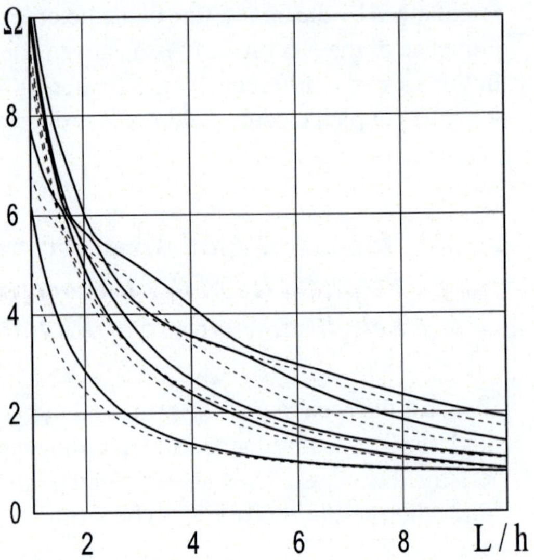

Fig. 2.7 Dependence of first natural frequencies cylinder relative length

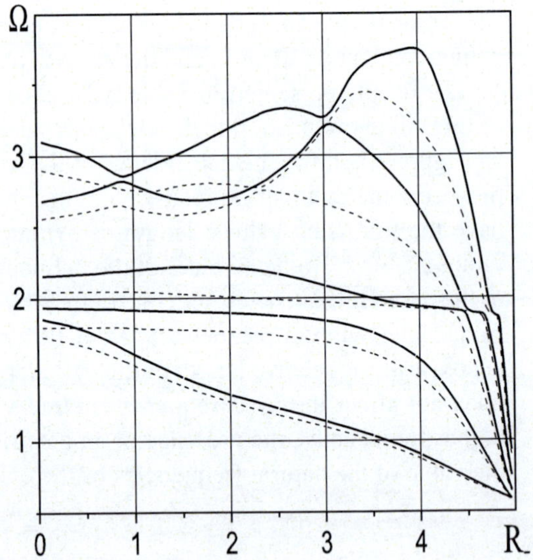

Fig. 2.8 Dependence of the first five natural frequencies cylinder on the relative inner radius

In Fig. 2.7, it can be seen that the influence of the piezoeffect leads to a rigidifying of the material, i.e., an increase in the value of the natural frequencies. In this case, in the determination of the first natural frequency, the influence of the piezoeffect can be neglected up to the relative length $L/h = 5$. For the second frequency, a noticeable influence of the piezoeffect is observed for fairly long cylinders ($L/h < 8$). For higher frequencies, this influence is pronounced for larger cylinders.

In Fig. 2.8, we show the dependencies of the first five natural frequencies on the relative inner radius of the R_-, for the fixed length of the cylinder $L/h = 5$ and external radius $R_+ = 5$. The change of the internal diameter is considered in the wide range from 0.05 to 4.95 units, i.e., from a practically solid cylinder to a very thin cylindrical shell. As material of the cylinder, the PZT-4 piezoceramic was chosen again. It follows from the analysis of the shown curves that with increase in the thickness of the cylinder, the natural frequencies of vibrations increase abruptly. It should be noted that for the first three frequencies, with increase in the thickness of the cylinder, the natural frequency of vibrations increases. For higher frequencies, after an abrupt increase in the frequency with increasing thickness, it decreases smoothly, which is not observed when the relative length of the cylinder increases (Fig. 2.7). As the relative length of the cylinder increases, the natural frequency of vibrations always increases. The influence of the piezoeffect for the second natural frequency and higher ones manifests itself even in fairly thin cylinders. Only in the determination of the first natural frequency, it can be neglected without loss of accuracy in the calculation of the frequency in the segment $R_{-} \geq 4$.

By analyzing the figures we note that only the first and third branches in Fig. 2.7 and the first two branches in Fig. 2.8 are relatively simple curves. For the dependencies of the higher natural frequencies, the structure of the spectrum is complicated. Segments with small changes in frequencies depending on the geometrical parameters (we call them plateau) with further approach of the values of frequencies (we call these points"points of attraction") are characteristic. Note that these "plateaus" in both Figs. 2.7 and 2.8 are located along some characteristic lines. For instance, in Fig. 2.7, it is the segment $1 \leq L/h \leq 3$ for the second, third, fourth, and fifth frequencies, the segment $4 \leq L/h \leq 6$ is for the fourth and fifth frequencies, and $L/h \approx 5$ is for the fourth and fifth natural frequencies. In Fig. 2.8 these are the domain $4 \leq R_{-} \leq 5$ for the third, fourth, and fifth frequencies, and segments $R_{-} \cong 3$ and $R_{-} \cong 1$ for the fourth and fifth frequencies. It can be assumed that the described"points of attraction" of the natural frequencies are "unfavorable" for the material because a small change in the frequency corresponds to a substantial reconstruction of the geometry of vibrations. However, this statement calls for additional investigations.

The problem of the free axisymmetric electroelastic vibrations of the hollow piezoceramic cylinders made of the functionally graded piezoceramic material polarized in the axially direction is considered. The material properties are assumed to change over the thickness by an law (2.28), (2.29). The physical and the mechanical characteristics of the material with index 0 have values on the outer surface of the cylinder:

$$c_{11}^0 = 13.9 \cdot 10^{10} \frac{\mathrm{N}}{\mathrm{m}^2}, c_{12}^0 = 7.43 \cdot 10^{10} \frac{\mathrm{N}}{\mathrm{m}^2}, c_{13}^0 = 7.78 \cdot 10^{10} \frac{\mathrm{N}}{\mathrm{m}^2},$$
$$c_{33}^0 = 11.5 \cdot 10^{10} \frac{\mathrm{N}}{\mathrm{m}^2}, c_{55}^0 = 2.56 \cdot 10^{10} \frac{\mathrm{N}}{\mathrm{m}^2}, e_{13}^0 = -5.2 \frac{\mathrm{C}}{\mathrm{m}^2},$$
$$e_{15}^0 = 12.7 \frac{\mathrm{C}}{\mathrm{m}^2}, e_{33}^0 = 15.1 \frac{\mathrm{C}}{\mathrm{m}^2}, \ \varepsilon_{11}^0/\varepsilon_0 = 730, \ \varepsilon_{33}^0/\varepsilon_0 = 635.$$

The change of the physical and the mechanical characteristics along the thickness of the cylinder on the example of the elastic module c_{11} is shown in Fig. 2.4.

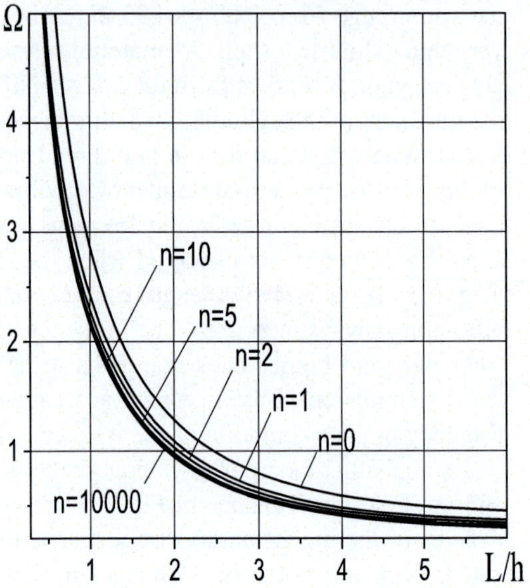

Fig. 2.9 Dependence of the first frequency of free vibrations of cylinder on relative length for different values of heterogeneity parameter

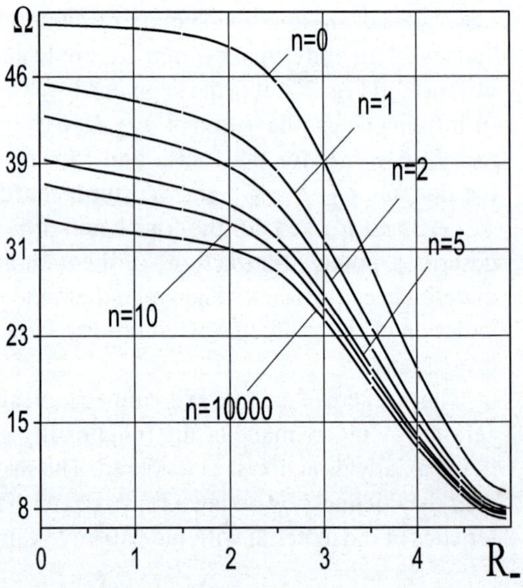

Fig. 2.10 Dependence of the first frequency of free vibrations of cylinder on the inner radius for different values of heterogeneity parameter

Figures 2.9 and 2.10 show the values of the vibrations frequencies of the piezoceramic cylinder at different values of the inhomogeneity parameter *n*. Figure 2.9 shows the dependence of the first frequency of the free vibrations of cylinder on relative length; Figure 2.10 shows the dependence of the first frequency of the free vibrations of cylinder on the inner radius. Note the significant increase of the value of the frequency for the short cylinders (Fig. 2.9) and high frequencies in case of a decrease in the inner radius of the cylinder (Fig. 2.10).

Table 2.4 Comparison of free vibration frequencies of a homogeneous and an inhomogeneous cylinder of finite length

No. of frequency	Inhomogeneous cylinder	Theory of effective moduli	Relative error, (%)	Without piezoeffect	Relative error, (%)
1	1.15	1.15	0.0	1.10	4.3
2	1.63	1.64	0.6	1.53	6.1
3	1.93	1.93	0.0	1.72	10.9
4	2.40	2.34	2.5	2.35	2.1
5	2.74	2.76	0.7	2.40	12.4

We will study the effect of the coupled electric field and the inhomogeneity parameter on the values of the natural vibrations frequencies of the cylinder. Let us consider the free vibration frequencies of the inhomogeneous and the homogeneous the cylinder with averaged characteristics calculated by the theory of effective moduli, when the module changes by the power law:

$$\mu = \frac{\mu^0}{R_+ - R_-} \int_{R_-}^{R_+} r^n \mathrm{d}r = \frac{\mu^0 R_-.}{2(n+1)h} \left(\left(\frac{R_+}{R_-} \right)^{n+1} - 1 \right). \tag{2.31}$$

In formula (2.31), we replace the corresponding the physical and the mechanical moduli of the material of the cylinder.

The values of the first five frequencies of the free vibrations of the inhomogeneous piezoceramic cylinders are compared with approximate values of the natural frequencies of the cylinder made of a homogeneous piezoceramics material with averaged characteristics calculated by the theory of effective moduli and with values of the natural frequencies of the inhomogeneity cylinders without (at absent, disregarding) piezoeffect (Table 2.4). The geometric parameters of the cylinder and the heterogeneity parameter of the cylinder material have the values $L = 5$, $R_+ = 5$, $n = 0.5$.

Based on the data in Table 2.4 note that the relative error in neglect of the inhomogeneity factor is less than the relative error in the case of neglect of the coupled electric field. We have the greatest relative error for the value of the fifth natural frequency (12.4%.) in the case of neglecting the coupled electric field. The greatest relative error at the value of the fourth natural frequency in the case of neglect of the inhomogeneity factor (2.5%) has a small value of the heterogeneity parameter ($n = 0.5$).

2.3.5 *Calculation of Asymmetric Vibrations of Hollow Piezoceramic Cylinder with Finite Length Based on Finite-Element Method*

Here, the calculation of the free vibrations of the final length piezoceramic cylinder with finite length in the case of the radial polarization based on FEM is carried out. Both end faces of the cylinders are rigidly fixed. The obtained calculation results make it possible to carry out a comparative analysis of the calculations of free vibration frequencies of a piezoceramic cylinder based on the analytical–numerical proposed method. Such a comparison can be a criterion for the reliability of the obtained calculation results based on the analytical–numerical proposed method.

For solving the axisymmetric electroelasticity problem by the finite-element method, we will use the Hamilton variation principle. In this case, the original functional is expressed by

$$J = \pi \int_{0}^{L} \int_{R_0-h}^{R_0+h} (T_1 S_1 + T_2 S_2 + T_3 S_3 + 2T_5 S_5 - \\ - E_1 D_1 - E_3 D_3 - \rho\omega^2 (u_1^2 + u_3^2)) r dr dz, \tag{2.32}$$

where L is the cylinder length, $R_0 - h$ and $R_0 + h$ are internal and external radiuses.

Suppose that the cylinder ends $z = 0$ and $z = 0$ are clamped, i.e., $u_r = u_z = 0$, and are covered by thin shortened electrodes, $\Phi = 0$. The lateral surfaces are free of external forces, so that $T_1 = T_5 = 0$, and a potential difference $\Phi = V_0 z$ is applied to them. In order to avoid passing of the frequencies corresponding to asymmetric over z modes, the function Φ can vary with.

By considering (2.1)–(2.3), (2.5), the equality (2.32) becomes

$$J = \pi \int_{0}^{L} \int_{R_0-h}^{R_0+h} \left\{ c_{33} \left(\frac{\partial u_1}{\partial r}\right)^2 + c_{11} \left(\frac{\partial u_3}{\partial z}\right)^2 + c_{55} \left(\frac{\partial u_1}{\partial z}\right)^2 + c_{55} \left(\frac{\partial u_3}{\partial r}\right)^2 - \right. \\ - \varepsilon_{33} \left(\frac{\partial \Phi}{\partial r}\right)^2 - \varepsilon_{11} \left(\frac{\partial \Phi}{\partial z}\right)^2 + c_{11} \left(\frac{u_1}{r}\right)^2 + 2c_{13} \frac{u_1}{r} \frac{\partial u_1}{\partial r} + 2c_{12} \frac{u_1}{r} \frac{\partial u_3}{\partial z} + \\ + 2e_{13} \frac{u_1}{r} \frac{\partial \Phi}{\partial r} + 2c_{13} \frac{\partial u_1}{\partial r} \frac{\partial u_3}{\partial z} + 2e_{33} \frac{\partial u_1}{\partial r} \frac{\partial \Phi}{\partial r} + 2e_{13} \frac{\partial u_3}{\partial z} \frac{\partial \Phi}{\partial r} + \\ \left. + 2e_{15} \frac{\partial u_1}{\partial z} \frac{\partial \Phi}{\partial z} + 2e_{15} \frac{\partial u_3}{\partial r} \frac{\partial \Phi}{\partial z} + 2c_{55} \frac{\partial u_1}{\partial z} \frac{\partial u_3}{\partial r} - \rho\omega^2 \left(u_1^2 + u_3^2\right) \right\} r dz dr. \tag{2.33}$$

To solve the problem, we will employ fourth-node rectangular finite elements [11]. Let the solution be

$$\Phi = \sum_{i=1}^{4} \Phi_i N_i,\ \ u_1 = \sum_{i=1}^{4} u_{1i} N_i,\ \ u_3 = \sum_{i=1}^{4} u_{3i} N_i, \tag{2.34}$$

where Φ_i, u_{1i}, and u_{3i} are the values of the unknown functions at the nodes of the finite element, N_i are the shape functions.

By substituting (2.34) into (2.33) with the condition $\delta J = 0$, we get the following system of equations for the unit area S_k

$$\begin{aligned}
&2\pi \iint\limits_{S_k} \{c_{33} r (u_{1i} N_{ri})\ N_{rj} + c_{55} r (u_{1i} N_{zi}) N_{zj} + \\
&+ \left(\frac{c_{11}}{r} - r\rho\omega^2\right) (u_{1i} N_i) N_j + c_{13}((u_{1i} N_{ri}) N_j + (u_{1i} N_i) N_{rj}) + \\
&+ c_{12}(u_{3i} N_{zi}) N_j + c_{55} r (u_{3i} N_{ri}) N_{zj} + + c_{13} r (u_{3i} N_{zi}) N_{rj} + \\
&+ e_{33} r (\Phi_i N_{ri}) N_{rj} + e_{15} r (\Phi_i N_{zi}) N_{zj} + e_{13}(\Phi_i N_{ri})\ N_j \big\}\, \mathrm{d}r \mathrm{d}z = 0, \\
&2\pi \iint\limits_{S_k} \{c_{12}(u_{1i} N_i)\ N_{zj} + c_{13} r (u_{1i} N_{ri}) N_{zj} + c_{55} r (u_{1i} N_{zi}) N_{rj} + \\
&+ c_{11} r (u_{3i} N_{zi}) N_{zj} + c_{55} r (u_{3i} N_{ri}) N_{rj} - r\rho\omega^2 (u_{3i} N_i) N_j + \\
&+ e_{13} r (\Phi_i N_{ri})\ N_{zj} + e_{15} r (\Phi_i N_{zi}) N_{rj} \big\}\, \mathrm{d}r \mathrm{d}z = 0, \\
&2\pi \iint\limits_{S_k} \{2 e_{13}(u_{1i} N_i)\ N_{rj} + e_{33} r (u_{1i} N_{ri}) N_{rj} + e_{15} r (u_{1i} N_{zi}) N_{zj} + \\
&+ e_{13} r (u_{3i} N_{zi}) N_{rj} + e_{15} r (u_{3i} N_{ri}) N_{zj} - \\
&- \varepsilon_{33} r (\Phi_i N_{ri}) N_{rj} - \varepsilon_{11} r (\Phi_i N_{zi})\ N_{zj} \big\}\, \mathrm{d}r \mathrm{d}z = 0.
\end{aligned} \tag{2.35}$$

In (2.35), it is assumed that the summation is carried out over the paired indices, e.g., $u_i N_i = \sum_{i=1}^{4} u_i N_i$. Here, the following notation is used:

$$N_{ri} = \frac{\partial N_i}{\partial r},\ N_{zi} = \frac{\partial N_i}{\partial z}.$$

Toper from the integration in (2.35), we use the Gauss quadrature method. The global system obtained by assembling, with the boundary conditions on the lateral surfaces of the cylinder, is inhomogeneous. The linear systems of algebraic equations obtained were solved with the Gauss method. Thus, as in the case of the spline-collocation method, we will determine the eigenfrequencies of the cylinder vibration using the step-by-step solving of the system of linear algebraic equations for the certain value of ω. The results are obtained.

Let us compare the results obtained for the cylinder of PZT-4 piezoceramic with different methods. The cylinder has the following input parameters: $R_- = R_0 - h = 3\,\mathrm{cm}$, $L = 10\,\mathrm{cm}$, $\rho = 7.5 \cdot 10^3\,\mathrm{kg/m^3}$, $c_{11} = 13.9 \cdot 10^{10}\,\mathrm{N/m^2}$, $c_{12} = 7.43 \cdot 10^{10}\,\mathrm{N/m^2}$, $c_{13} = 7.78 \cdot 10^{10}\,\mathrm{N/m^2}$, $c_{33} = 11.5 \cdot 10^{10}\,\mathrm{N/m^2}$, $c_{55} = 2.56 \cdot 10^{10}$

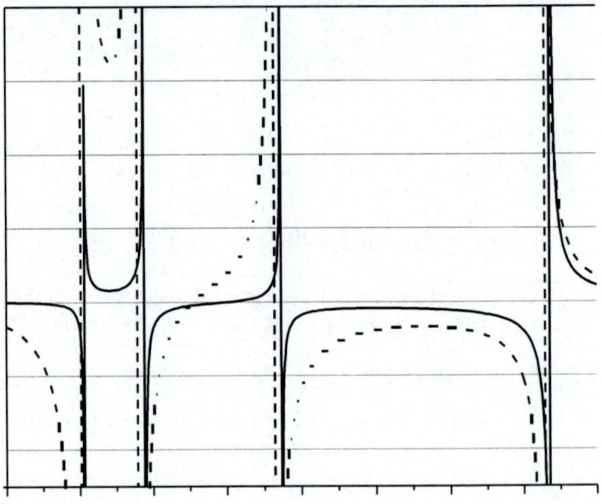

Fig. 2.11 Frequency-dependent distributions of the displacement u_1 (continuous line:FEM, dashed line: SCM)

N/m^2, $e_{13} = -5.2\,\text{C/m}^2$, $e_{15} = 12.7\,\text{C/m}^2$, $e_{33} = 15.1\,\text{C/m}^2$, $\varepsilon_{11}/\varepsilon_0 = 730$, $\varepsilon_{33}/\varepsilon_0 = 635$, $V_0 = 1$; and ε_0 is the vacuum permittivity.

These dimensionless quantities were used in the calculations: $\tilde{u} = u/h, \tilde{z} = z/h$, $\tilde{u}_i = u_i/h$, $\tilde{\Phi} = \Phi\sqrt{\varepsilon_0}/\left(h\sqrt{\lambda}\right)$, $\Omega = \omega h\sqrt{\rho/h}$, $\tilde{c}_{ij} = c_{ij}/\lambda$, $\tilde{\varepsilon}_{ij} = \varepsilon_{ij}/\varepsilon_0$, where $\lambda = 10^{10}\,\text{N/m}^2$.

Figure 2.11 shows how the first four frequencies are determined by using as an example the resonance of the displacements u. The displacement distributions obtained with the finite-element method are shown by solid line, while those obtained with the spline-collocation method are shown by dotted lines. In this case, by employing the spline-collocation method, we have chosen 20 collocation points ($N = 19$) and 2000 of 0.1×0.1 square elements.

Six frequencies at different levels of approximation, i.e., for 16, 20, and 24 collocation points in the spline-collocation method and K being equal to 500, 2000, 8000 in the case of the finite-element method, are collected in Table 2.5. The linear dimensions of the element at $K = 500$ are doubled, while at $K = 8000$ they decrease in comparison with $K = 2000$.

As can be seen from the plots and Table 2.11, the results obtained with both methods agree with high accuracy. Values of the frequencies obtained with spline-collocation method at $N = 19$ and 23 differ insignificantly as in the case of the finite-element method at $K = 2000$ and 8000. The difference in the results obtained by these methods does not exceed 0.8%.

By using the spline-collocation method with discrete orthogonalization in combination with the finite-element method, we have studied the vibrations of a radially polarized piezoceramic cylinder. The values of the frequencies determined with different methods are compared. As it is shown, they agree with high accuracy.

Table 2.5 Values of the frequencies obtained with different methods

Ω	SCM			FEM		
	$N = 15$	$N = 19$	$N = 23$	$K = 500$	$K = 2000$	$K = 8000$
Ω_1	0.8	0.802	0.803	0.807	0.806	0.806
Ω_2	0.875	0.879	0.881	0.891	0.889	0.888
Ω_3	1.062	1.065	1.066	1.076	1.074	1.073
Ω_4	1.431	1.428	1.428	1.441	1.436	1.434
Ω_5	1.841	1.846	1.849	1.866	1.862	1.861
Ω_6	1.999	1.99	1.988	2.002	1.992	1.989

2.3.6 *Torsional Free Vibrations of the Piezoceramic Cylinder with Finite Length*

2.3.6.1 Resolving System and Solving Method

The closed system of equations describing the torsional axisymmetric free vibrations of the piezoceramic cylinder of the finite length in cylindrical coordinates system (r, θ, z) has the following form:

Equations of motion:

$$\frac{\partial T_6}{\partial r} + \frac{2}{r}T_6 + \frac{\partial T_4}{\partial z} + \rho\omega^2 u_2 = 0; \tag{2.36}$$

Equations of electrostatics:

$$\frac{\partial D_1}{\partial r} + \frac{1}{r}D_1 + \frac{\partial D_3}{\partial z} = 0,\ E_r = -\frac{\partial \Phi}{\partial r}E_z = -\frac{\partial \Phi}{\partial z}; \tag{2.37}$$

Kinematic relations:

$$S_6 = \frac{1}{2}\left(\frac{\partial u_2}{\partial r} - \frac{u_2}{r}\right);\ S_4 = \frac{1}{2}\frac{\partial u_2}{\partial z}. \tag{2.38}$$

The constitutive equations for the piezoceramic material polarized in the circular direction have the form:

$$\begin{aligned} T_6 &= 2c_{55}S_6 - e_{15}E_1,\ T_4 = 2c_{55}S_4 - e_{15}E_3, \\ D_1 &= 2e_{15}S_6 + \varepsilon_{11}E_1,\ D_3 = 2e_{15}S_4 + \varepsilon_{11}E_3. \end{aligned} \tag{2.39}$$

We set the following boundary conditions on the surfaces of the cylinder: the lateral surfaces $r = R_+$ and $r = R_-$ are free from external forces, i.e.:

$$T_6 = 0 \tag{2.40}$$

And covered with short-circuited thin electrode, i.e., $\tilde{\Phi} = 0$, the ends of the cylinder (at $z = \pm L/2$) are rigidly fixed:

$$u_2 = 0 \tag{2.41}$$

And covered with shirt circuited thin electrode, i.e:

$$D_3 = 0, \tag{2.42}$$

where R_+ is the outer radius of the cylinder; R_- is inner radius of the cylinder; L is the length of the cylinder.

After the same transformations of the Eqs. (2.36) and (2.37), we the get system of the equations respect to $\frac{\partial^2 \Phi}{\partial r^2}$ and $\frac{\partial^2 u_2}{\partial r^2}$:

$$\frac{\partial^2 \Phi}{\partial r^2} = -\frac{\partial^2 \Phi}{\partial z^2} - \left(1 + \frac{\tilde{e}_{15}^2}{\delta}\right)\frac{1}{r}\frac{\partial \Phi}{\partial r} + \left(\frac{\tilde{c}_{55}}{r^2} - \Omega^2\right)\frac{\tilde{e}_{15}}{\delta}u_2 - \frac{\tilde{c}_{55}\tilde{e}_{15}}{\delta}\frac{1}{r}\frac{\partial u_2}{\partial z} \tag{2.43}$$

$$\frac{\partial^2 u_2}{\partial r^2} = -\frac{\tilde{e}_{15}\tilde{\varepsilon}_{11}}{\delta}\frac{1}{r}\frac{\partial \Phi}{\partial r} + \left(\frac{\tilde{c}_{55}}{r^2} - \Omega^2\right)\frac{\tilde{\varepsilon}_{11}}{\delta}u_2 - \frac{\partial^2 u_2}{\partial z^2} - \frac{\tilde{c}_{55}\tilde{\varepsilon}_{11}}{\delta}\frac{1}{r}\frac{\partial u_2}{\partial r}. \tag{2.44}$$

Here, we denote δ and define dimensionless values:

$$\delta = \tilde{c}_{55}\tilde{\varepsilon}_{11} + \tilde{e}_{15}^2,\ \Omega = \omega h\sqrt{\frac{\rho}{\lambda}},\ \tilde{c}_{ij} = \frac{c_{ij}}{\lambda},\ \tilde{e}_{ij} = \frac{e_{ij}}{\sqrt{\varepsilon_0 \lambda}},\ \tilde{\varepsilon}_i = \frac{\varepsilon_i}{\varepsilon_0}. \tag{2.45}$$

ω is the circular frequency; $\lambda = 10^{10}\,\mathrm{N/m^2}$, ε_0 is the dielectric constant of the vacuum, ρ is the density of the material, and h is half of the thickness of the cylinder.

To reduce the differential equations in the partial derivatives (2.43) and (2.44) to systems of the ordinary differential equations, we will use the spline-collocation method. We will now search for functions $\tilde{u}_r\,(r, z)$, $\tilde{u}_z\,(r, z)$, $\Phi\,(r, z)$, $u_\theta\,(r, z)$ of the form:

$$\tilde{\Phi}\,(r, z) = \sum_{i=0}^{N} \Phi_i\,(x)\,\psi_i\,(z)\,,\ \tilde{u}_2\,(r, z) = \sum_{i=0}^{N} v_i\,(x)\,\psi_i\,(z)\,, \tag{2.46}$$

where $x = (r - R_0)/h$, $R_0 = \frac{R_+ + R_-}{2}$, $\Phi_i\,(x), v_i\,(x)$ are the unknown functions of the variable x, $\psi_i\,(z)\big(i = \overline{0,\ N}\big)$ are linear combinations of B-splines on a uniform mesh Δ; $-L/2 = z_0 < z_1 < \cdots < z_n = L/2$.

Considering the boundary conditions on the ends of the cylinder (at $z = -L/2$ and at $z = L/2$), we can see that the system involves derivatives of the solution vector that are not higher than second order. Therefore, we can restrict ourselves to an approximation by third-order spline-functions

If we introduce the notation:

$$\begin{aligned}
&\Psi = [\psi_i(\xi_k)], k,\ i = \overline{0,\ N}, j = 1,\ 2; \Phi = [\Phi_0,\ \Phi_1,\ \dots,\ \Phi_N]^{\mathrm{T}},\\
&v = [v_0,\ v_1,\ \dots,\ v_N]^{\mathrm{T}}, (k,\ l) \in \left\{(k,\ l)|\, k,\ l = \overline{1,\ 4}\right\},\\
&a_{kl}^T = \left\{a_{kl}\left(x,\ \xi_0,\ \Omega^2\right),\ a_{kl}\left(x,\ \xi_1,\ \Omega^2\right),\ \dots,\ a_{kl}\left(x,\ \xi_N,\ \Omega^2\right)\right\},
\end{aligned} \tag{2.47}$$

we transform the system (2.43) and (2.44) into a system of $4(N+1)$ linear differential equations for the functions v, $\tilde{v}$, Φ, $\tilde{\Phi}$,

$$\begin{aligned}
&\frac{\mathrm{d}\Phi}{\mathrm{d}x} = \tilde{\Phi},\ \frac{\mathrm{d}\tilde{\Phi}}{\mathrm{d}x} = \Psi^{-1}\left[a_{11}\Psi''\Phi + a_{12}\Psi\tilde{\Phi} + a_{13}\Psi v + a_{14}\Psi\tilde{v}\right],\\
&\frac{\mathrm{d}v}{\mathrm{d}x} = \tilde{v},\ \frac{\mathrm{d}\tilde{v}}{\mathrm{d}x} = \Psi^{-1}\left[a_{21}\Psi\tilde{\Phi} + \left(a_{22}\Psi + a_{23}\Psi''\right)v + a_{24}\Psi\tilde{v}\right],
\end{aligned} \tag{2.48}$$

here,

$$\begin{aligned}
&\bar{a}_{11} = -1; \bar{a}_{12} = -\left(1 - \frac{\tilde{e}_{15}^2}{\delta}\right)\frac{1}{x}; \bar{a}_{13} = \frac{\tilde{e}_{15}}{\delta}\left(\frac{\tilde{c}_{55}}{x^2} - \Omega^2\right); \bar{a}_{14} = -\frac{\tilde{e}_{15}\tilde{c}_{55}}{\delta}\frac{1}{x},\\
&\bar{a}_{22} = -\frac{\tilde{e}_{15}\tilde{S}_{11}}{\Delta}\frac{1}{x}; \bar{a}_{23} = \frac{\tilde{S}_{11}}{\Delta}\left(\frac{\tilde{c}_{55}}{x^2} - \Omega^2\right); \bar{a}_{24} = -1; \bar{a}_{25} = -\frac{\tilde{c}_{55}\tilde{S}_{11}}{\Delta}\frac{1}{x}.
\end{aligned}$$

If one defines the symbols $\frac{1}{\chi'} = \frac{\Psi'}{\Psi}$ and $\frac{1}{\chi''} = \frac{\Psi''}{\Psi}$, this system can be represented as follows:

$$\frac{\mathrm{d}}{\mathrm{d}x}\begin{Vmatrix}\Phi\\ \tilde{\Phi}\\ v\\ \tilde{v}\end{Vmatrix} = \begin{Vmatrix}
0 & 1 & 0 & 0\\
-\frac{1}{\chi''} & -\left(1 + \frac{\tilde{e}_{15}^2}{\delta}\right)\frac{1}{x} & \frac{\tilde{e}_{15}}{\delta}\left(\frac{\tilde{c}_{55}}{x^2} - \Omega^2\right) & -\frac{\tilde{c}_{55}\tilde{e}_{15}}{\Delta}\frac{1}{x}\\
0 & 0 & 0 & 1\\
0 & -\frac{\tilde{e}_{15}\tilde{\varepsilon}_{11}}{\delta}\frac{1}{x} & \frac{\tilde{\varepsilon}_{11}}{\delta}\left(\frac{\tilde{c}_{55}}{x^2} - \Omega^2\right) - \frac{1}{\chi''} & -\frac{\tilde{c}_{55}\tilde{\varepsilon}_{11}}{\delta}\frac{1}{x}
\end{Vmatrix}\begin{Vmatrix}\Phi\\ \tilde{\Phi}\\ v\\ \tilde{v}\end{Vmatrix}. \tag{2.49}$$

The boundary conditions take the form:

$$\tilde{c}_{55}\left(v + \tilde{v}\right) + \tilde{e}_{15}\tilde{\Phi} = 0,\ \tilde{c}_{55}\left(-v + \tilde{v}\right) + \tilde{e}_{15}\tilde{\Phi} = 0,\ \Phi|_{x=-1} = -V_0,\ \Phi|_{x=1} = V_0. \tag{2.50}$$

2.3.6.2 Numerical Analysis of Torsional Free Vibration Frequencies of a Piezoceramic Cylinder of Finite Length

To assess the accuracy of the proposed method, we compared (Table 2.1) the dimensionless vibration frequencies of a homogeneous piezoceramic cylinder simply supported at the edges, which were obtained by using this approach for a different number of collocation points, with the analytical solution [15]. The cylinder is made of the piezoceramic PZT-4. The geometric parameters of the cylinder have the following values: length of cylinder $L = 10$, inner radius $R_{-} = 3$, outer radius $R_{+} = 5$ ($\varepsilon = 0, 25$). A comparative analysis of the first six eigenvalues of the calculated based on the spline-collocation method and the analytical method is presented in Table 2.6.

The results presented in Table 2.6 indicate convergence and sufficient accuracy of the spline-collocation method that was used.

Figure 2.12 shows the dependence of the first five free vibration frequencies of the piezoceramic cylinder made from PZT-4 on relative length. The geometric parameters of the cylinder have the following values: length of the cylinder $L = 10$, inner radius $R_{-} = 3$, outer radius $R_{+} = 5$. Both end faces of the cylinders are rigidly fixed. We show the values of the natural frequencies with piezoeffect by solid lines and the values of the natural frequencies without piezoeffect ($e_{ij} = 0$) by dashed lines. The natural frequencies decrease as the relative length of the cylinder increases. The electric field does practically not affect the first frequency of free vibrations of the cylinder. However, at higher frequencies, the presence of an electric field has a more significant effect on the value of the natural frequencies of the cylinder. It is seen from Fig. 2.12 that the influence of the piezoeffect leads to an increase in the rigidity of the material, i.e., to an increase in the value of natural frequencies. In this case, in the determination of the first natural frequency, the influence of the piezoeffect can be neglected up to a relative length of $L/h = 5$. For the second frequency, the noticeable influence of the piezoeffect is observed for fairly long cylinders with $L/h < 8$ for higher frequencies, this influence is noticeable for longer cylinders.

Table 2.6 Comparative analysis of cylinder free vibration frequencies based on numerical and analytical approaches

No. of frequency	Spline-collocation method, N=24	Spline-collocation method, N=30	Analytical solution
1	0.4958	0.5050	0.5050
2	1.0311	1.0321	1.0322
3	1.5922	1.5902	1.5900
4	2.1760	2.1786	2.1788
5	2.7859	2.7879	2.7882
6	2.8130	2.8230	2.8224

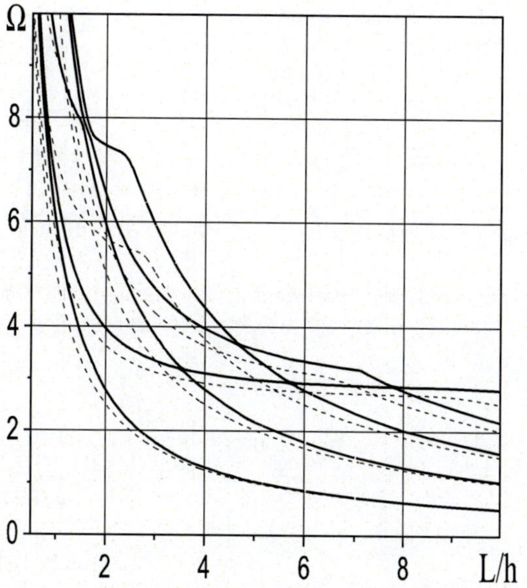

Fig. 2.12 Dependence of the first frequency of free vibrations of cylinder on relative length at different values

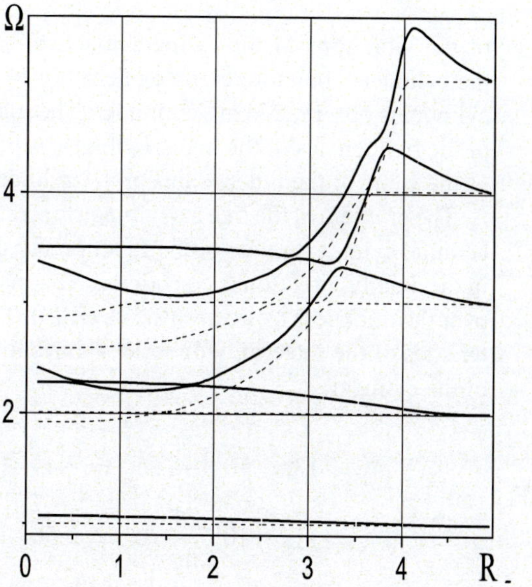

Fig. 2.13 Dependence of the first frequency of free vibrations of cylinder on the inner radius

Figure 2.13 shows a dependence of the first five natural frequencies from the internal radius of the cylinder (R_{-}), whose length and external radius diameter, in this case, remain fixed: $R_{+} = 5$ and $L = 5$. The values of natural frequencies with regard to the piezoeffect are shown by solid lines, and their values without regard to the piezoeffect ($e_{ij} = 0$) are shown by dotted lines. We consider the change in the internal radius in a wide range, namely, from 0.05 to 4.95 dimensionless units. For the cylinder material, we also chose PZT-4 piezoceramic. The first two frequencies

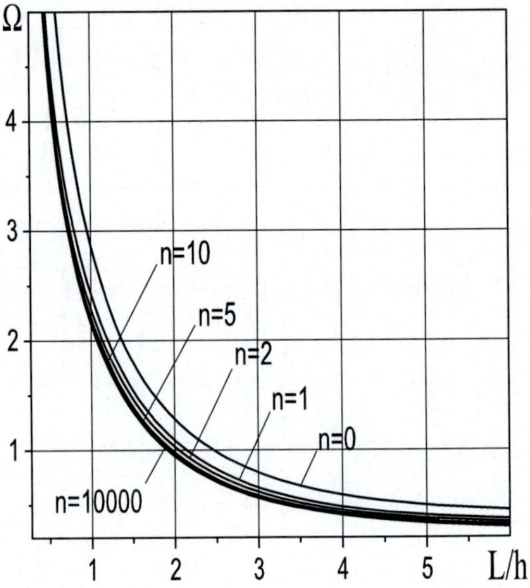

Fig. 2.14 Dependence of the first frequency of free vibrations of cylinder on relative length at different values of heterogeneity parameter

of the vibrations of the cylinder practically do not depend on the inner radius of the cylinder. The more complex behavior of the values of the free vibrations of the cylinder is observed in higher modes. The influence of the electric field on the values of the frequencies of the natural vibrations of the cylinder is observed for a decrease of the value of the inner radius of the cylinder and at high modes of the vibrations.

The problem of the free axisymmetric electroelastic vibrations of the hollow piezoceramic cylinders made of the functionally graded piezoceramic material polarized in the axially direction is considered. The material properties are assumed to change over the thickness by an law (2.28), (2.29). The physical and the mechanical characteristics of the material with index 0 have the following values on the outer surface of the cylinder:

$$c_{11}^0 = 13.9 \cdot 10^{10}\frac{\mathrm{N}}{\mathrm{m}^2},\ c_{12}^0 = 7.43 \cdot 10^{10}\frac{\mathrm{N}}{\mathrm{m}^2},\ c_{13}^0 = 7.78 \cdot 10^{10}\frac{\mathrm{N}}{\mathrm{m}^2},$$
$$c_{33}^0 = 11.5 \cdot 10^{10}\frac{\mathrm{N}}{\mathrm{m}^2},\ c_{55}^0 = 2.56 \cdot 10^{10}\frac{\mathrm{N}}{\mathrm{m}^2},\ e_{13}^0 = -5.2\frac{\mathrm{C}}{\mathrm{m}^2},$$
$$e_{15}^0 = 12.7\frac{\mathrm{C}}{\mathrm{m}^2},\ e_{33}^0 = 15,1\frac{\mathrm{C}}{\mathrm{m}^2},\ \varepsilon_{11}^0/\varepsilon_0 = 730,\ \varepsilon_{33}^0/\varepsilon_0 = 635.$$

The values of the vibration frequencies of the piezoceramic cylinder at different values of the inhomogeneity parameter n are shown in Figs. 2.14 and 2.15. Figure 2.9 shows the dependence of the first frequency of the free vibrations of cylinder on relative length. In Fig. 2.10, the dependence of the first frequency of the free vibrations of cylinder on the inner radius is shown. Note the significant increase of the value of the frequency for the short cylinders (Fig. 2.14) and high frequencies in case of a decrease in the inner radius of the cylinder (Fig. 2.15).

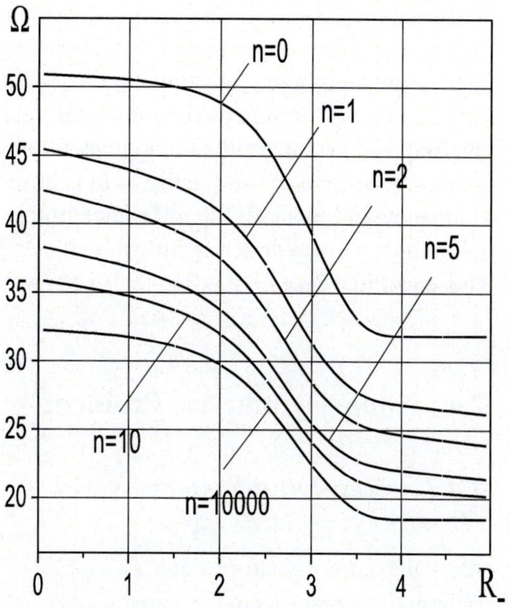

Fig. 2.15 Dependence of the first five free vibration frequencies on the inner radius of the cylinder for different values of heterogeneity parameter

Table 2.7 Comparison of free vibration frequencies of homogeneous and inhomogeneous cylinder of the finite length

No. of frequency	Inhomogeneous cylinder	Theory of effective moduli	Relative error, %	Without piezoeffect	Relative error, %
1	0.73	0.73	0.0	0.72	1.4
2	1.57	1.50	4.5	1.44	8.3
3	1.62	1.58	2.5	1.57	3.1
4	2.17	1.99	8.3	2.15	0.9
5	2.43	2.46	1.2	2.23	8.2

The values of the first five frequencies of the free vibrations of the inhomogeneous piezoceramic cylinders are compared with approximate values of the natural frequencies of the cylinder made of a homogeneous piezoceramics material with averaged characteristics calculated by the theory of effective moduli and with values of the natural frequencies of the inhomogeneity cylinders without piezoeffect (Table 2.7). The geometric parameters of the cylinder and the heterogeneity parameter of the cylinder material have the values $L = 5$, $R_+ = 5$, $n = 1, 0$. For the material of the cylinder, we also chose a PZT-4 piezoceramic.

Table 2.7 shows the values of the first five frequencies of free vibrations of the inhomogeneous piezoceramic cylinder and compares the analysis with the eigenfrequency values for a homogeneous cylinder by using the formulas of the theory of effective modules and comparison with the frequency value when neglecting the coupled electric field.

Based on the data from Table 2.7, note that the influence of the inhomogeneous structure and the coupled electric field on the values of the free vibration frequencies of the cylinder is practically the same. The exception is the effect of a nonhomogeneity in the cylinder structure on the fourth natural frequency. In this case, the difference in the frequency value for the homogeneous and inhomogeneous cylinder is 8.3%. The greatest influence of the electric field is observed for the value of the second natural frequency. In this case, the difference in the frequency value for homogeneous and inhomogeneous cylinders with coupled electric field and without piezoeffect and is also equal to 8.3%.

2.4 Nonaxisymmetric Problem

2.4.1 Resolving Systems

After the same transformation (2.1)–(2.6), we get the resolving system of partial differential equations in the general case with variable coefficients.

In the case of axial polarization, it reads

$$
\begin{aligned}
\frac{\partial^2 u_1}{\partial r^2} &= \frac{1}{\tilde{c}_{11}}\left(\frac{\tilde{c}_{11}}{r^2} - \Omega^2\right) u_1 - \frac{\tilde{c}_{66}}{\tilde{c}_{11}} \frac{1}{r^2} \frac{\partial^2 u_1}{\partial \theta^2} - \frac{\tilde{c}_{55}}{\tilde{c}_{11}} \frac{\partial^2 u_1}{\partial z^2} - \frac{1}{r} \frac{\partial u_1}{\partial r} - \\
&- \frac{\tilde{c}_{12} + \tilde{c}_{66}}{\tilde{c}_{11}} \frac{1}{r} \frac{\partial^2 u_2}{\partial r \partial \theta} - \frac{\tilde{c}_{11} + \tilde{c}_{66}}{\tilde{c}_{11}} \frac{1}{r^2} \frac{\partial u_2}{\partial \theta} - \frac{\tilde{c}_{13} + \tilde{c}_{55}}{\tilde{c}_{11}} \frac{\partial^2 u_3}{\partial r \partial z} - \frac{\tilde{e}_{13} + \tilde{e}_{15}}{\tilde{c}_{11}} \frac{\partial^2 \Phi}{\partial r \partial z}, \\
\frac{\partial^2 u_2}{\partial r^2} &= -\frac{\tilde{c}_{11} + \tilde{c}^{E}_{66}}{\tilde{c}_{66}} \frac{1}{r^2} \frac{\partial u_1}{\partial \theta} - \frac{\tilde{c}_{12} + \tilde{c}_{66}}{\tilde{c}_{66}} \frac{1}{r} \frac{\partial^2 u_1}{\partial r \partial \theta} + \frac{1}{\tilde{c}_{66}}\left(\frac{\tilde{c}_{66}}{r^2} - \Omega^2\right) u_2 - \\
&- \frac{\tilde{c}_{11}}{\tilde{c}_{66}} \frac{1}{r^2} \frac{\partial^2 u_2}{\partial \theta^2} - \frac{\tilde{c}_{55}}{\tilde{c}_{66}} \frac{\partial^2 u_2}{\partial z^2} - \frac{1}{r} \frac{\partial u_2}{\partial r} - \frac{\tilde{c}_{13} + \tilde{c}_{55}}{\tilde{c}_{66}} \frac{1}{r} \frac{\partial^2 u_3}{\partial \theta \partial z} - \frac{\tilde{e}_{13} + \tilde{e}_{15}}{\tilde{c}_{66}} \frac{1}{r} \frac{\partial^2 \Phi}{\partial \theta \partial z}, \\
\frac{\partial^2 u_3}{\partial r^2} &= -\left(1 + \frac{\delta_1}{\delta}\right) \frac{1}{r} \frac{\partial u_1}{\partial z} - \left(1 + \frac{\delta_1}{\delta}\right) \frac{\partial^2 u_1}{\partial r \partial z} - \left(1 + \frac{\delta_1}{\delta}\right) \frac{1}{r} \frac{\partial^2 u_2}{\partial \theta \partial z} - \\
&- \frac{\tilde{\varepsilon}_{11} \Omega^2}{\delta} u_3 - \frac{1}{r^2} \frac{\partial^2 u_3}{\partial \theta^2} - \frac{\delta_2}{\delta} \frac{\partial^2 u_3}{\partial z^2} - \frac{1}{r} \frac{\partial u_3}{\partial r} - \frac{\delta_3}{\delta} \frac{\partial^2 \Phi}{\partial z^2}, \\
\frac{\partial^2 \Phi}{\partial r^2} &= -\frac{\delta_4}{\delta} \frac{1}{r} \frac{\partial u_1}{\partial z} - \frac{\delta_4}{\delta} \frac{\partial^2 u_1}{\partial r \partial z} - \frac{\delta_4}{\delta} \frac{1}{r} \frac{\partial^2 u_2}{\partial \theta \partial z} - \frac{\tilde{e}_{15}}{\delta} \Omega^2 u_3 + \\
&+ \frac{\delta_5}{\delta} \frac{\partial^2 u_3}{\partial z^2} - \frac{1}{r} \frac{\partial \Phi}{\partial r} - \frac{1}{r^2} \frac{\partial^2 \Phi}{\partial \theta^2} - \frac{\tilde{c}_{55} \tilde{\varepsilon}_{33}}{\delta} \frac{\partial^2 \Phi}{\partial z^2},
\end{aligned}
\tag{2.51}
$$

where the following symbols were used:

$$\delta = c_{55}\varepsilon_{11} + e_{15}^2,\ \delta_1 = c_{13}\varepsilon_{11} + e_{13}e_{15},\ \delta_2 = c_{33}\varepsilon_{11} + e_{15}e_{33},$$
$$\delta_3 = e_{15}\varepsilon_{33} - e_{33}\varepsilon_{11},\ \delta_4 = c_{13}e_{15} - c_{55}e_{13},\ \delta_5 = c_{55}e_{33} - c_{33}e_{15}. \quad (2.52)$$

The resolving system in the case of the radial polarization:

$$\begin{aligned}
\frac{\partial^2 u_1}{\partial r^2} &= \frac{\tilde{\varepsilon}_{33}}{\delta}\left(\frac{\tilde{c}_{11}}{r^2} - \Omega^2\right)u_1 - \frac{\delta_1}{\delta}\frac{1}{r^2}\frac{\partial^2 u_1}{\partial \theta^2} - \frac{\delta_1}{\delta}\frac{\partial^2 u_1}{\partial z^2} \\
&- \left(1 + \frac{\tilde{e}_{13}\tilde{e}_{33}}{\delta}\right)\frac{1}{r}\frac{\partial u_1}{\partial r} - \\
&- \frac{\delta_1 + \tilde{c}_{11}\tilde{\varepsilon}_{33}}{\delta}\frac{1}{r^2}\frac{\partial u_2}{\partial \theta} - \frac{\delta_1 + \delta_2}{\delta}\frac{1}{r}\frac{\partial^2 u_2}{\partial r \partial \theta} - \frac{\delta_2 - \tilde{c}_{12}\tilde{\varepsilon}_{33}}{\delta}\frac{1}{r}\frac{\partial u_3}{\partial z} - \\
&- \frac{\delta_1 + \delta_2}{\delta}\frac{\partial^2 u_3}{\partial r \partial z} + \frac{\delta_3}{\delta}\frac{1}{r^2}\frac{\partial^2 \Phi}{\partial \theta^2} - \frac{\delta_3}{\delta}\frac{\partial^2 \Phi}{\partial z^2} + \frac{\tilde{e}_{13}\tilde{\varepsilon}_{33}}{\delta}\frac{1}{r}\frac{\partial \Phi}{\partial r}, \\
\frac{\partial^2 u_2}{\partial r^2} &= -\frac{\tilde{c}_{11} + \tilde{c}_{55}}{\tilde{c}_{55}}\frac{1}{r^2}\frac{\partial u_1}{\partial \theta} - \frac{\tilde{c}_{13} + \tilde{c}_{55}}{\tilde{c}_{55}}\frac{1}{r}\frac{\partial^2 u_1}{\partial r \partial \theta} - \frac{\tilde{c}_{11}}{\tilde{c}_{55}}\frac{1}{r^2}\frac{\partial^2 u_2}{\partial \theta^2} - \\
&- \frac{\tilde{c}_{66}}{\tilde{c}_{55}}\frac{\partial^2 u_2}{\partial z^2} - \frac{b}{r}\frac{\partial u_2}{\partial r} + \left(\frac{1}{r^2} - \frac{\Omega^2}{\tilde{c}_{55}}\right)u_2 - \frac{\tilde{c}_{12} + \tilde{c}_{66}}{\tilde{c}_{55}}\frac{1}{r}\frac{\partial^2 u_3}{\partial \theta \partial z} - \\
&- \frac{\tilde{e}_{15}}{\tilde{c}_{55}}\frac{1}{r^2}\frac{\partial \Phi}{\partial \theta} - \frac{\tilde{e}_{13} + \tilde{e}_{15}}{\tilde{c}_{55}}\frac{1}{r}\frac{\partial^2 \Phi}{\partial r \partial \theta}, \qquad (2.53)\\
\frac{\partial^2 u_3}{\partial r^2} &= -\frac{\tilde{c}_{12} + \tilde{c}_{55}}{\tilde{c}_{55}}\frac{1}{r}\frac{\partial u_1}{\partial z} - \frac{\tilde{c}_{13} + \tilde{c}_{55}}{\tilde{c}_{55}}\frac{\partial^2 u_1}{\partial r \partial z} - \frac{\tilde{c}_{12} + \tilde{c}_{66}}{\tilde{c}_{55}}\frac{1}{r}\frac{\partial^2 u_2}{\partial \theta \partial z} + \frac{\Omega^2}{c_{55}}u_3 - \\
&- \frac{\tilde{c}_{66}}{\tilde{c}_{55}}\frac{1}{r^2}\frac{\partial^2 u_3}{\partial \theta^2} - \frac{\tilde{c}_{11}}{\tilde{c}_{55}}\frac{\partial^2 u_3}{\partial z^2} - \frac{1}{r}\frac{\partial u_3}{\partial r} - \frac{\tilde{e}_{15}}{\tilde{c}_{55}}\frac{1}{r}\frac{\partial \Phi}{\partial z} - \frac{\tilde{e}_{13} + \tilde{e}_{15}}{\tilde{c}_{55}}\frac{\partial^2 \Phi}{\partial r \partial z}, \\
\frac{\partial^2 \Phi}{\partial r^2} &= \frac{\tilde{e}_{33}}{\delta}\left(\frac{\tilde{c}_{11}}{r^2} - \Omega^2\right)u_1 - \frac{\delta_4}{\delta}\frac{1}{r^2}\frac{\partial^2 u_1}{\partial \theta^2} - \frac{\delta_4}{\delta}\frac{\partial^2 u_1}{\partial z^2} - \frac{\tilde{c}_{33}\tilde{e}_{13}}{\delta}\frac{1}{r}\frac{\partial u_1}{\partial r} - \\
&- \frac{\delta_7 + \tilde{c}_{11}\tilde{e}_{33}}{\delta}\frac{1}{r^2}\frac{\partial u_2}{\partial \theta} - \frac{\delta_4 + \delta_5}{\delta}\frac{1}{r}\frac{\partial^2 u_2}{\partial r \partial \theta} - \frac{\delta_5 - \tilde{c}_{12}\tilde{e}_{33}}{\delta}\frac{1}{r}\frac{\partial u_3}{\partial z} - \\
&- \frac{\delta_4 + \delta_5}{\delta}\frac{\partial^2 u_3}{\partial r \partial z} - \frac{\delta_6}{\delta}\frac{1}{r^2}\frac{\partial^2 \Phi}{\partial \theta^2} - \frac{\delta_6}{\delta}\frac{\partial^2 \Phi}{\partial z^2} - \left(1 - \frac{\tilde{e}_{13}\tilde{e}_{33}}{\delta}\right)\frac{1}{r}\frac{\partial \Phi}{\partial r},
\end{aligned}$$

with the following symbols:

$$\delta = \tilde{c}_{33}\tilde{\varepsilon}_{33} + \tilde{e}_{33}^2,\ \delta_1 = \tilde{c}_{55}\tilde{\varepsilon}_{33} + \tilde{e}_{15}\tilde{e}_{33},\ \delta_2 = \tilde{c}_{13}\tilde{\varepsilon}_{33} + \tilde{e}_{13}\tilde{e}_{33},$$
$$\delta_3 = \tilde{e}_{33}\tilde{\varepsilon}_{11} - \tilde{e}_{15}\varepsilon_{33},\ \delta_4 = \tilde{c}_{55}\tilde{e}_{33} - \tilde{c}_{33}\tilde{e}_{15},\ \delta_6 = \tilde{c}_{33}\tilde{\varepsilon}_{11} + \tilde{e}_{15}\tilde{e}_{33}. \quad (2.54)$$

The resolving system in the case of circumferential polarization reads

$$\begin{aligned}
\frac{\partial^2 u_1}{\partial r^2} &= \frac{1}{\tilde{c}_{11}}\left(\frac{\tilde{c}_{33}}{r^2} - \Omega^2\right)u_1 - \frac{\tilde{c}_{55}}{\tilde{c}_{11}}\frac{1}{r^2}\frac{\partial^2 u_1}{\partial \theta^2} - \frac{\tilde{c}_{66}}{\tilde{c}_{11}}\frac{\partial^2 u_1}{\partial z^2} - \frac{1}{r}\frac{\partial u_1}{\partial r} - \frac{\tilde{c}_{33}+\tilde{c}_{55}}{\tilde{c}_{11}}\frac{1}{r^2}\frac{\partial u_2}{\partial \theta} - \\
&\quad - \frac{\tilde{c}_{13}+\tilde{c}_{55}}{\tilde{c}_{11}}\frac{1}{r}\frac{\partial^2 u_2}{\partial r \partial \theta} - \frac{\tilde{c}_{12}-\tilde{c}_{13}}{\tilde{c}_{11}}\frac{1}{r}\frac{\partial u_3}{\partial z} - \frac{\tilde{c}_{12}+\tilde{c}_{66}}{\tilde{c}_{11}}\frac{\partial^2 u_3}{\partial r \partial z} + \frac{\tilde{e}_{33}}{\tilde{c}_{11}}\frac{1}{r^2}\frac{\partial \Phi}{\partial \theta} - \frac{\tilde{e}_{13}+\tilde{e}_{15}}{\tilde{c}_{11}}\frac{1}{r}\frac{\partial^2 \Phi}{\partial r \partial \theta}, \\
\frac{\partial^2 u_2}{\partial r^2} &= -\frac{\delta_1 + \tilde{c}_{55}\tilde{\varepsilon}_{11}}{\delta}\frac{1}{r^2}\frac{\partial u_1}{\partial \theta} - \left(1 + \frac{\delta_2}{\delta}\right)\frac{1}{r}\frac{\partial^2 u_1}{\partial r \partial \theta} + \frac{\tilde{\varepsilon}_{11}}{\delta}\left(\frac{\tilde{c}_{55}}{r^2} - \Omega^2\right)u_2 - \\
&\quad - \frac{\delta_1}{\delta}\frac{1}{r^2}\frac{\partial^2 u_2}{\partial \theta^2} - \frac{\partial^2 u_2}{\partial z^2} - \frac{1}{r}\frac{\partial u_2}{\partial r} - \left(1 + \frac{\delta_2}{\delta}\right)\frac{1}{r}\frac{\partial^2 u_3}{\partial \theta \partial z} - \frac{\delta_3}{\delta}\frac{1}{r^2}\frac{\partial^2 \Phi}{\partial \theta^2}, \\
\frac{\partial^2 u_3}{\partial r^2} &= -\frac{\tilde{c}_{13}+\tilde{c}_{66}}{\tilde{c}_{66}}\frac{1}{r}\frac{\partial u_1}{\partial z} - \frac{\tilde{c}_{12}+\tilde{c}_{66}}{\tilde{c}_{66}}\frac{\partial^2 u_1}{\partial r \partial z} - \frac{\tilde{c}_{13}+\tilde{c}_{55}}{\tilde{c}_{66}}\frac{1}{r}\frac{\partial^2 u_2}{\partial \theta \partial z} - \frac{\Omega^2}{\tilde{c}_{66}}u_3 - \\
&\quad - \frac{\tilde{c}_{55}}{\tilde{c}_{66}}\frac{1}{r^2}\frac{\partial^2 u_3}{\partial \theta^2} - \frac{\tilde{c}_{11}}{\tilde{c}_{66}}\frac{\partial^2 u_3}{\partial z^2} - \frac{1}{r}\frac{\partial u_3}{\partial r} - \frac{\tilde{e}_{13}+\tilde{e}_{15}}{\tilde{c}_{66}}\frac{\partial^2 \Phi}{\partial z^2},
\end{aligned} \tag{2.55}$$

with the following symbols:

$$\begin{aligned}
&\delta = \tilde{c}_{55}\tilde{\varepsilon}_{11} + \tilde{e}_{15}^2,\ \delta_1 = \tilde{c}_{33}\tilde{\varepsilon}_{11} + \tilde{e}_{15}\tilde{e}_{33},\ \delta_2 = \tilde{c}_{13}\tilde{\varepsilon}_{11} + \tilde{e}_{13}\tilde{e}_{15}, \\
&\delta_3 = \tilde{e}_{33}\tilde{\varepsilon}_{11} - \tilde{e}_{15}\tilde{\varepsilon}_{33},\ \delta_4 = \tilde{c}_{33}\tilde{e}_{15} - \tilde{c}_{55}\tilde{e}_{33}, \\
&\delta_5 = \tilde{c}_{13}\tilde{e}_{15} - \tilde{c}_{55}\tilde{e}_{13},\ \delta_6 = \tilde{c}_{55}\tilde{\varepsilon}_{33} + \tilde{e}_{15}\tilde{e}_{33},
\end{aligned} \tag{2.56}$$

and dimensionless values:

$$\varepsilon = \frac{h}{R_0},\ \Omega = \omega h\sqrt{\frac{\rho}{\lambda}},\ \tilde{c}_{ij} = \frac{c_{ij}}{\lambda},\ \tilde{e}_{ij} = \frac{e_{ij}}{\sqrt{\varepsilon_0 \lambda}},\ \tilde{\varepsilon}_{ij} = \frac{\varepsilon_{ij}}{\varepsilon_0}, \tag{2.57}$$

where R_0 is the radius of the middle surface of the cylinder; h is the half-thickness of the cylinder; ω is the circular frequency; $\lambda = 10^{10}\text{N/m}^2$, and ε_0 is the vacuum permittivity.

2.4.2 *Solution Method for the Problem*

The basic resolving systems of the three-dimensional differential equations of the theory of electric elasticity (2.51)–(2.57) in partial derivatives in the general case with variable coefficients reduce to two-dimensional differential equations. We will use the method of separation of variables. The cylinder is a closed body in the circumferential direction. Therefore, we represent the components of the resolution vectors in the form of the standing waves in the circumferential direction. The nature of these waves depends on the preliminary polarization of piezoceramics.

For the piezoceramics polarized in axial and radial directions, the components of the resolution vector are given in the form:

$$u_1 = hU_1(r,z)\cos m\theta;\ u_2 = hU_2(r,z)\sin m\theta;$$
$$u_3 = hU_3(r,z)\cos m\theta;\ \Phi = h\sqrt{\frac{\lambda}{\varepsilon_0}}U_4(r,z)\cos m\theta. \tag{2.58}$$

In the case of the circumferential polarization, the components of the resolving vector have the form:

$$u_1 = hU_1(r,z)\cos m\theta;\ u_2 = hU_2(r,z)\sin m\theta;$$
$$u_3 = hU_3(r,z)\cos m\theta;\ \Phi = h\sqrt{\frac{\lambda}{\varepsilon_0}}U_4(r,z)\sin m\theta. \tag{2.59}$$

The systems of differential equations (2.51) in the case of the axial polarization of the material of the cylinder based on the presentation (2.58) will have the form:

$$\begin{aligned}
\frac{\partial^2 U_1}{\partial r^2} &= \frac{1}{\tilde{c}_{11}}\left(\frac{\tilde{c}_{11}+m^2\tilde{c}_{66}}{r^2}-\Omega^2\right)U_1-\frac{\tilde{c}_{55}}{\tilde{c}_{11}}\frac{\partial^2 U_1}{\partial z^2}-\frac{1}{r}\frac{\partial U_1}{\partial r}-\\
&-\frac{\tilde{c}_{11}+\tilde{c}_{66}}{\tilde{c}_{11}}\frac{m}{r^2}U_2-\frac{\tilde{c}_{12}+\tilde{c}_{66}}{\tilde{c}_{11}}\frac{m}{r}\frac{\partial U_2}{\partial r}-\frac{\tilde{c}_{11}+\tilde{c}_{66}}{\tilde{c}_{11}}\frac{\partial^2 U_3}{\partial r\partial z}-\frac{\tilde{e}_{13}+\tilde{e}_{15}}{\tilde{c}_{11}}\frac{\partial^2 U_4}{\partial r\partial z},\\
\frac{\partial^2 U_2}{\partial r^2} &= \frac{\tilde{c}_{11}+\tilde{c}_{66}^{E}}{\tilde{c}_{66}}\frac{m}{r^2}U_1+\frac{\tilde{c}_{12}+\tilde{c}_{66}}{\tilde{c}_{66}}\frac{m}{r}\frac{\partial U_1}{\partial r}+\frac{1}{\tilde{c}_{66}}\left(\frac{\tilde{c}_{66}+m^2\tilde{c}_{11}}{r^2}-\Omega^2\right)U_2-\\
&-\frac{\tilde{c}_{55}}{\tilde{c}_{66}}\frac{\partial^2 U_2}{\partial z^2}-\frac{1}{r}\frac{\partial U_2}{\partial r}+\frac{\tilde{c}_{13}+\tilde{c}_{55}}{\tilde{c}_{66}}\frac{m}{r}\frac{\partial U_3}{\partial z}+\frac{\tilde{e}_{13}+\tilde{e}_{15}}{\tilde{c}_{66}}\frac{m}{r}\frac{\partial U_4}{\partial z},\\
\frac{\partial^2 U_3}{\partial r^2} &= -\left(1+\frac{\delta_1}{\delta}\right)\frac{1}{r}\frac{\partial U_1}{\partial z}-\left(1+\frac{\delta_1}{\delta}\right)\frac{\partial^2 U_1}{\partial r\partial z}-\left(1+\frac{\delta_1}{\delta}\right)\frac{m}{r}\frac{\partial U_2}{\partial z}+\\
&+\frac{1}{\delta}\left(\frac{m^2\delta}{r^2}-\tilde{\varepsilon}_{11}\Omega^2\right)U_3-\frac{\delta_2}{\delta}\frac{\partial^2 U_3}{\partial z^2}-\frac{1}{r}\frac{\partial U_3}{\partial r}-\frac{\delta_3}{\delta}\frac{\partial^2 U_4}{\partial z^2},\\
\frac{\partial^2 U_4}{\partial r^2} &= -\frac{\delta_4}{\delta}\frac{1}{r}\frac{\partial U_1}{\partial z}-\frac{\delta_4}{\delta}\frac{\partial^2 U_1}{\partial r\partial z}-\frac{(\tilde{c}_{13}+\tilde{c}_{55})\tilde{e}_{15}}{\delta}\frac{m}{r}\frac{\partial U_2}{\partial z}-\frac{\tilde{e}_{15}\Omega^2}{\delta}U_3-\\
&-\frac{\delta_5}{\delta}\frac{\partial^2 U_3}{\partial z^2}+\frac{m^2}{r^2}U_4-\frac{\tilde{c}_{55}\tilde{\varepsilon}_{33}}{\delta}\frac{\partial^2 U_4}{\partial z^2}-\frac{1}{r}\frac{\partial U_4}{\partial r}.
\end{aligned} \tag{2.60}$$

The systems of differential equations (2.53) in the case of the radial polarization of the material of the cylinder based on the presentation (2.58) read the following:

$$
\begin{aligned}
\frac{\partial^2 U_1}{\partial r^2} &= \frac{1}{\delta}\left(\frac{\tilde{c}_{11}\tilde{\varepsilon}_{33}+m^2\delta_1}{r^2}-\Omega^2\right)U_1-\frac{\delta_1}{\delta}\frac{\partial^2 U_1}{\partial z^2}-\left(1+\frac{\tilde{e}_{13}\tilde{e}_{33}}{\delta}\right)\frac{1}{r}\frac{\partial U_1}{\partial r}-\\
&-\frac{\delta_1+\tilde{c}_{11}\tilde{\varepsilon}_{33}}{\delta}\frac{m}{r^2}U_2-\frac{\delta_1+\delta_2}{\delta}\frac{m}{r}\frac{\partial U_2}{\partial r}-\frac{\delta_2-\tilde{c}_{12}\tilde{\varepsilon}_{33}}{\delta}\frac{1}{r}\frac{\partial U_3}{\partial z}-\\
&-\frac{\delta_1+\delta_2}{\delta}\frac{\partial^2 U_3}{\partial r\partial z}+\frac{\delta_3}{\delta}\frac{m^2}{r^2}U_4-\frac{\delta_3}{\delta}\frac{\partial^2 U_4}{\partial z^2}+\frac{\tilde{e}_{13}\tilde{\varepsilon}_{33}}{\delta}\frac{1}{r}\frac{\partial U_4}{\partial r},\\
\frac{\partial^2 U_2}{\partial r^2} &= \frac{\tilde{c}_{11}+\tilde{c}_{55}}{\tilde{c}_{55}}\frac{m}{r^2}U_1+\frac{\tilde{c}_{13}+\tilde{c}_{55}}{\tilde{c}_{55}}\frac{m}{r}\frac{\partial U_1}{\partial r}+\frac{1}{\tilde{c}_{55}}\left(\frac{\tilde{c}_{55}+m^2\tilde{c}_{11}}{r^2}-\Omega^2\right)U_2-\\
&-\frac{\tilde{c}_{66}}{\tilde{c}_{55}}\frac{\partial^2 U_2}{\partial z^2}-\frac{1}{r}\frac{\partial U_2}{\partial r}+\frac{\tilde{c}_{12}+\tilde{c}_{66}}{\tilde{c}_{55}}\frac{m}{r}\frac{\partial U_3}{\partial z}+\frac{\tilde{e}_{15}}{\tilde{c}_{55}}\frac{m}{r^2}U_4+\frac{\tilde{e}_{13}+\tilde{e}_{15}}{\tilde{c}_{55}}\frac{m}{r}\frac{\partial U_4}{\partial r},\\
\frac{\partial^2 U_3}{\partial r^2} &= -\frac{\tilde{c}_{12}+\tilde{c}_{55}}{\tilde{c}_{55}}\frac{1}{r}\frac{\partial U_1}{\partial z}-\frac{\tilde{c}_{13}+\tilde{c}_{55}}{\tilde{c}_{55}}\frac{\partial^2 U_1}{\partial r\partial z}-\frac{\tilde{c}_{12}+\tilde{c}_{66}}{\tilde{c}_{55}}\frac{m}{r}\frac{\partial U_2}{\partial z}+ \qquad (2.61)\\
&+\frac{1}{\tilde{c}_{55}}\left(\frac{m^2\tilde{c}_{66}}{r^2}-\Omega^2\right)U_3-\frac{\tilde{c}_{11}}{\tilde{c}_{55}}\frac{\partial^2 U_3}{\partial z^2}-\frac{1}{r}\frac{\partial U_3}{\partial r}-\frac{\tilde{e}_{15}}{\tilde{c}_{55}}\frac{1}{r}\frac{\partial U_4}{\partial z}-\frac{\tilde{e}_{13}+\tilde{e}_{15}}{\tilde{c}_{55}}\frac{\partial^2 U_4}{\partial r\partial z};\\
\frac{\partial^2 U_4}{\partial r^2} &= \frac{1}{\delta}\left(\frac{\tilde{c}_{11}\tilde{e}_{33}+m^2\delta_4}{r^2}-\tilde{e}_{33}\Omega^2\right)U_1-\frac{\delta_4}{\delta}\frac{\partial^2 U_1}{\partial z^2}+\frac{\tilde{c}_{33}\tilde{e}_{13}}{\delta}\frac{1}{r}\frac{\partial U_1}{\partial r}-\\
&-\frac{\delta_4+\tilde{c}_{11}\tilde{e}_{33}}{\delta}\frac{m}{r^2}U_2-\frac{\delta_4+\delta_5}{\delta}\frac{m}{r}\frac{\partial U_2}{\partial r}-\frac{\delta_5-\tilde{c}_{12}\tilde{e}_{33}}{\delta}\frac{1}{r}\frac{\partial U_3}{\partial z}-\\
&-\frac{\delta_4+\delta_5}{\delta}\frac{\partial^2 U_3}{\partial r\partial z}+\frac{\delta_6}{\delta}\frac{m^2}{r^2}U_4-\frac{\delta_6}{\delta}\frac{\partial^2 U_4}{\partial z^2}--\left(1-\frac{\tilde{e}_{13}\tilde{e}_{33}}{\delta}\right)\frac{1}{r}\frac{\partial U_4}{\partial r}.
\end{aligned}
$$

The systems of differential equations (2.55) in the case of the circumferential polarization of the material of the cylinder based on the presentation (2.59) are given by

$$
\begin{aligned}
\frac{\partial^2 U_1}{\partial r^2} &= \frac{1}{\tilde{c}_{11}}\left(\frac{\tilde{c}_{33}+m^2\tilde{c}_{55}}{r^2}-\Omega^2\right)U_1-\frac{\tilde{c}_{66}}{\tilde{c}_{11}}\frac{\partial^2 U_1}{\partial z^2}-\frac{1}{r}\frac{\partial U_1}{\partial r}-\\
&-\frac{\tilde{c}_{33}+\tilde{c}_{55}}{\tilde{c}_{11}}\frac{m}{r^2}U_2-\frac{\tilde{c}_{13}+\tilde{c}_{55}}{\tilde{c}_{11}}\frac{m}{r}\frac{\partial U_2}{\partial r}-\frac{\tilde{c}_{12}-\tilde{c}_{13}}{\tilde{c}_{11}}\frac{1}{r}\frac{\partial U_3}{\partial z}-\\
&-\frac{\tilde{c}_{12}+\tilde{c}_{66}}{\tilde{c}_{11}}\frac{\partial^2 U_3}{\partial r\partial z}+\frac{\tilde{e}_{33}}{\tilde{c}_{11}}\frac{m^2}{r^2}U_4-\frac{\tilde{e}_{13}+\tilde{e}_{15}}{\tilde{c}_{11}}\frac{m}{r}\frac{\partial U_4}{\partial r},\\
\frac{\partial^2 U_2}{\partial r^2} &= \frac{(\delta_1+\tilde{c}_{55}\tilde{\varepsilon}_{11})}{\delta}\frac{m}{r^2}U_1+\left(1+\frac{\delta_2}{\delta}\right)\frac{m}{r}\frac{\partial U_1}{\partial r}+\frac{1}{\delta}\left(\frac{\tilde{c}_{55}\tilde{\varepsilon}_{11}+m^2\delta_1}{r^2}-\tilde{\varepsilon}_{11}\Omega^2\right)U_2-\\
&-\frac{\partial^2 U_2}{\partial z^2}-\frac{1}{r}\frac{\partial U_2}{\partial r}+\left(1+\frac{\delta_2}{\delta}\right)\frac{m}{r}\frac{\partial U_3}{\partial z}+\frac{\delta_3}{\delta}\frac{m^2}{r^2}U_4, \qquad (2.62)\\
\frac{\partial^2 U_3}{\partial r^2} &= -\frac{\tilde{c}_{13}+\tilde{c}_{66}}{\tilde{c}_{66}}\frac{1}{r}\frac{\partial U_1}{\partial z}-\frac{\tilde{c}_{12}+\tilde{c}_{66}}{\tilde{c}_{66}}\frac{\partial^2 U_1}{\partial r\partial z}-\frac{\tilde{c}_{13}+\tilde{c}_{55}}{\tilde{c}_{66}}\frac{m}{r}\frac{\partial U_2}{\partial z}+\\
&+\frac{1}{\tilde{c}_{66}}\left(\frac{m^2\tilde{c}_{55}}{r^2}-\Omega^2\right)U_3-\frac{\tilde{c}_{11}}{\tilde{c}_{66}}\frac{\partial^2 U_3}{\partial z^2}-\frac{1}{r}\frac{\partial U_3}{\partial r}-\frac{\tilde{e}_{13}+\tilde{e}_{15}}{\tilde{c}_{66}}\frac{\partial^2 U_4}{\partial z^2},\\
\frac{\partial^2 U_4}{\partial r^2} &= \frac{\delta_4+\tilde{c}_{55}\tilde{e}_{15}}{\delta}\frac{m}{r^2}U_1+\frac{\delta_5}{\delta}\frac{m}{r}\frac{\partial U_1}{\partial r}+\frac{1}{\delta}\left(\frac{\tilde{c}_{55}\tilde{e}_{15}+m^2\delta_4}{r^2}-\tilde{e}_{15}\Omega^2\right)U_2-\\
&-\frac{\delta_4}{\delta}\frac{1}{r}\frac{\partial U_2}{\partial r}+\frac{m\delta_5}{\delta}\frac{1}{r}\frac{\partial U_3}{\partial z}+\frac{\delta_6}{\delta}\frac{m^2}{r^2}U_4-\frac{\partial^2 U_4}{\partial z^2}-\frac{1}{r}\frac{\partial U_4}{\partial r}.
\end{aligned}
$$

To reduce the differential equations in the partial derivatives (2.60) and (2.62) to systems of ordinary differential equations, we will use the spline-collocation method. We will now search for the unknown functions $U_1(r,z)$, $U_2(r,z)$, $U_3(r,z)$, $U_4(r,z)$ for different types of polarization of the piezoceramic material of the cylinder in the form:

Axially polarization:

$$U_1(r,z) = \sum_{i=0}^{N} u_{1i}(x)\,\psi_{1i}(z)\,,\ U_2(r,z) = \sum_{i=0}^{N} u_{2i}(x)\,\psi_{1i}(z)\,,$$
$$U_3(r,z) = \sum_{i=0}^{N} u_{3i}(x)\,\psi_{2i}(z)\,,\ U_4(r,z) = \sum_{i=0}^{N} u_{4i}(x)\,\psi_{2i}(z)\,. \tag{2.63}$$

Radial and circumferential polarization:

$$U_1(r,z) = \sum_{i=0}^{N} u_{1i}(x)\,\psi_{1i}(z)\,,\ U_2(r,z) = \sum_{i=0}^{N} u_{2i}(x)\,\psi_{1i}(z)\,,$$

$$U_3(r,z) = \sum_{i=0}^{N} u_{3i}(x)\,\psi_{2i}(z)\,;\ U_4(r,z) = \sum_{i=0}^{N} u_{4i}(x)\,\psi_{1i}(z)\,, \tag{2.64}$$

where $x = (r - R_0)/h, u_{1i}(x)$, $u_{2i}(x)$, $u_{3i}(x)$, $u_{4i}(x)$ are the unknown functions with respect to the variable x, $\psi_{ji}(z)$ $(j = 1, 2; i = 0, 1, \ldots, N)$ are linear combinations of B-splines on the uniform mesh $\Delta : -L/2 = z_0 < z_1 < \cdots < z_n = L/2$. By considering the boundary conditions on the ends of the cylinder (at $z = -L/2$ and at $z = L/2$), we can see that the system involves derivatives of the solution vector that are not higher than second order. Therefore, we can restrict ourselves to an approximation by third-order spline-functions:

$$B_3^i = \frac{1}{6}\begin{cases} 0; & -\infty < z < z_{i-2} \\ \tilde{z}^3; & \tilde{z}_{i-2} \le \tilde{z} < \tilde{z}_{i-1} \\ -3\tilde{z}^2 + 3\tilde{z}^2 + 3\tilde{z} + 1; & z_{i-1} \le z < z_i \\ 3\tilde{z}^3 - 6\tilde{z}^2 + 4; & z_i \le z < z_{i+1} \\ (1 - \tilde{z})^3; & z_{i+1} \le z < z_{i+2} \\ 0; & \tilde{z}_{i+2} \le z < \infty, \end{cases} \tag{2.65}$$

where $\tilde{z} = \dfrac{(z - z_k)}{h_z}$ on the interval: $[z_k; z_{k+1}]$. $k = \overline{i-2, i+1}$; $i = \overline{-1, N+1}$; $h_z = z_{k+1} - z_k = \text{const}$.

At the same time, the functions ψ_{ji} will be written in this way:

(a) if the corresponding vector component of the solution is equal to zero, then

$$\psi_{j0}(z) = -4B_3^{-1}(z) + B_3^0(z),\ \psi_{j1}(z) = -4B_3^{-1}(z) - \frac{1}{2}B_3^0(z) + B_3^1(z),$$
$$\psi_{ji}(z) = B_3^i(z),\ \left(i = \overline{2, N-2}\right). \tag{2.66}$$

(b) if the derivative of the corresponding vector component of the solution with respect to z is equal to zero, then

$$\psi_{j0}(z) = B_3^0(z),\ \psi_{j1}(z) = B_3^{-1}(z) - \frac{1}{2}B_3^0(z) + B_3^1(z),$$
$$\psi_{ji}(z) = B_3^i(z),\ \left(i = \overline{2, N-2}\right). \tag{2.67}$$

Analogous formulae hold for $\psi_{jN-1}(z)$ and $\psi_{jN}(z)$.

Let us substitute the solution (2.63) and (2.64) in Eqs. (2.60)–(2.62) and demand that they be satisfied at the set collocation points $\xi_k \in [-L/2, L/2], k = 0, \ldots, N$.

We consider the case where the number of nodes of the mesh is even, i.e., $N = 2n + 1$ $(n \geq 3)$ (is taken into account z_0), at the collocation points choose the min the following way: $\xi_{2i} \in \left[z_{2i}, z_{2i+1}\right]$, $\xi_{2i=1} \in \left[z_{2i}, z_{2i+1}\right]$, $\left(i = \overline{1, n}\right)$. Then, on the segment $[z_{2i}, z_{2i+1}]$, we have two collocation nodes and no collocation node on the neighboring $[z_{2i+1}, z_{2i+2}]$ segments. In each of the segments $[z_{2i+1}, z_{2i+2}]$, we choose collocation points in the following way: $\xi_{2i} = z_{2i} + w_1 h_z$, $\xi_{2i+1} = z_{2i} + w_2 h_z$, $i = 1, \ldots, n$, where w_1 and w_2 are the roots of the Legendre polynomial of the second order on the segment [0, 1], which are equal to $w_1 = \frac{1}{2} - \frac{\sqrt{3}}{6}$, $w_2 = \frac{1}{2} + \frac{\sqrt{3}}{6}$. This choice of the collocation points is optimal and substantially increases the degree of accuracy of the approximation.

If we introduce the notation:

$$\Psi_j = [\psi_{ji}(\xi_k)],\ k, i = \overline{0, N},\ j = 1, 2,\ u_1 = [u_{10}, u_{11}, \ldots, u_{1N}]^T,$$
$$u_2 = [u_{20}, u_{21}, \ldots, u_{2N}]^T,\ u_3 = [u_{30}, u_{31}, \ldots, u_{3N}]^T,$$
$$u_4 = [u_{40}, u_{41}, \ldots, u_{4N}]^T,\ (k, l) \in \left\{(k, l)\ |k, l = \overline{1, 8}\right\},$$
$$\bar{a}_{kl}^T = \left\{a_{kl}\left(x, \xi_0, \Omega^2\right), a_{kl}(x, \xi_1, \Omega^2), \ldots, a_{kl}(x, \xi_N, \Omega^2)\right\}.$$

The system (2.60) is transformed into the following system of the ordinary differential equations of the high order 8 $(N + 1)$ for the functions u_1, $\tilde{u}_1$, u_2, $\tilde{u}_2$, u_3, $\tilde{u}_3$, u_4, $\tilde{u}_4$:

$$
\begin{aligned}
&\frac{\mathrm{d}u_1}{\mathrm{d}x} = \tilde{u}_1, \frac{\mathrm{d}u_2}{\mathrm{d}x} = \tilde{u}_2, \frac{\mathrm{d}u_3}{\mathrm{d}x} = \tilde{u}_3, \frac{\mathrm{d}u_4}{\mathrm{d}x} = \tilde{u}_4,\\
&\frac{\mathrm{d}\tilde{u}_1}{\mathrm{d}x} = \Psi_1^{-1}\left[\left(a_{11}\Psi_1 + a_{12}\Psi_1''\right)u_1 + a_{13}\Psi_1\tilde{u}_1 + \right.\\
&\qquad \left. + a_{14}\Psi_1 u_2 + a_{15}\Psi_1\tilde{u}_2 + a_{16}\Psi_2'\tilde{u}_3 + a_{17}\Psi_2'\tilde{u}_4\right],\\
&\frac{\mathrm{d}\tilde{u}_2}{\mathrm{d}x} = \Psi_1^{-1}\left[a_{21}\Psi_1 u_1 + a_{22}\Psi_1\tilde{u}_1 + \left(a_{23}\Psi_1 + a_{24}\Psi_1''\right)u_2 + \right.\\
&\qquad \left. + a_{25}\Psi_1\tilde{u}_2 + a_{26}\Psi_2' u_3 + a_{27}\Psi_2' u_4\right], \qquad (2.68)\\
&\frac{\mathrm{d}\tilde{u}_3}{\mathrm{d}x} = \Psi_2^{-1}\left[a_{31}\Psi_1' u_1 + a_{32}\Psi_1'\tilde{u}_1 + a_{33}\Psi_1' u_2 + \right.\\
&\qquad \left. + \left(a_{34}\Psi_2 + a_{35}\Psi_2''\right)u_3 + a_{36}\Psi_2\tilde{u}_3 + a_{37}\Psi_2'' u_4\right],\\
&\frac{\mathrm{d}\tilde{u}_4}{\mathrm{d}x} = \Psi_2^{-1}\left[a_{41}\Psi_1' u_1 + a_{42}\Psi_1'\tilde{u}_1 + a_{43}\Psi_1' u_2 + \right.\\
&\qquad \left. + \left(a_{44}\Psi_2 + a_{45}\Psi_2''\right)u_3 + \left(a_{46}\Psi_2 + a_{47}\Psi_2''\right)u_4 + a_{48}\Psi_2\tilde{u}_4\right],
\end{aligned}
$$

where

$$
\begin{aligned}
&a_{11} = \frac{\tilde{c}_{11} + m^2\tilde{c}_{66}}{\tilde{c}_{11}x^2} - \frac{\Omega^2}{\tilde{c}_{11}}, a_{12} = -\frac{\tilde{c}_{55}}{\tilde{c}_{11}}, a_{13} = -\frac{1}{x}, a_{14} = -\frac{m\left(\tilde{c}_{11} + \tilde{c}_{66}\right)}{\tilde{c}_{11}x^2},\\
&a_{15} = -\frac{m\left(\tilde{c}_{12} + \tilde{c}_{66}\right)}{\tilde{c}_{11}x}, a_{16} = -\frac{\tilde{c}_{11} + \tilde{c}_{66}}{\tilde{c}_{11}}, a_{17} = -\frac{\tilde{e}_{13} + \tilde{e}_{15}}{\tilde{c}_{11}}, a_{21} = \frac{m\left(\tilde{c}_{11} + \tilde{c}_{66}\right)}{\tilde{c}_{66}x^2},\\
&a_{22} = \frac{m\left(\tilde{c}_{12} + \tilde{c}_{66}\right)}{\tilde{c}_{66}x}, a_{23} = \left(\frac{\tilde{c}_{66} + m^2\tilde{c}_{11}}{x^2} - \Omega^2\right)\frac{1}{\tilde{c}_{66}}, a_{24} = -\frac{\tilde{c}_{55}}{\tilde{c}_{66}}, a_{25} = -\frac{1}{x},\\
&a_{26} = \frac{\tilde{c}_{13} + \tilde{c}_{55}}{\tilde{c}_{66}}\frac{m}{x}, a_{27} = \frac{\tilde{e}_{13} + \tilde{e}_{15}}{\tilde{c}_{66}}\frac{m}{x}, a_{31} = -\left(1 + \frac{\delta_1}{\delta}\right)\frac{1}{x}, a_{32} = -\left(1 + \frac{\delta_1}{\delta}\right),\\
&a_{33} = -\left(1 + \frac{\delta_1}{\delta}\right)\frac{m}{x}, a_{34} = \left(\frac{m^2\delta}{x^2} - \tilde{\varepsilon}_{11}\Omega^2\right)\frac{1}{\delta}, a_{35} = -\frac{\delta_2}{\delta}, a_{36} = -\frac{1}{x}, a_{37} = -\frac{\delta_3}{\delta},\\
&a_{41} = -\frac{\delta_4}{\delta}\frac{1}{x}, a_{42} = -\frac{\delta_4}{\delta}, a_{43} = -\frac{m\left(\tilde{c}_{13} + \tilde{c}_{55}\right)\tilde{e}_{15}}{\delta x}, a_{44} = -\frac{\tilde{e}_{15}\Omega^2}{\delta}, a_{45} = -\frac{\delta_5}{\delta},\\
&a_{46} = \frac{m^2}{x^2}, a_{47} = -\frac{\tilde{c}_{55}\tilde{\varepsilon}_{33}}{\delta}, a_{48} = -\frac{1}{x}
\end{aligned}
$$

and the system (2.61) is transformed in the following systems of the ordinary differential equations of the high order $8(N+1)$ for the functions u_1, $\tilde{u}_1$, u_2, $\tilde{u}_2$, u_3, $\tilde{u}_3$, u_4, $\tilde{u}_4$:

$$
\begin{aligned}
&\frac{du_1}{dx}=\tilde{u}_1,\ \frac{du_2}{dx}=\tilde{u}_2,\ \frac{du_3}{dx}=\tilde{u}_3,\ \frac{du_4}{dx}=\tilde{u}_4,\\
&\frac{d\tilde{u}_1}{dx}=\Psi_1^{-1}\left[\left(a_{11}\Psi_1+a_{12}\Psi_1''\right)u_1+a_{13}\Psi_1\tilde{u}_1+a_{14}\Psi_1u_2+a_{15}\Psi_1\tilde{u}_2+\right.\\
&\quad\left.+a_{16}\Psi_2'u_3+a_{17}\Psi_2'\tilde{u}_3+\left(a_{18}\Psi_1+a_{19}\Psi_1''\right)u_4+\bar{a}_{110}\Psi_1\tilde{u}_4\right],\\
&\frac{d\tilde{u}_2}{dx}=\Psi_1^{-1}\left[a_{21}\Psi_1u_1+a_{22}\Psi_1\tilde{u}_1+\left(a_{23}\Psi_1+a_{24}\Psi_1''\right)u_2+\right.\\
&\quad\left.+a_{25}\Psi_1\tilde{u}_2+a_{26}\Psi_2'u_3+a_{27}\Psi_1u_4+a_{28}\Psi_1\tilde{u}_4\right],\\
&\frac{d\tilde{u}_3}{dx}=\Psi_2^{-1}\left[a_{31}\Psi_1'u_1+a_{32}\Psi_1'\tilde{u}_1+a_{33}\Psi_1'u_2+\left(a_{34}\Psi_2+a_{35}\Psi_2''\right)u_3+\right.\\
&\quad\left.+a_{36}\Psi_2\tilde{u}_3+a_{37}\Psi_1'u_4+a_{38}\Psi_1'\tilde{u}_4\right],\\
&\frac{d\tilde{u}_4}{dx}=\Psi_1^{-1}\left[\left(a_{41}\Psi_1+a_{42}\Psi_1''\right)u_1+a_{43}\Psi_1\tilde{u}_1+a_{44}\Psi_1u_2+a_{45}\Psi_1\tilde{u}_2+\right.\\
&\quad\left.+a_{46}\Psi_2'u_3+a_{47}\Psi_2'\tilde{u}_3+\left(a_{48}\Psi_1+a_{49}\Psi_1''\right)u_4+a_{410}\Psi_1\tilde{u}_4\right],
\end{aligned}
\tag{2.69}
$$

where

$$
\begin{aligned}
&a_{11}=\frac{\tilde{c}_{11}\tilde{\varepsilon}_{33}+m^2\delta_1}{\delta x^2}-\frac{\tilde{\varepsilon}_{33}\Omega^2}{\delta},\ a_{12}=-\frac{\delta_1}{\delta},\ a_{13}=-\left(\frac{1}{x}+\frac{\tilde{e}_{13}\tilde{e}_{33}}{\delta x}\right),\\
&a_{14}=-\frac{m\left(\delta_1+\tilde{c}_{11}\tilde{\varepsilon}_{33}\right)}{\delta x^2},\ a_{15}=-\frac{m\left(\delta_1+\delta_2\right)}{\delta x},\ a_{16}=-\frac{\delta_2-\tilde{c}_{12}\tilde{\varepsilon}_{33}}{\delta x},\\
&a_{17}=-\frac{\delta_1+\delta_2}{\delta},\ a_{18}=\frac{m^2\delta_3}{\delta x^2},\ a_{19}=-\frac{\delta_3}{\delta},\ a_{110}=\frac{\tilde{e}_{13}\tilde{\varepsilon}_{33}}{\delta x},\ a_{21}=\frac{m\left(\tilde{c}_{11}+\tilde{c}_{55}\right)}{\tilde{c}_{55}x^2},\\
&a_{22}=\frac{m\left(\tilde{c}_{13}+\tilde{c}_{55}\right)}{\tilde{c}_{55}x},\ a_{23}=\frac{\tilde{c}_{55}+m^2\tilde{c}_{11}}{\tilde{c}_{55}x^2}-\frac{\Omega^2}{\tilde{c}_{55}},\ a_{24}=-\frac{\tilde{c}_{66}}{\tilde{c}_{55}},\ a_{25}=-\frac{1}{x},\\
&a_{26}=\frac{m\left(\tilde{c}_{12}+\tilde{c}_{66}\right)}{\tilde{c}_{55}x},\ a_{27}=\frac{m\tilde{e}_{15}}{\tilde{c}_{55}x^2},\ a_{28}=\frac{m\left(\tilde{e}_{13}+\tilde{e}_{15}\right)}{\tilde{c}_{55}x},\ a_{31}=-\frac{\tilde{c}_{12}+\tilde{c}_{55}}{\tilde{c}_{55}x},\\
&a_{32}=-\frac{\tilde{c}_{13}+\tilde{c}_{55}}{\tilde{c}_{55}},\ a_{33}=-\frac{m\left(\tilde{c}_{12}+\tilde{c}_{66}\right)}{\tilde{c}_{55}x},\ a_{34}=\frac{m^2\tilde{c}_{66}}{\tilde{c}_{55}x^2}-\frac{\Omega^2}{\tilde{c}_{55}},\ a_{35}=-\frac{\tilde{c}_{11}}{\tilde{c}_{55}},\\
&a_{36}=-\frac{1}{x},\ a_{37}=-\frac{\tilde{e}_{15}}{\tilde{c}_{55}x},\ a_{38}=-\frac{\tilde{e}_{13}+\tilde{e}_{15}}{\tilde{c}_{55}},\ a_{41}=\frac{\tilde{c}_{11}\tilde{e}_{33}+m^2\delta_4}{\delta x^2}-\frac{\tilde{e}_{33}\Omega^2}{\delta},\\
&a_{42}=-\frac{\delta_4}{\delta},\ a_{43}=\frac{\tilde{c}_{33}\tilde{e}_{33}}{\delta x},\ a_{44}=-\frac{m\left(\delta_4+\tilde{c}_{11}\tilde{e}_{33}\right)}{\delta x^2},\ a_{45}=-\frac{m\left(\delta_4+\delta_5\right)}{\delta x},\\
&a_{46}=-\frac{\delta_5-\tilde{c}_{12}\tilde{e}_{33}}{\delta x},\ a_{47}=-\frac{\delta_4+\delta_5}{\delta},\ a_{48}=\frac{m^2\delta_6}{\delta x^2},\\
&a_{49}=-\frac{\delta_6}{\delta},\ a_{410}=-\left(\frac{1}{x}-\frac{\tilde{e}_{13}\tilde{e}_{33}}{\delta x}\right).
\end{aligned}
$$

The system (2.62) is transformed in the following systems of the ordinary equations of the high order -8 $(N+1)$ for the functions $u_1, \tilde{u}_1, u_2, \tilde{u}_2, u_3, \tilde{u}_3, u_4, \tilde{u}_4$:

$$\frac{\mathrm{d}u_1}{\mathrm{d}x} = \tilde{u}_1,\ \frac{\mathrm{d}u_2}{\mathrm{d}x} = \tilde{u}_2,\ \frac{\mathrm{d}u_3}{\mathrm{d}x} = \tilde{u}_3,\ \frac{\mathrm{d}u_4}{\mathrm{d}x} = \tilde{u}_4,$$

$$\frac{\mathrm{d}\tilde{u}_1}{\mathrm{d}x} = \Psi_1^{-1}\left[\left(a_{11}\Psi_1 + a_{12}\Psi_1''\right)u_1 + a_{13}\Psi_1\tilde{u}_1 + a_{14}\Psi_1 u_2 + a_{15}\Psi_1\tilde{u}_2 + \right.$$
$$\left. + \ a_{16}\Psi_2' u_3 + a_{17}\Psi_2'\tilde{u}_3 + a_{18}\Psi_1 u_4 + \bar{a}_{19}\Psi_1\tilde{u}_4\right],$$

$$\frac{\mathrm{d}\tilde{u}_2}{\mathrm{d}x} = \Psi_1^{-1}\left[a_{21}\Psi_1 u_1 + a_{22}\Psi_1\tilde{u}_1 + \left(a_{23}\Psi_1 + a_{24}\Psi_1''\right)u_2 + \right. \tag{2.70}$$
$$\left. + \ a_{25}\Psi_1\tilde{u}_2 + a_{26}\Psi_2' u_3 + a_{27}\Psi_1 u_4\right], \tag{2.71}$$

$$\frac{\mathrm{d}\tilde{u}_3}{\mathrm{d}x} = \Psi_2^{-1}\left[a_{31}\Psi_1' u_1 + a_{32}\Psi_1'\tilde{u}_1 + a_{33}\Psi_1' u_2 + \left(a_{34}\Psi_2 + a_{35}\Psi_2''\right)u_3 + \right.$$
$$\left. + \ a_{36}\Psi_2\tilde{u}_3 + a_{37}\Psi_1'' u_4\right],$$

$$\frac{\mathrm{d}\tilde{u}_4}{\mathrm{d}x} = \Psi_1^{-1}\left[a_{41}\Psi_1 u_1 + a_{42}\Psi_1\tilde{u}_1 + a_{43}\Psi_1 u_2 + a_{44}\Psi_1\tilde{u}_2 + a_{45}\Psi_2' u_3 + \right.$$
$$\left. + \ \left(a_{46}\Psi_1 + a_{47}\Psi_1''\right)u_4 + a_{48}\Psi_1\tilde{u}_4\right], \tag{2.72}$$

where

$$a_{11} = \frac{\tilde{c}_{33} + m^2\tilde{c}_{55}}{\tilde{c}_{11}x^2} - \frac{\Omega^2}{\tilde{c}_{11}},\ a_{12} = -\frac{\tilde{c}_{66}}{\tilde{c}_{11}},\ a_{13} = -\frac{1}{x},\ a_{15} = -\frac{m\left(\tilde{c}_{13} + \tilde{c}_{55}\right)}{x\tilde{c}_{11}},$$

$$a_{16} = \frac{\tilde{c}_{13} - \tilde{c}_{12}}{x\tilde{c}_{11}},\ a_{17} = -\frac{\tilde{c}_{12} + \tilde{c}_{66}}{\tilde{c}_{11}},\ a_{18} = \frac{m^2 e_{33}}{x^2 c_{11}},\ a_{19} = -\frac{m\left(\tilde{e}_{13} + \tilde{e}_{15}\right)}{x\tilde{c}_{11}},$$

$$a_{21} = \frac{m\left(\delta_1 + \tilde{c}_{55}\tilde{\varepsilon}_{11}\right)}{\delta x^2},\ a_{22} = \left(1 + \frac{\delta_2}{\delta}\right)\frac{m}{x},\ a_{23} = \frac{\tilde{c}_{55}\tilde{\varepsilon}_{11} + m^2\delta_1}{\delta x^2} - \frac{\tilde{\varepsilon}_{11}\Omega^2}{\delta},$$

$$a_{24} = -1,\ a_{25} = -\frac{1}{x},\ a_{26} = \left(1 + \frac{\delta_2}{\delta}\right)\frac{m}{x},\ a_{27} = \frac{m^2\delta_4}{\delta x^2},\ a_{31} = -\frac{\tilde{c}_{13} + \tilde{c}_{66}}{x\tilde{c}_{66}}{}_1,$$

$$a_{32} = -\frac{\tilde{c}_{12} + \tilde{c}_{66}}{\tilde{c}_{66}}{}_1,\ a_{33} = -\frac{m\left(\tilde{c}_{13} + \tilde{c}_{55}\right)}{x\tilde{c}_{66}},\ a_{34} = \frac{m^2\tilde{c}_{55}}{\tilde{c}_{66}x^2} - \frac{\Omega^2}{\tilde{c}_{66}},\ a_{35} = -\frac{\tilde{c}_{11}}{\tilde{c}_{66}},$$

$$a_{36} = -\frac{1}{x},\ a_{37} = -\frac{\tilde{e}_{13} + \tilde{e}_{15}}{\tilde{c}_{66}},\ a_{41} = \frac{m\left(\delta_4 + \tilde{c}_{55}\tilde{e}_{15}\right)}{\delta x^2},\ a_{42} = \frac{m\delta_4}{\delta x},$$

$$a_{43} = \frac{\tilde{c}_{55}\tilde{e}_{15} + m^2\delta_4}{\delta x^2} - \frac{\tilde{e}_{15}\Omega^2}{\delta},\ a_{44} = -\frac{\delta_4}{\delta x},\ a_{45} = \frac{m\delta_5}{\delta x},\ a_{46} = \frac{m^2\delta_6}{\delta x^2},$$

$$a_{47} = -1,\ a_{48} = -\frac{1}{x}.$$

These systems (2.68)–(2.70) can be reduced in the form:

$$\frac{\mathrm{d}R}{\mathrm{d}x} = A\left(x, \Omega\right)R, \tag{2.73}$$

where

$$R = \{u_{10}, u_{11}, \ldots, u_{1N}, \tilde{u}_{10}, \tilde{u}_{11}, \ldots, \tilde{u}_{1N}, u_{20}, u_{21}, \ldots, u_{2N}, \tilde{u}_{20}, \tilde{u}_{21}, \ldots, \tilde{u}_{2N},$$
$$u_{30}, u_{31}, \ldots, u_{3N}, \tilde{u}_{30}, \tilde{u}_{31}, \ldots, \tilde{u}_{3N}, u_{40}, u_{41}, \ldots, u_{4N}, \tilde{u}_{40}, \tilde{u}_{41}, \ldots, \tilde{u}_{4N}\}^T .$$

Here, the following symbols are used:

$$\frac{1}{\chi'_{11}} = \frac{\Psi'_1}{\Psi_1}, \frac{1}{\chi'_{12}} = \frac{\Psi'_1}{\Psi_2}, \frac{1}{\chi'_{21}} = \frac{\Psi'_2}{\Psi_1}, \frac{1}{\chi'_{22}} = \frac{\Psi'_2}{\Psi_2}, \frac{1}{\chi''_{11}} = \frac{\Psi''_1}{\Psi_1},$$
$$\frac{1}{\chi''_{12}} = \frac{\Psi''_1}{\Psi_2}, \frac{1}{\chi''_{21}} = \frac{\Psi''_2}{\Psi_1}, \frac{1}{\chi''_{22}} = \frac{\Psi''_2}{\Psi_2}, \text{ and also :}$$

Axial polarization:

$$\alpha_1 = \frac{\tilde{c}_{11} + m^2\tilde{c}_{66}}{\tilde{c}_{11}x^2} - \frac{\tilde{c}_{55}}{\tilde{c}_{11}\chi''_{11}}, \alpha_2 = \frac{1}{\tilde{c}_{11}}, \beta_1 = \frac{\tilde{c}_{66} + m^2\tilde{c}_{11}}{x^2\tilde{c}_{66}} - \frac{\tilde{c}_{55}}{\tilde{c}_{66}\chi''_{11}},$$
$$\beta_2 = \frac{1}{\tilde{c}_{66}}, \beta_2 = \frac{1}{\tilde{c}_{66}}, \gamma_1 = \frac{m^2\delta}{x^2} + \frac{\delta_2}{\delta\chi'_{22}}, \gamma_2 = \frac{\tilde{\varepsilon}_{11}}{\delta}, \tau_1 = -\frac{\delta_5}{\delta\chi''_{22}}, \tau_2 = \frac{\tilde{e}_{15}}{\delta};$$

Radial polarization:

$$\alpha_1 = \frac{\tilde{c}_{11}\tilde{\varepsilon}_{33} + m^2\delta_1}{\delta x^2} - \frac{\delta_1}{\delta\chi''_{11}}, \alpha_2 = \frac{\tilde{\varepsilon}_{33}}{\delta}, \beta_1 = \frac{\tilde{c}_{55} + m^2\tilde{c}_{11}}{x^2\tilde{c}_{55}} - \frac{\tilde{c}_{66}}{\tilde{c}_{55}\chi''_{11}}, \beta_2 = \frac{1}{\tilde{c}_{55}},$$
$$\gamma_1 = \frac{m^2\tilde{c}_{66}}{\tilde{c}_{55}x^2} - \frac{\tilde{c}_{11}}{\tilde{c}_{55}\chi'_{22}}, \gamma_2 = \frac{1}{\tilde{c}_{55}}, \tau_1 = \frac{\tilde{c}_{11}\tilde{e}_{33} + m^2\delta_4}{\delta x^2} - \frac{\delta_4}{\delta\chi''_{11}}, \tau_2 = \frac{\tilde{e}_{33}}{\delta};$$

Circumferential polarization:

$$\alpha_1 = \frac{\tilde{c}_{33} + m^2\tilde{c}_{55}}{x^2\tilde{c}_{11}} - \frac{\tilde{c}_{66}}{\tilde{c}_{11}\chi''_{11}}, \alpha_2 = \frac{1}{\tilde{c}_{11}}, \beta_1 = \frac{\tilde{c}_{55}\tilde{\varepsilon}_{11} + m^2\delta_3}{\delta x^2} - \frac{1}{\chi''_{11}}, \beta_2 = \frac{\tilde{\varepsilon}_{11}}{\delta},$$
$$\gamma_1 = \frac{m^2\tilde{c}_{55}}{\tilde{c}_{66}x^2} - \frac{\tilde{c}_{11}}{\tilde{c}_{66}\chi''_{22}}, \gamma_2 = \frac{1}{\tilde{c}_{66}}, \tau_1 = \frac{\tilde{c}_{55}\tilde{e}_{15} + m^2\delta_4}{\delta x^2}, \tau_2 = \frac{\tilde{e}_{15}}{\delta}.$$

Then, matrix A in the case of the system (2.68) is given by

$$\left\| \begin{array}{cccccccc}
0 & 1 & 0 & 0 & 0 & 0 & 0 & 0 \\
\alpha_1 - \alpha_2\Omega^2 & -\frac{1}{x} & -\frac{m(\tilde{c}_{11}+\tilde{c}_{66})}{\tilde{c}_{11}x^2} & -\frac{m(\tilde{c}_{12}+\tilde{c}_{66})}{\tilde{c}_{11}x} & \frac{\tilde{c}_{13}-\tilde{c}_{12}}{x\tilde{c}_{11}\chi'_{21}} & -\frac{\tilde{c}_{12}+\tilde{c}_{66}}{\tilde{c}_{11}\chi'_{21}} & \frac{m^2 e_{33}}{x^2 c_{11}} & -\frac{m(\tilde{e}_{13}+\tilde{e}_{15})}{x\tilde{c}_{11}} \\
0 & 0 & 0 & 1 & 0 & 0 & 0 & 0 \\
\frac{m(\tilde{c}_{11}+\tilde{c}_{66})}{\tilde{c}_{66}x^2} & \frac{m(\tilde{c}_{12}+\tilde{c}_{66})}{\tilde{c}_{66}x} & \beta_1 - \beta_2\Omega^2 & -\frac{1}{x} & \frac{m(\tilde{c}_{13}+\tilde{c}_{55})}{x\tilde{c}_{66}\chi'_{21}} & 0 & \frac{m(\tilde{e}_{13}+\tilde{e}_{15})}{x\tilde{c}_{66}\chi'_{21}} & 0 \\
0 & 0 & 0 & 0 & 0 & 1 & 0 & 0 \\
-\frac{(\delta+\delta_1)}{\delta x\chi'_{12}} & -\frac{(\delta+\delta_1)}{\delta\chi'_{12}} & -\frac{m(\delta+\delta_1)}{\delta x\chi'_{12}} & 0 & \gamma_1 - \gamma_2\Omega^2 & -\frac{1}{x} & -\frac{\delta_3}{\delta\chi''_{22}} & 0 \\
0 & 0 & 0 & 0 & 0 & 0 & 0 & 1 \\
-\frac{\delta_4}{\delta x\chi'_{12}} & -\frac{\delta_4}{\delta\chi'_{12}} & -\frac{m\tilde{e}_{15}(\tilde{c}_{13}+\tilde{c}_{55})}{\delta x\chi'_{12}} & 0 & \tau_1 - \tau_2\Omega^2 & 0 & \frac{m^2}{x^2} - \frac{\tilde{c}_{55}\tilde{\varepsilon}_{33}}{\Delta\chi''_{22}} & -\frac{1}{x}
\end{array} \right\| .$$

The boundary conditions will have the form:

$$B_1 R(-1) = 0,\; B_2 R(1) = 0. \tag{2.74}$$

In the case of the system (2.68), the matrices B_1 and B_2 are given by

$$B_1 = \begin{pmatrix} -\tilde{c}_{12}\Psi_1 & \tilde{c}_{11}\Psi_1 & -A_{12}m\Psi_1 & 0 & \tilde{c}_{13}\Psi_2' & 0 & \tilde{e}_{13}\Psi_2' & 0 \\ -m\Psi_1 & 0 & \Psi_1 & \Psi_1 & 0 & 0 & 0 & 0 \\ \tilde{c}_{55}\Psi_1' & 0 & 0 & 0 & 0 & \tilde{c}_{55}\Psi_2 & 0 & \tilde{e}_{15}\Psi_2 \\ 0 & 0 & 0 & 0 & 0 & 0 & -1 & 0 \end{pmatrix},$$

$$B_2 = \begin{pmatrix} \tilde{c}_{12}\Psi_1 & \tilde{c}_{11}\Psi_1 & A_{12}m\Psi_1 & 0 & \tilde{c}_{13}\Psi_2' & 0 & \tilde{e}_{13}\Psi_2' & 0 \\ m\Psi_1 & 0 & -\Psi_1 & \Psi_1 & 0 & 0 & 0 & 0 \\ \tilde{c}_{55}\Psi_1' & 0 & 0 & 0 & 0 & \tilde{c}_{55}\Psi_2 & 0 & \tilde{e}_{15}\Psi_2 \\ 0 & 0 & 0 & 0 & 0 & 0 & 1 & 0 \end{pmatrix}.$$

Then, matrix A in the case of the system (2.69) is given by

$$\begin{Vmatrix} 0 & 1 & 0 & 0 & 0 & 0 & 0 & 0 \\ \alpha_1 - \alpha_2\Omega^2 & -\frac{1}{x} - \frac{\tilde{e}_{13}\tilde{e}_{33}}{\Delta x} & -\frac{m\Delta_2}{\Delta x^2} & -\frac{m\Delta_3}{x\Delta} & -\frac{\Delta_4}{\Delta x \chi_{21}'} & -\frac{\Delta_2}{\Delta\chi_{21}'} & \frac{m^2\Delta_5}{\Delta x^2} - \frac{\Delta_5}{\Delta\chi_{11}'} & -\frac{\tilde{e}_{13}\tilde{S}_{33}}{\Delta}\frac{1}{x} \\ 0 & 0 & 0 & 1 & 0 & 0 & 0 & 0 \\ \frac{m(\tilde{c}_{11}+\tilde{c}_{55})}{\tilde{c}_{55}x^2} & \frac{m(\tilde{c}_{13}+\tilde{c}_{55})}{\tilde{c}_{55}x} & \beta_1 - \beta_2\Omega^2 & -\frac{1}{x} & \frac{m(\tilde{c}_{12}+\tilde{c}_{66})}{x\tilde{c}_{55}\chi_{21}'} & 0 & \frac{\tilde{e}_{15}}{\tilde{c}_{55}}\frac{m}{x^2} & \frac{m(\tilde{e}_{13}+\tilde{e}_{15})}{\tilde{c}_{55}x} \\ 0 & 0 & 0 & 0 & 0 & 1 & 0 & 0 \\ -\frac{(\tilde{c}_{12}+\tilde{c}_{55})}{\tilde{c}_{55}x\chi_{12}'} & -\frac{(\tilde{c}_{13}+\tilde{c}_{55})}{\tilde{c}_{55}\chi_{12}'} & -\frac{m(\tilde{c}_{12}+\tilde{c}_{66})}{\tilde{c}_{55}x\chi_{12}'} & 0 & \gamma_1 - \gamma_2\Omega^2 & -\frac{1}{x} & -\frac{\tilde{e}_{15}}{\tilde{c}_{55}x\chi_{12}'} & -\frac{\tilde{e}_{13}+\tilde{e}_{15}}{\tilde{c}_{55}\chi_{12}'} \\ 0 & 0 & 0 & 0 & 0 & 0 & 0 & 1 \\ \delta_1 - \delta_2\Omega^2 & -\frac{\tilde{c}_{33}\tilde{e}_{33}}{\Delta}\frac{1}{x} & -\frac{m\Delta_7}{x^2\Delta} & -\frac{m\Delta_8}{x\Delta} & -\frac{\Delta_9}{x\Delta\chi_{21}'} & 0 & \frac{m^2\delta_6}{\delta x^2} - \frac{\delta_6}{\delta\chi_{11}'} & \left(\frac{\tilde{e}_{13}\tilde{e}_{33}}{\delta} - 1\right)\frac{1}{x} \end{Vmatrix}.$$

In the case of the system (2.69), the matrices B_1 and B_2 (2.74) are given by

$$B_1 = \begin{Vmatrix} -\tilde{c}_{13}\Psi_1 & \tilde{c}_{33}\Psi_1 & -\tilde{c}_{13}m\Psi_1 & 0 & \tilde{c}_{13}\Psi_2' & 0 & 0 & \tilde{e}_{33}\Psi_1 \\ \tilde{c}_{55}m\Psi_1 & 0 & \tilde{c}_{55}\Psi_1 & \tilde{c}_{55}\Psi_1 & 0 & 0 & e_{15}m\Psi_1 & 0 \\ \tilde{c}_{55}\Psi_1' & 0 & 0 & 0 & 0 & \tilde{c}_{55}\Psi_2 & e_{15}\Psi_1' & 0 \\ 0 & 0 & 0 & 0 & 0 & 0 & -1 & 0 \end{Vmatrix},$$

$$B_2 = \begin{Vmatrix} \tilde{c}_{13}\Psi_1 & \tilde{c}_{33}\Psi_1 & \tilde{c}_{13}m\Psi_1 & 0 & \tilde{c}_{13}\Psi_2' & 0 & 0 & \tilde{e}_{33}\Psi_1 \\ -\tilde{c}_{55}m\Psi_1 & 0 & -\tilde{c}_{55}\Psi_1 & \tilde{c}_{55}\Psi_1 & 0 & 0 & -e_{15}m\Psi_1 & 0 \\ \tilde{c}_{55}\Psi_1' & 0 & 0 & 0 & 0 & \tilde{c}_{55}\Psi_2 & e_{15}\Psi_1' & 0 \\ 0 & 0 & 0 & 0 & 0 & 0 & 1 & 0 \end{Vmatrix}.$$

Then, in the case of the system (2.70), matrix A is given by

$$\left\| \begin{matrix} 0 & 1 & 0 & 0 & 0 & 0 & 0 & 0 \\ \alpha_1 - \alpha_2 \Omega^2 & -\frac{1}{x} & -\frac{m\left(\tilde{c}_{33} + \tilde{c}_{55}\right)}{x^2 \tilde{c}_{11}} & -\frac{m\left(\tilde{c}_{13} + \tilde{c}_{55}\right)}{x \tilde{c}_{11}} & \frac{\tilde{c}_{13} - \tilde{c}_{12}}{x \tilde{c}_{11} \chi'_{21}} & -\frac{\tilde{c}_{12} + \tilde{c}_{66}}{\tilde{c}_{11} \chi'_{21}} & \frac{m^2 e_{33}}{x^2 c_{11}} & \frac{m\left(\tilde{e}_{13} + \tilde{e}_{15}\right)}{x \tilde{c}_{11}} \\ 0 & 0 & 0 & 1 & 0 & 0 & 0 & 0 \\ \frac{m\left(\delta_1 + \tilde{c}_{55} \tilde{\varepsilon}_{11}\right)}{\delta x^2} & \left(1 + \frac{\delta_2}{\delta}\right) \frac{m}{x} & \beta_1 - \beta_2 \Omega^2 & -\frac{1}{x} & \left(1 + \frac{\delta_2}{\delta}\right) \frac{m}{x \chi'_{21}} & 0 & \frac{m^2 \delta_4}{\delta x^2} & 0 \\ 0 & 0 & 0 & 0 & 0 & 1 & 0 & 0 \\ -\frac{\tilde{c}_{13} + \tilde{c}_{66}}{x \tilde{c}_{66} \chi'_{12}} & -\frac{\tilde{c}_{12} + \tilde{c}_{66}}{\tilde{c}_{66} \chi'_{12}} & -\frac{m\left(\tilde{c}_{13} + \tilde{c}_{55}\right)}{x \tilde{c}_{66} \chi'_{12}} & 0 & \gamma_1 - \gamma_2 \Omega^2 & -\frac{1}{x} & -\frac{\tilde{e}_{13} + \tilde{e}_{15}}{\tilde{c}_{66} \chi''_{12}} & 0 \\ 0 & 0 & 0 & 0 & 0 & 0 & 0 & 1 \\ \frac{m\left(\delta_4 + \tilde{c}_{55} \tilde{e}_{15}\right)}{\delta x^2} & \frac{m \delta_4}{\delta x} & \tau_1 - \tau_2 \Omega^2 & -\frac{\delta_4}{\delta x} & \frac{m \delta_5}{\delta x \chi'_{21}} & 0 & \frac{m^2 \delta_6}{\delta x^2} - \frac{1}{\chi''_{11}} & -\frac{1}{x} \end{matrix} \right\|.$$

In the case of the system (2.70), the matrices B_1 and B_2 (2.74) are given by

$$B_1 = \left\| \begin{matrix} -\tilde{c}_{13}\Psi_1 & \tilde{c}_{11}\Psi_1 & -\tilde{c}_{13}m\Psi_1 & 0 & \tilde{c}_{12}\Psi'_2 & 0 & \tilde{e}_{13}m\Psi_1 & 0 \\ \tilde{c}_{55}m\Psi_1 & 0 & \tilde{c}_{55}\Psi_1 & \tilde{c}_{55}\Psi_1 & 0 & 0 & 0 & e_{15}\Psi_1 \\ \Psi'_1 & 0 & 0 & 0 & 0 & \Psi_2 & 0 & 0 \\ 0 & 0 & 0 & 0 & 0 & 0 & -1 & 0 \end{matrix} \right\|,$$

$$B_2 = \left\| \begin{matrix} \tilde{c}_{13}\Psi_1 & \tilde{c}_{11}\Psi_1 & \tilde{c}_{13}m\Psi_1 & 0 & \tilde{c}_{12}\Psi'_2 & 0 & \tilde{e}_{13}m\Psi_1 & 0 \\ -\tilde{c}_{55}m\Psi_1 & 0 & -\tilde{c}_{55}\Psi_1 & \tilde{c}_{55}\Psi_1 & 0 & 0 & 0 & e_{15}\Psi_1 \\ \Psi'_1 & 0 & 0 & 0 & 0 & \Psi_2 & 0 & 0 \\ 0 & 0 & 0 & 0 & 0 & 0 & 1 & 0 \end{matrix} \right\|.$$

The solution of the above boundary problems for eigenvalues for the systems of the ordinary differential equations is made by a stable numerical method of discrete orthogonalization in combination with the step-by-step search method.

2.4.3 *Numerical Analysis of Nonaxisymmetric Free Vibration Frequencies of a Piezoceramic Cylinder with Finite Length in the Case of Axial Polarization*

2.4.3.1 Homogeneous Piezoceramic Cylinder

To assess the accuracy of the calculation result of the studied problems based on the proposed method, we compared (Table 2.8) the dimensionless vibration frequencies of a homogeneous piezoceramic cylinder simply supported at the edges, which were obtained by using this approach for a different number of N collocation points with the analytical solution [17]. The cylinder was made of the piezoceramic PZT-4. The geometric parameters of the cylinder have the following values: length of the cylinder $L = 10$, inner radius $R_0 - h = 3$, outer radius $R_0 + h = 5$, $(\varepsilon = 0, 25)$. The comparative analysis of the first six eigenvalues of the calculated based on the spline-collocation method and the analytical method is presented in Table 2.8.

The results presented in Table 2.8 indicate convergence and sufficient accuracy of the spline-collocation method used.

Table 2.8 Comparative analysis of cylinder free vibration frequencies based on numerical and analytical approaches

No. of frequency	Spline-collocation method, $N = 24$	Spline-collocation method $N = 30$	Analytical method
1	0.1501	0.1412	0.1411
2	0.2743	0.2739	0.2738
3	0.3774	0.3767	0.3764
4	0.4358	0.4339	0.4338
5	0.5081	0.5072	0.5071
6	0.6057	0.6048	0.6045

Figure 2.16 shows the dependence of the first five free vibration frequencies of the piezoceramic cylinder on relative length. The geometric parameters of the cylinder have the following values length of the cylinder $L = 10$, inner radius $R_0 - h = 3$, outer radius $R_0 + h = 5$. Both of the cylinders are rigidly fixed. We show the values of the natural frequencies with piezoeffect by solid lines and the values of the natural frequencies without piezoeffect ($e_{ij} = 0$) by dashed lines. The natural frequencies decrease as the relative length of the cylinder increases. The electric field practically does not affect the first frequency of free vibrations of the cylinder. However, at higher frequencies, the presence of an electric field has a more significant effect on the value of the natural frequencies of the cylinder.

Figure 2.17 shows the dependence of the first five natural frequencies on the cylinder inner radius, $R_0 - h$ and here its length ($L = 5$) and the external radius $R_0 + h = 5$ remain fixed. We consider the change in the inner radius from 0.05 to 4.5 dimensionless units. The cylinder material is PZT-4 piezoceramics as before. The solid and dashed lines show the frequencies of natural vibrations of piezoceramic and elastic($e_{ij} = 0$) cylinders, respectively. For the first three eigenfrequencies, an increase in eigenfrequency values is observed with an increase in cylinder radius, in the case of higher frequencies, the nature of the spectrum becomes more complicated, intervals appear where there is a slight dependence of eigenfrequency values on the considered geometric parameter of the cylinder. It is worth noting that the influence of piezoelectric effect on the spectrum of natural frequencies is insignificant only for the first four frequencies. For higher frequencies, we observe a more substantial influence of the piezoelectric effect on the natural frequencies of nonaxisymmetric vibrations.

2.4.3.2 Inhomogeneous Piezoceramic Cylinder

The problem of the free nonaxisymmetric electroelastic vibrations of the hollow piezoceramic cylinders made of the functionally graded piezoceramic material polarized in the axial direction is considered.The material properties are assumed to change

Fig. 2.16 Dependence of the first five natural frequencies on the cylinder on relative length

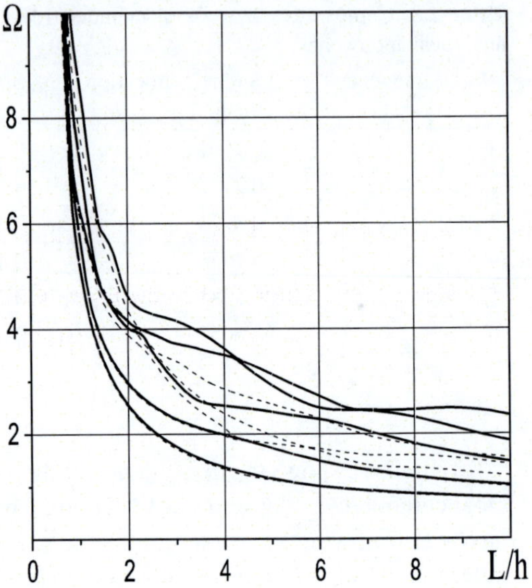

Fig. 2.17 Dependence of the first five natural frequencies on the cylinder inner radius

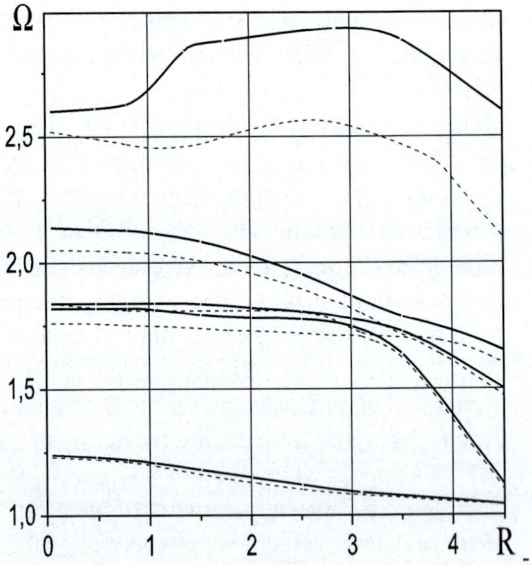

over the thickness by an law (2.28), (2.29). The physical and the mechanical characteristics of the material with index 0 that have values on the outer surface $R_0 + h = 5$ of the cylinder have the values:

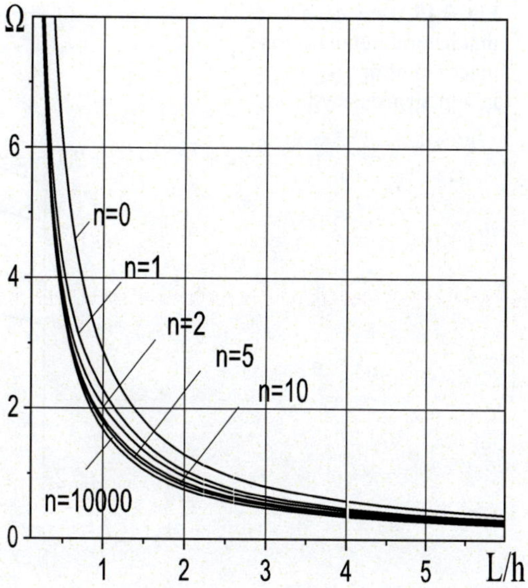

Fig. 2.18 Dependence of first natural frequency on relative length of inhomogeneous cylinder

$$
\begin{aligned}
&c^0_{11} = 13.9 \cdot 10^{10}\frac{\mathrm{N}}{\mathrm{m}^2}, c^0_{12} = 7.43 \cdot 10^{10}\frac{\mathrm{N}}{\mathrm{m}^2}, c^0_{13} = 7.78 \cdot 10^{10}\frac{\mathrm{N}}{\mathrm{m}^2},\\
&c^0_{33} = 11.5 \cdot 10^{10}\frac{\mathrm{N}}{\mathrm{m}^2}, c^0_{55} = 2.56 \cdot 10^{10}\frac{\mathrm{N}}{\mathrm{m}^2}, e^0_{13} = -5, 2\frac{\mathrm{C}}{\mathrm{m}^2},\\
&e^0_{15} = 12.7\frac{\mathrm{C}}{\mathrm{m}^2}, e^0_{33} = 15.1\frac{\mathrm{C}}{\mathrm{m}^2}, \varepsilon^0_{11}/\varepsilon_0 = 730,\ \varepsilon^0_{33}/\varepsilon_0 = 635.
\end{aligned}
\tag{2.75}
$$

Figure 2.18 shows the dependence of the first natural frequency on the relative length of the inhomogeneous cylinder for different values of the inhomogeneity n ($m = 1$).

Figure 2.19 shows the dependence of the first natural frequency on the inner radius $R_0 - h$ of the inhomogeneous cylinder for the different values of the inhomogeneity parameter n, the outer radius remains at the constant value ($R_0 + h = 5$) and the value of the cylinder length is $L = 5$ ($m = 1$).

Note the significant increase of the value of the frequency for the short cylinders (Fig. 2.18) and high frequencies in case of a decrease in the inner radius of the cylinder (Fig. 2.19).

We will study the effect of the coupled electric field and the inhomogeneity parameter on the values of the natural vibration frequencies of the cylinder. Let us consider the free vibration frequencies of the inhomogeneous and the homogeneous the cylinder with averaged with averaged characteristics calculated by the theory of effective moduli, when the moduli change by a power law:

$$
\mu = \frac{\mu_0}{2h_-}\int\limits_{R_0-h}^{R_0+h} e^{nr}\,\mathrm{d}r = \frac{\mu_0}{2nh}e^{n(R_0-h)}\left(e^{2nh} - 1\right). \tag{2.76}
$$

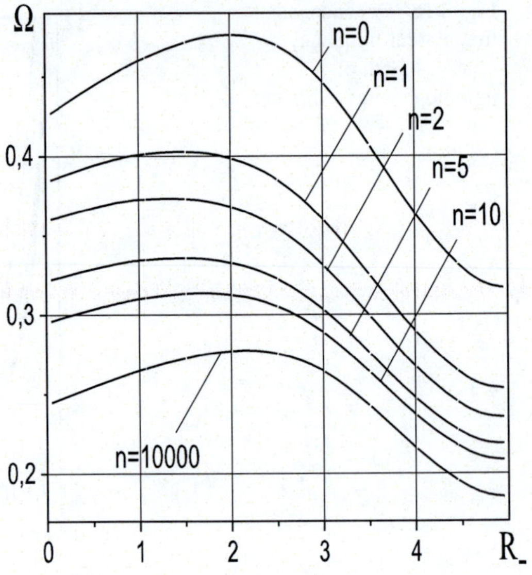

Fig. 2.19 Dependence of first natural frequency on inner radius of inhomogeneous cylinder

Table 2.9 Comparison of free vibration frequencies of homogeneous and inhomogeneous cylinder

No. of frequency	Inhomogeneous cylinder	The theory of effective moduli	Relative error, %	Without piezoeffect	Relative error, %
1	1.15	1.16	0.9	1.14	0.9
2	2.18	2.18	0.0	1.79	17.9
3	2.31	2.35	1.8	2.04	11.7
4	2.82	2.64	6.4	2.19	22.3
5	2.98	3.04	2.0	2.84	4.7

The values of the first five frequencies of the free vibrations of the inhomogeneous piezoceramic cylinders are compared with approximate values of the natural frequencies of the cylinder made of a homogeneous piezoceramics material with averaged characteristics calculated by the theory of effective moduli and with values of the natural frequencies of the inhomogeneity cylinders without piezoeffect (Table 2.9). The geometric parameters of the cylinder and the heterogeneity parameter of the cylinder material have values: $R_0 = 4, h = 1, n = 1, 5, m = 1$. As material of the cylinder, we also chose PZT-4 piezoceramic with the preliminary axial polarization.

Based on the data in Table 2.9, note that the relative error when neglecting the inhomogeneity factor is less than the relative error in the case of neglecting the coupled electric field. We have the greatest relative error in the value of the third natural frequency (22.3%) for the case of neglecting the coupled electric field. The greatest relative error that occurs for the value of the fourth natural frequency in

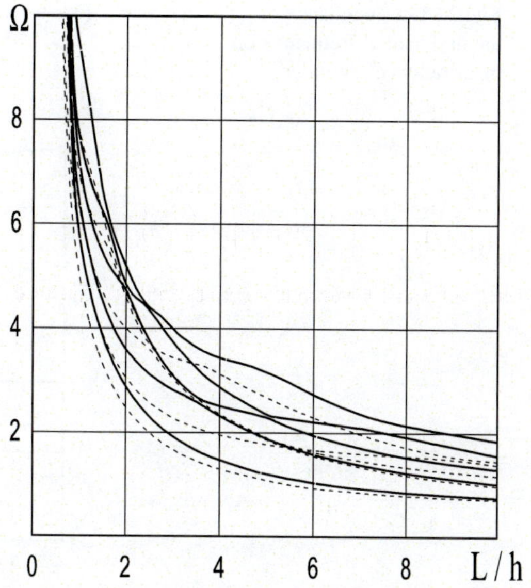

Fig. 2.20 Dependence of the first natural frequency on the relative length of cylinder

the case of neglecting the inhomogeneity factor (6.4%) has a small value of the heterogeneity parameter ($n = 1, 5$).

2.4.4 Numerical Analysis of Nonaxisymmetric Free Vibration Frequencies of the Piezoceramic Cylinder with Finite Length in the Case of Radial Polarization

2.4.4.1 Homogeneous Cylinder

Figure 2.20 shows the dependence of the first five free vibration frequencies of the piezoceramic cylinder on the relative length L/h. For the cylinder material, we also chose a PZT-4 piezoceramic. The geometric parameters of the cylinder have the following values: inner radius $R_0 - h = 3$, outer radius $R_0 + h = 5$. Both end faces of the cylinders are rigidly fixed. We show the values of the natural frequencies with piezoeffect by solid lines and the values of the natural frequencies without piezoeffect ($e_{ij} = 0$) by dashed lines. From the results above, it can be seen that the influence of the piezoelectric effect leads to a "tightening" of the material, that is, to an increase in the value of the natural frequencies. In this case, when determining the first and second natural frequencies, the influence of the piezoelectric effect can be neglected for all the lengths of the cylinder. For higher frequencies, this effect is noticeable for longer cylinders. The natural frequencies decrease as the relative length of the cylinder increases.

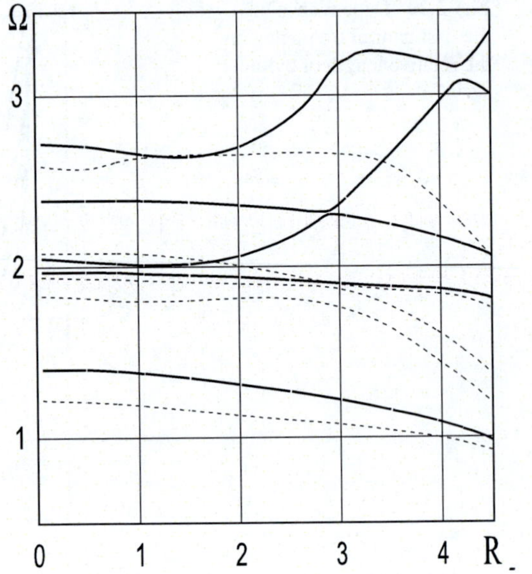

Fig. 2.21 Dependence of the first natural frequency on inner radius of cylinder

Figure 2.21 shows the dependence of the first five free vibration frequencies of the piezoceramic cylinder on the inner radius of the cylinder $R_0 - h$. Note that the cylinder length ($L = 5$) and the outer diameter ($R_0 + h = 5$) are fixed. The analysis of the presented results shows that the effect of the piezoelectric is significant in the range under consideration. Its smallest influence is observed only for the first natural frequency. For higher frequencies, a significant changes in the behavior of the spectrum of natural frequencies of the vibrations of the cylinder is observed.

Figure 2.22 shows the dependence of the first six frequencies of the natural vibrations of the cylinder on the number m of half-waves in the circumferential direction. Although the intermediate eigenfrequency values between the integer values of the number of half-waves do not have a physical meaning, they are connected by a line for greater clarity of the dependence. A solid line indicates values of the frequencies of the natural vibrations for the case of piezoceramic cylinder, a dashed one stands for the case of the elastic cylinder ($e_{ij} = 0$). Based on the presented data it can be seen that the smallest influence of the piezoelectric effect can be observed on the first two frequencies for small values of m. With an increase in the number of half-waves in the circumferential direction, an increase in the effect of the electric field on the spectrum of the natural frequencies is observed.

2.4.4.2 Inhomogeneous Cylinder

The problem of the free nonaxisymmetric electroelastic vibrations of the hollow piezoceramic cylinders made of the functionally graded piezoceramic material polarized in the radial direction is considered. The material properties are assumed to

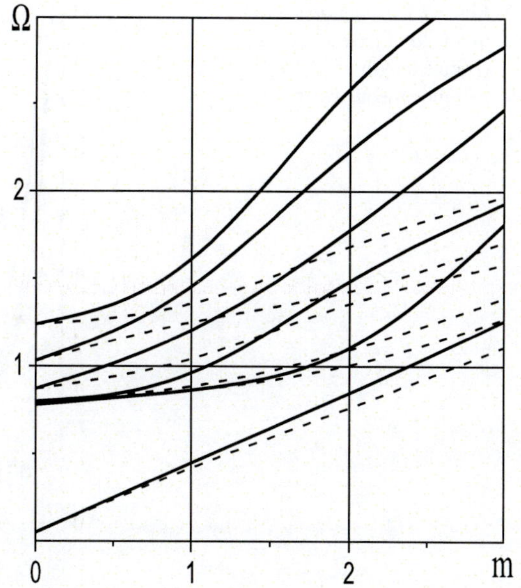

Fig. 2.22 Dependence of the first six frequencies of the natural vibrations of the cylinder on the number of half-waves in the circumferential direction

change over the thickness by an law (2.28), (2.29). The physical and the mechanical characteristics of the material with index 0 have values on the outer surface $R_0 + h = 5$ of the cylinder have the value: (2.75).

The nature of the change in the physical and mechanical characteristics of the material is shown in Fig. 2.4 using the example of the elastic module c_{11}.

Figure 2.22 shows the dependence of the first five free vibration frequencies of the piezoceramic cylinder on relative length L/h. As a material of the cylinder, we also chose PZT-4 piezoceramic. The geometric parameters of the cylinder have the following values: inner radius $R_0 - h = 3$, outer radius $R_0 + h = 5$. Both end faces of the cylinders are rigidly fixed. The number of half-waves in the circumferential direction is $m = 1$. As can be seen from Fig. 2.22, the smaller the value of the inhomogeneity parameter n the greater the natural frequency.

Figure 2.3 shows how the change in the internal radius of the cylinder affects the value of the first natural oscillation frequency at different parameter values, while the external radius of the cylinder remains constant ($R_0 + h = 5$). The length of the cylinder is $L = 5$, the number of half-waves in the circumferential direction is $m = 1$. Here we consider a change of the inner radius of the cylinder in an interval from 0.005 to 4.95 dimensionless units or, in other words, a geometry ranging from a solid cylinder to thin cylindrical shells. By analyzing Fig. 2.23, we can conclude that greater the natural frequencies correspond to smaller values of n. At the same time, the influence of the value of considerably higher in the case of a thick cylinder (Fig. 2.24).

We will study the effect of the coupled electric field and the inhomogeneity parameter on the values of the natural vibration frequencies of the cylinder. Let us consider the free vibration frequencies of the inhomogeneous and homogeneous cylinder with

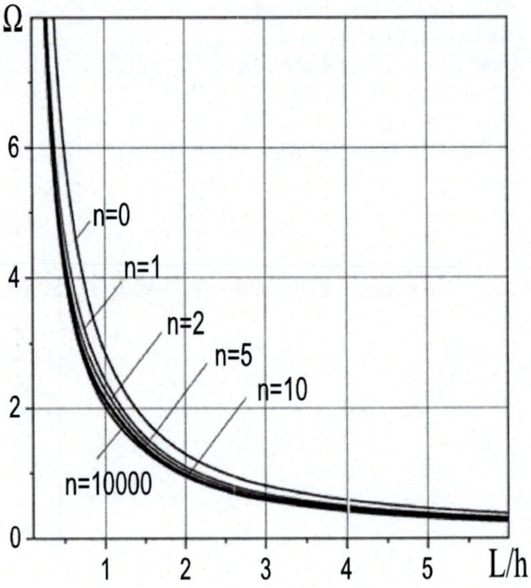

Fig. 2.23 Dependence of first natural frequency on relative length of inhomogeneous cylinder

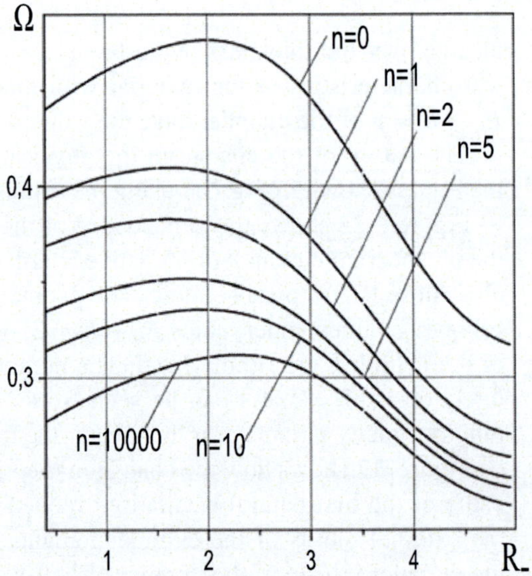

Fig. 2.24 Dependence of first natural frequency on inner radius of inhomogeneous cylinder

averaged characteristics calculated by the theory of effective moduli, when the moduli change by the power law (2.76).

The values of the first five frequencies of the free vibrations of the inhomogeneous piezoceramic cylinders are compared with approximate values of the natural frequencies of the cylinder made of a homogeneous piezoceramics material with averaged characteristics calculated by the theory of effective moduli and with values of the natural frequencies of the inhomogeneity cylinders without piezoeffect (Table

Table 2.10 Comparison of free vibration frequencies of homogeneous and inhomogeneous cylinder

No. of frequency	Inhomogeneous cylinder	The theory of effective moduli	Relative error, %	Without piezoeffect	Relative error, %
1	1.31	1.31	0.0	1.21	7.6
2	2.16	2.10	2.8	1.83	15.3
3	2.29	2.30	0.4	2.22	3.1
4	2.69	2.68	0.4	2.31	14.1
5	3.08	3.11	1.0	2.74	11.0

2.10). The geometric parameters of the cylinder and the heterogeneity parameter of the cylinder material have the values: $L = 5$, $R_0 + h = 5$, $R_0 - h = 3n = 1, 5$, $m = 1$. For the material of the cylinder, we also chose a PZT-4 piezoceramic with the preliminary radial polarization.

Based on the date of Table 2.10, note that the relative error in neglect of the inhomogeneity factor is less than the relative error in the case of neglect of the coupled electric field. We have the greatest relative error for the value of the second natural frequency (15.3 %) in the case of neglecting the coupled electric field; the greatest relative error for the value of the second natural frequency in the case of neglecting the inhomogeneity factor (2.8 %); for a small value of the heterogeneity parameter ($n = 1, 5$).

2.4.5 *Numerical Analysis of Nonaxisymmetric Free Vibration Frequencies of the Piezoceramic Cylinder with Finite Length in the Case of Circumferential Polarization*

2.4.5.1 Homogeneous Cylinder

The problem of the free vibrations of the homogeneous piezoceramic cylinder with finite length with circumferential preliminary polarization is considered. Figure 2.25 shows the dependence of the first five free vibration frequencies of the piezoceramic cylinder on the relative length L/h. As the material of the cylinder, we chose PZT-4 piezoceramic. The geometric parameters of the cylinder have the following values: inner radius $R_0 - h = 3$, outer radius $R_0 + h = 5$. Both end faces of the cylinders are rigidly fixed. We show the values of the natural frequencies with piezoeffect by solid lines and the values of the natural frequencies without piezoeffect ($e_{ij} = 0$) by dashed lines. From the above results, it can be seen that the influence of the piezo-electric effect leads to a "tightening" of the material, that is, to an increase in the value of the natural frequencies. In this case, when determining the first and second natural frequencies, the influence of the piezoelectric effect can be neglected for all

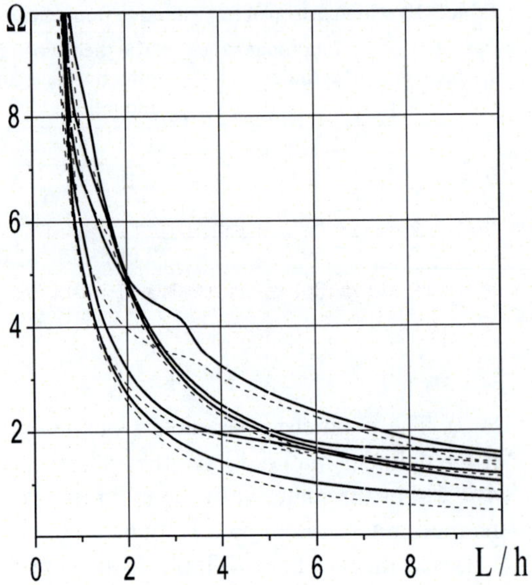

Fig. 2.25 Dependence of first natural frequency on relative length of cylinder

the lengths of the cylinder. For higher frequencies, this effect is noticeable for longer cylinders. The natural frequencies decrease as the relative length of the cylinder increases. For the natural frequencies, starting from the second, note that the intervals at which the natural frequency slightly depends on the geometric parameters of the cylinder ("plateau") and the convergence point of frequencies (not crossing). This behavior of the frequencies is observed starting from the second natural frequency of the vibrations of the cylinder. Obviously, note the behavior of the values of the frequencies, problems arise to the reliable functioning of the corresponding cylindrically-shaped structural elements. In this case, a slight change of the natural vibration frequency corresponds to a significant change in the vibration shape. Of course, the conclusions above require additional research.

Figure 2.26 shows the dependence of the first five free vibration frequencies of the piezoceramic cylinder on the inner radius of the cylinder $R_0 - h$. Note here that the cylinder length ($L = 5$) and outer diameter ($R_0 + h = 5$) are fixed. The number of half-waves in the circumferential direction is as follows $m = 1$. An analysis of the presented results shows that the effect of the piezoelectric is significant in the range under consideration. Its smallest influence is observed only for the first natural frequency. We can see significant changes in the behavior of the spectrum of natural frequencies of the vibrations of the cylinder for the higher frequencies. For the first natural frequency, there is an increase of the frequency with an increase in the radius of the cylinder. In the case of the higher frequencies, the character of the behavior of the free frequencies of the vibration of the cylinder becomes more complicated. Intervals appear where there is a slight dependence of the natural frequency values on the considered the geometric parameters of the cylinder.

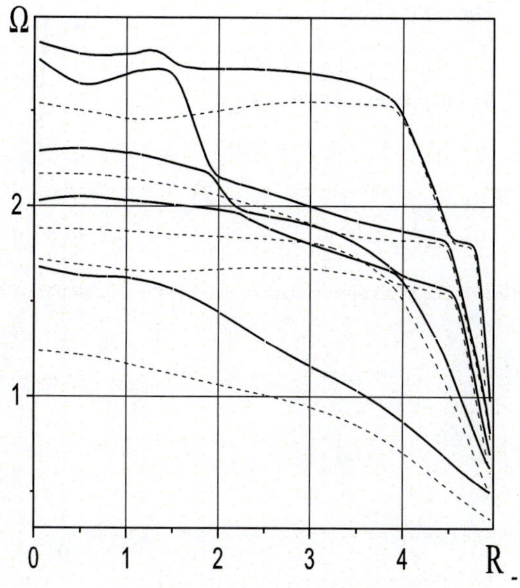

Fig. 2.26 Dependence of first natural frequency on inner radius of cylinder

2.4.5.2 Inhomogeneous Cylinder

The problem of the free nonaxisymmetric electroelastic vibrations of the hollow piezoceramic cylinders made of the functionally graded piezoceramic material polarized in the circumferential direction is considered. The material properties are assumed to change over the thickness by an law (2.28), (2.29). The physical and the mechanical characteristics of the material with index 0 have values on the outer surface $R_0 + h = 5$ of the cylinder have the value: (2.75).

The nature of the change in the physical and mechanical characteristics of the material is shown in Fig. 2.4 using the example of the elastic modulus c_{11}

Figure 2.27 shows the dependence of the first five free vibration frequencies of the piezoceramic cylinder on relative length L/h. For the material of the cylinder, we chose a PZT-4 piezoceramic. The geometric parameters of the cylinder have the following values: inner radius $R_0 - h = 3$, outer radius $R_0 + h = 5$. Both endfaces of the cylinders are rigidly fixed. The number of half-waves in the circumferential direction is $m = 1$. A significant difference is observed in the values of the first vibration frequency of the inhomogeneous piezoceramic cylinder in the case of the radial and the circumferential polarization, when the relative length of the cylinder decreases.

Figure 2.28 shows how the change in the internal radius of the cylinder affects the value of the first natural oscillation frequency at different parameter values, while the external radius of the cylinder remains constant ($R_0 + h = 5$). The length of the cylinder is $L = 2, 5$, the number of half-waves in the circumferential direction $m = 1$. Here, we consider a change of the inner radius of the cylinder an interval from 0.005 to 4.95 dimensionless units or, in other words, a geometry ranging from

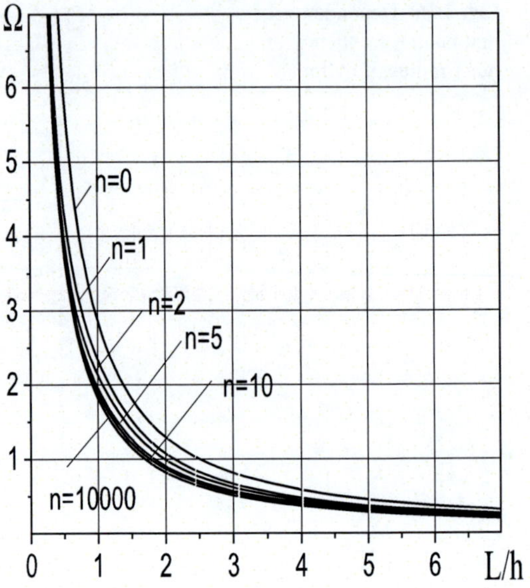

Fig. 2.27 Dependence of first natural frequency on relative length of inhomogeneous cylinder

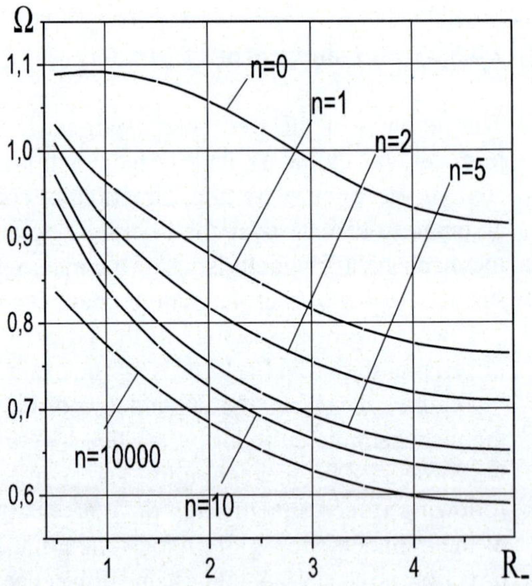

Fig. 2.28 Dependence of first natural frequency on inner radius of inhomogeneous cylinder

a solid cylinder to thin cylindrical shells. By analyzing Fig. 2.28, we can conclude that greater natural frequencies correspond to smaller values of n. The influence of the inhomogeneity parameter of the material of the cylinder on the values of the natural frequencies is significant when the cylinder radius decreases. In this case, there is also a more significant frequency value for the radial and the circumferential polarization of the material of the cylinder.

We will study the effect of the coupled electric field and the inhomogeneity parameter on the values of the natural vibration frequencies of the cylinder. Let us consider

Table 2.11 Comparison of free vibration frequencies of homogeneous and inhomogeneous cylinder

No. of frequency	Inhomogeneous cylinder	The theory of effective moduli	Relative error, %	Without piezoeffect	Relative error, %
1	1.22	1.22	0.0	1.16	4.9
2	1.30	1.28	1.5	1.26	3.1
3	1.49	1.56	4.7	1.42	4.7
4	1.90	1.82	4.2	1.63	14.2
5	2.01	2.05	2.0	1.83	9.0

the free vibration frequencies of the inhomogeneous and the homogeneous cylinder with averaged characteristics calculated by the theory of effective moduli, when the moduli change by the power law (2.76)

The values of the first five frequencies of the free vibrations of the inhomogeneous piezoceramic cylinders are compared with approximate values of the natural frequencies of the cylinder made of a homogeneous piezoceramics material with averaged characteristics calculated by the theory of effective moduli and with values of the natural frequencies of the inhomogeneity cylinders without piezoeffect (Table 2.10). The geometric parameters of the cylinder and the heterogeneity parameter of the cylinder material have the values: $L = 5$, $R_0 + h = 5$, $R_0 - h = 3n = 1, 5$, $m = 1$. For the material of the cylinder, we chose a PZT-4 piezoceramic with the preliminary circumferential polarization.

Based on Table 2.10 data note that the relative error in neglect of the inhomogeneity factor is less than the relative error in the case of neglecting the coupled electric field. We have the greatest relative error at the value of the second natural frequency (14,2%.) for the case of neglecting the coupled electric field. The greatest relative error at the value of the third natural frequency in the case of neglect of the inhomogeneity factor (4,7 %) has a small value of the heterogeneity parameter ($n = 1.5$).

References

1. Alibeigloo A, Kani AM (2010) 3D free vibration analysis of laminated cylindrical shell integrated piezoelectric layers using the differential quadrature method. Appl Math Model 34(12):4123–4137
2. Dai HL, Hong L, Fu YM, Xiao X (2010) Analytical solution for electromagnetothermoelastic behaviors of functionally graded piezoelectric hollow cylinder. Appl Math Modeling 34(2):343–357
3. Ebenezer DD, Ramesh R (2012) Exact analysis of axially polarized piezoelectric ceramic cylinders with certain uniform boundary conditions. Int J Mech Appl 2(5):74–80
4. Grigorenko AY, Efimova TL, Loza IA (2010) Free vibrations of axially polarized piezoceramic hollow cylinders of finite length. Int Appl Mech 46(2):625–623

5. Grigorenko AY, Müller WH, Grigorenko YM, Vlaikov GG (2016) Recent developments in anisotropic heterogeneous shell theory. General theory and applications of classical theory, vol. I, Springer, Berlin
6. Grigorenko AY, Müller WH, Wille R, Loza IA (2012) Nonaxisymmetric vibrations of radially polarized hollow cylinders made of functionally gradient piezoelectric materials. Continuum Mech Thermodyn 24(4):515–524
7. Grigorenko YM, Grigorenko AY, Vlaikov GG (2009) Problems of mechanics for anisotropic inhomogeneous shells on basis of different models. Kyiv Akademperiodika
8. Heyliger PR, Ramirez G (2000) Free vibrations of laminated circular piezoelectric plates and discs. J Sound Vib 229(4):935–956
9. Hussein M, Heyliger PR (1996) Discrete layer analysis of axisymmetric vibrations of laminated piezoelectric cylinders. J Sounds Vib 192(5):995–1013
10. Kharouf N, Heyliger PR (1996) Axisymmetric free vibrations of homogeneous and laminated piezoelectric cylinders. J Sound Vib 174(4):539–561
11. Lewis RWC, Bowen CR, Dent ACE, Jona K (2019) Finite element and experimental analysis of the vibration response of radially poled piezoceramic cylinders. Ferroelectrics 389(1):95–106
12. Loza IA (1984) Axisymmetric acoustoelectrical wave propagation in a hollow circularly polarized cylindrical waveguide. Soviet Appl Mech 20(12):1103–106
13. Loza IA (1985) Propagation of nonaxisymmetric waves in hollow piezoceramic cylinder with radial polarization. Soviet Appl Mech 21(1):22–27
14. Loza IA (2011) Free vibrations of piezoceramic hollow cylinders with radial polarization. J Math Sci 174(3):295–302
15. Loza IA (2012) Torsional vibrations of piezoceramic hollow cylinders with circular polarization. J Math Sci 180(2):146–152
16. Rabbani V, Bahari A, Hodaei M, Maghoul P, Wu N (2019) Three-dimensional free vibration analysis of triclinic piezoelectric hollow cylinder. Compos Part B Eng 158(1):352–363
17. Shul'ga NA, Grigorenko AY, Loza IA (1984) Propagation of nonaxisymmetric acoustoelectric waves in a hollow cylinder. Soviet Appl Mech 20(6):517–521

Chapter 3
Electric Elastic Waves in Layered Inhomogeneous and Continuously Inhomogeneous Piezoceramic Cylinders

Abstract The propagation of axisymmetric and nonaxisymmetric electroelastic waves in hollow inhomogeneous piezoceramic cylinders based on 3D electroelasticity are considered. The elastic and electric properties of the material vary in the radial direction. Two variants of materials are considered: piecewise constant properties of the material (layered structures with metal and dielectric layers) and continuously varying properties (functionally gradient piezoelectric materials—FGPM). Free and forced motions are investigated. In the case of free motion, the surfaces of the cylinder are not loaded and free from electrodes, insulation or short-circuited by electrodes. Two variants of boundary conditions are considered in the case of forced motions: electric excitation—when an electrostatic potential with an alternating sign is applied to the external cylindrical surface; and mechanical excitation—when a pressure with an alternating sign is applied to the external cylindrical surface. An efficient numerical–analytical method to solving this problem is proposed. Components of the elasticity tensor, mechanical and electric displacement vector, electrostatic potential, and stress tensor are presented in the form of standing circumferential waves and by running waves in an axial direction. The three-dimensional system of resolving equation is reduced to a boundary-value problem described by a system of inhomogeneous ordinary differential equations. In the case of free motion, this system represents a differential eigenvalue problem. The discrete-orthogonalization method and a step-by-step search approach method is used to solve the problem. In the case of forced motions, a similar procedure is followed and the problem is solved by discrete-orthogonalization methods. Different variants of polarized piezoceramic materials are considered. The influence of the mechanical and electric parameters of the material on the kinematic (mechanical displacement and electrostatic potential) and dynamic (mechanical stress and electric displacement) characteristics are analyzed. As before, significant attention is paid to the validation of the reliability of the results obtained by numerical calculations.

Keywords 3D electroelasticity theory · Piezoceramic cylinders · Wave propagation · Inhomogeneous materials · Numerical methods

A. Ya. Grigorenko et al., *Selected Problems in the Elastodynamics of Piezoceramic Bodies*, Advanced Structured Materials 154,
https://doi.org/10.1007/978-3-030-74199-0_3

3.1 Introduction

Circular cylindrical piezoceramic waveguides are widely used in acoustoelectronics, which indicates the importance of studies on wave processes in piezoceramic bodies. Acoustoelectric waves propagating in homogeneous cylindrical waveguides are addressed in many publications. The axisymmetric problem for solid and hollow waveguides was studied in [13, 18, 24], etc., and the nonaxisymmetric problem in [14, 18, 20–25], etc. To solve the problem under consideration, we will use the effective numerical analytic approach applied in [5, 6] to the solution of similar problems for elastic bodies.

For layered cylinders, not only the boundary conditions, but also the interface conditions must be satisfied, which increases the order of the systems of equations. Wave processes in inhomogeneous structures were studied in [1, 6, 7, 9, 15, 21, 26], etc. This problem was solved in [15] by expanding the unknown functions into power series. However, the difficulties in implementing this method hinder obtaining detailed quantitative data. To solve the problem under consideration, we will use the effective numerical analytic approach applied in [10] to the solution of similar problems for elastic bodies.

Allowing for the inhomogeneity of the cylinder material complicates the problem even more, while precisely inhomogeneous piezoelectric materials (bimorphs) are used in many devices. At present, functionally gradient piezoelectric materials, which combine advantages of the bimorphs and free-interface materials, showing different coefficients of thermal expansion, find ever widening application.The attempt to allow for continually varying properties of a material leads to the situation when the material moduli are not constants but functions with respect to one coordinate [2, 4, 8, 11, 12], etc. This hinders application of many numerical methods.

In what follows, we will consider an axisymmetric problem on the propagation of forced acoustoelectric waves in a hollow cylinder with nonuniform thickness made of functionally gradient piezoceramics polarized in a radial direction. The first sets of the solutions of partial differential equations (Lame's equations) in cylindrical coordinates were obtained in studies of Pochhammer [19] and Cree [3].To solve the above problem, we will employ the efficient numerical–analytical method with which we will analyze the kinematics of acoustoelectric waves propagating along the cylinder axis. The influence of inhomogeneity on the kinematic characteristics of propagating waves will be studied as well.

3.2 Nonaxisymmetric Electroelastic Waves Propagation in Layered Piezoceramic Cylinders

3.2.1 Problem Statement. Basic Equations for Hollow Cylinders

Consider a hollow layered cylinder with unchangeable piezoceramic or metal layers. A nonaxisymmetric acoustoelectric wave propagates along the axis of the cylinder. The closed system of equations describing this problem in cylindrical coordinates system $(r,\ \theta,\ z)$ consists of [11]

(i) the equations of nonaxisymmetric motion of the ith layer:

$$\begin{aligned} &\frac{\partial T_1^i}{\partial r} + \frac{1}{r}\left(\frac{\partial T_6^i}{\partial \theta} + T_1^i - T_2^i\right) + \frac{\partial T_5^i}{\partial z} + \rho\omega^2 u_1^i = 0, \\ &\frac{\partial T_6^i}{\partial r} + \frac{1}{r}\left(\frac{\partial T_2^i}{\partial \theta} + 2T_6^i\right) + \frac{\partial T_4^i}{\partial z} + \rho\omega^2 u_2^i = 0, \\ &\frac{\partial T_5^i}{\partial r} + \frac{1}{r}\left(\frac{\partial T_4^i}{\partial \theta} + T_5^i\right) + \frac{\partial T_3^i}{\partial z} + \rho\omega^2 u_3^i = 0; \end{aligned} \tag{3.1}$$

(ii) the equations of electrostatics for the ith layer:

$$E_2^i = -\frac{\partial \Phi^i}{\partial r},\ E_2^i = -\frac{1}{r}\frac{\partial \Phi^i}{\partial \theta},\ E_3^i = -\frac{\partial \Phi^i}{\partial z},\ \frac{\partial D_1^i}{\partial r} + \frac{1}{r}D_1^i + \frac{1}{r}\frac{\partial D_2^i}{\partial \theta} + \frac{\partial D_3^i}{\partial z} = 0; \tag{3.2}$$

(iii) the kinematic equations for the ith layer:

$$\begin{aligned} &S_1^i = \frac{\partial u_1^i}{\partial r},\ S_2^i = \frac{1}{r}\frac{\partial u_2^i}{\partial \theta} + \frac{u_1^i}{r},\ S_3^i = \frac{\partial u_3^i}{\partial z},\ 2S_6^i = \frac{1}{r}\frac{\partial u_1^i}{\partial \theta} + \frac{\partial u_2^i}{\partial r} - \frac{u_2^i}{r}, \\ &2S_5^i = \frac{\partial u_3^i}{\partial r} + \frac{\partial u_1^i}{\partial z},\ 2S_4^i = \frac{\partial u_2^i}{\partial z} + \frac{1}{r}\frac{\partial u_3^i}{\partial \theta}. \end{aligned} \tag{3.3}$$

The constitutive equations for the ith piezoceramic layer are:
Axial polarization:

$$\begin{aligned} &T_1^i = c_{11}^i S_1^i + c_{12}^i S_2^i + c_{13}^i S_3^i - e_{13}^i S_3^i, \\ &T_2^i = c_{12}^i S_1^i + c_{11}^i S_2^i + c_{13}^i S_3^i - e_{13}^i E_3^i, \\ &T_3^i = c_{13}^i S_1^i + c_{13}^i S_2^i + c_{33}^i S_3^i - e_{33}^i E_3^i, \\ &T_5^i = 2c_{55}^i S_5^i - e_{15}^i E_1^i,\ T_4^i = 2c_{55}^i S_4^i - e_{15}^i E_2^i, \\ &T_6^i = 2c_{66}^i S_6^i,\ D_3^i = e_{13}^i S_1^i + e_{13}^i S_2^i + e_{33}^i S_3^i + \varepsilon_{33}^i E_3^i, \\ &D_1^i = 2e_{15}^i S_5^i + \varepsilon_{11}^i E_1^i,\ D_2^i = 2e_{15}^i S_4^i + \varepsilon_{11}^i E_2^i; \end{aligned} \tag{3.4}$$

Radial polarization:

$$\begin{aligned}
T_1^i &= c_{33}^i S_1^i + c_{13}^i S_2^i + c_{13}^i S_3^i - e_{33}^i E_1^i, \\
T_2^i &= c_{13}^i S_1^i + c_{11}^i S_2^i + c_{12}^i S_3^i - e_{13}^i E_1^i, \\
T_3^i &= c_{13}^i S_1^i + c_{12}^i S_2^i + c_{11}^i S_3^i - e_{13}^i E_1^i, \\
T_6^i &= 2c_{55}^i S_6^i - e_{15}^i E_2^i,\ T_5^i = 2c_{55}^i S_5^i - e_{15}^i E_3^i, \\
T_4^i &= 2c_{66}^i S_4^i,\ D_1^i = e_{33}^i S_1^i + e_{13}^i S_2^i + e_{13}^i S_3^i + \varepsilon_{33}^i E_1^i, \\
D_2^i &= 2e_{15}^i S_4^i + \varepsilon_{11}^i E_2^i,\ D_3^i = 2e_{15}^i S_5^i + \varepsilon_{11}^i E_3^i;
\end{aligned} \tag{3.5}$$

Circumferential polarization:

$$\begin{aligned}
T_1^i &= c_{11}^i S_1^i + c_{13}^i S_2^i + c_{12}^i S_3^i - e_{13}^i E_2^i, \\
T_2^i &= c_{13}^i S_1^i + c_{33}^i S_2^i + c_{13}^i S_3^i - e_{33}^i E_2^i, \\
T_3^i &= c_{12}^i S_1^i + c_{13}^i S_2^i + c_{11}^i S_3^i - e_{13}^i E_2^i, \\
T_6^i &= 2c_{55}^i S_6^i - e_{15}^i E_1^i,\ T_4^i = 2c_{55}^i S_4^i - e_{15}^i E_3^i, \\
T_5^i &= 2c_{66}^i S_5^i,\ D_2^i = e_{13}^i S_1^i + e_{33}^i S_2^i + e_{13}^i S_3^i + \varepsilon_{33}^i E_2^i, \\
D_1^i &= 2e_{15}^i \varepsilon_6^i + S_{11}^i E_1^i,\ D_3^i = 2e_{15}^i S_4^i + \varepsilon_{11}^i E_3^i.
\end{aligned} \tag{3.6}$$

The constitutive equations for the metal material for the ith layer are described as

$$\begin{aligned}
T_1^i &= \frac{(1-\nu^i)E^i}{(1+\nu^i)(1-2\nu^i)} S_1^i + \frac{\nu^i E^i}{(1+\nu^i)(1-2\nu^i)} S_2^i + \frac{\nu^i E^i}{(1+\nu^i)(1-2\nu^i)} S_3^i, \\
T_2^i &= \frac{\nu^i E^i}{(1+\nu^i)(1-2\nu^i)} S_1^i + \frac{(1-\nu^i)E^i}{(1+\nu^i)(1-2\nu^i)} S_2 + \frac{\nu^i E^i}{(1+\nu^i)(1-2\nu^i)} S_3, \\
T_3^i &= \frac{\nu^i E^i}{(1+\nu^i)(1-2\nu^i)} S_1^i + \frac{\nu^i E^i}{(1+\nu^i)(1-2\nu^i)} S_2^i + \frac{(1-\nu^i)E^i}{(1+\nu^i)(1-2\nu^i)} T_3^i, \\
T_5^i &= 2\frac{E^i}{2(1+\nu^i)} S_5^i,\ T_6^i = 2\frac{E^i}{2(1+\nu^i)} S_6^i,\ T_4^i = 2\frac{E^i}{2(1+\nu^i)} S_4^i.
\end{aligned} \tag{3.7}$$

Hereafter the superscript "i" will be omitted.

The boundary conditions on the lateral surfaces of the piezoceramic sphere are free of external mechanical forces

$$T_1(R_0 \pm h, \theta, z) = 0,\ T_6(R_0 \pm h, \theta, z) = 0,\ T_5(R_0 \pm h, \theta, z) = 0. \tag{3.8}$$

We will consider the two variants of the electrical boundary conditions on the lateral surfaces:

(i) free of electrodes:

$$D_1(R_0 \pm h, \theta, z) = 0; \tag{3.9}$$

(ii) covered with thin short electrodes:

$$\Phi(R_0 \pm h, \theta, z) = 0, \tag{3.10}$$

where R_0—the radius of the sphere mid-surface, h—half the thickness of the cylinder.

In the piecewise nonhomogeneity of the cylinder at the interface of the layers, we will prescribe mechanical and electrical conditions, respectively. On the surface of the discontinuity on properties one conditions of the rigid mechanical contact without slipping and tearing off. Therefore, we have the conditions

$$\begin{gathered} T_1^i = T_1^{i+1}, T_6^i = T_6^{i+1}, T_5^i = T_5^{i+1}, \\ u_1^i = u_1^{i+1}, u_2^i = u_2^{i+1}, u_3^i = u_3^{i+1}. \end{gathered} \tag{3.11}$$

The conditions for the electrical values on the contact surface of the layers depend on the material type. If the ith and $(i+1)$th layers are piezoelectric or dielectric materials then the conditions (3.11) must be supplemented by the condition of continuity of the normal component vector of electrical induction

$$D_1^i = D_1^{i+1}$$

If the ith layers are piezoelectric and the $(i+1)$th are conductor layers, we get the next boundary conditions (1.178)

$$\Phi^i = 0, \Phi^i = \Phi^{i+1}, D_1^i = D_1^{i+1}.$$

Hereafter the superscript "i" will be omitted.

For the unknown functions, we choose the vector-function in terms of which the boundary and interface conditions are formulated.

The resolving vector-function of the mixed form is

$$R = \{T_1, T_6, T_5, \Phi, u_1, u_2, u_3, D_1\}. \tag{3.12}$$

The different types of the preliminary polarization of the piezoceramic cylinder for the component of the vector-function R after identical transformation will be as follows:

axial polarization:

$$\frac{\partial T_1}{\partial r} = \frac{1}{r}\left(\frac{c_{12}}{c_{11}} - 1\right) T_1 - \frac{1}{r}\frac{\partial T_6}{\partial \theta} - \frac{\partial T_5}{\partial z} + \frac{\Delta_5}{r c_{11}}\frac{\partial \Phi}{\partial z} +$$

$$
\begin{aligned}
&+\left(\frac{\Delta_4}{r^2c_{11}}-\rho\frac{\partial^2}{\partial t^2}\right)u_1+\frac{1}{r^2}\left(\frac{\Delta_4}{c_{11}}+c_{66}\right)\frac{\partial u_2}{\partial\theta}+\frac{\Delta_1}{rc_{11}}\frac{\partial u_3}{\partial z},\\
\frac{\partial T_6}{\partial r}&=-\frac{c_{12}}{rc_{11}}\frac{\partial T_1}{\partial\theta}-\frac{2}{r}T_6-\left(\frac{\Delta_5}{c_{11}}+e_{15}\right)\frac{1}{r}\frac{\partial^2\Phi}{\partial\theta\partial z}-\frac{\Delta_4}{r^2c_{11}}\frac{\partial u_1}{\partial\theta}-\\
&-\left(\frac{\Delta_4}{r^2c_{11}}\frac{\partial^2}{\partial\theta^2}+c_{55}\frac{\partial^2}{\partial z^2}-\rho\frac{\partial^2}{\partial t^2}\right)u_2-\left(\frac{\Delta_1}{c_{11}}+c_{55}\right)\frac{1}{r}\frac{\partial^2u_3}{\partial\theta\partial z},\\
\frac{\partial T_5}{\partial r}&=-\frac{c_{13}}{c_{11}}\frac{\partial T_1}{\partial z}-\frac{1}{r}T_5+\left(\frac{\Delta_3}{c_{11}}\frac{\partial^2}{\partial z^2}-\frac{e_{15}}{r^2}\frac{\partial^2}{\partial\theta^2}\right)\Phi-\frac{\Delta_1}{rc_{11}}\frac{\partial u_1}{\partial z}-\\
&-\left(\frac{\Delta_1}{c_{11}}+c_{55}\right)\frac{1}{r}\frac{\partial^2u_2}{\partial\theta\partial z}-\left(\frac{c_{55}}{r^2}\frac{\partial^2}{\partial\theta^2}-\frac{\Delta_2}{c_{11}}\frac{\partial^2}{\partial z^2}-\rho\frac{\partial^2}{\partial t^2}\right)u_3,\\
\frac{\partial\Phi}{\partial r}&=\frac{e_{15}}{\Delta}T_5-\frac{c_{55}}{\Delta}D_1,\ \frac{\partial u_1}{\partial r}=\frac{1}{c_{11}}T_1-\frac{e_{13}}{c_{11}}\frac{\partial\Phi}{\partial z}-\frac{c_{12}}{rc_{11}}u_1-\frac{c_{12}}{rc_{11}}\frac{\partial u_2}{\partial\theta}-\frac{c_{13}}{c_{11}}\frac{\partial u_3}{\partial z},\\
\frac{\partial u_2}{\partial r}&=\frac{1}{c_{66}}T_6-\frac{1}{r}\frac{\partial u_1}{\partial\theta}+\frac{1}{r}u_2,\ \frac{\partial u_3}{\partial r}=\frac{\varepsilon_{11}}{\Delta}T_5-\frac{\partial u_1}{\partial z}+\frac{e_{15}}{\Delta}D_1,\\
\frac{\partial D_1}{\partial r}&=-\frac{e_{13}}{c_{11}}\frac{\partial T_1}{\partial z}+\left(\frac{\varepsilon_{11}}{r^2}\frac{\partial^2}{\partial\theta^2}+\frac{\Delta_6}{c_{11}}\frac{\partial^2}{\partial z^2}\right)\Phi-\frac{\Delta_5}{c_{11}}\frac{1}{r}\frac{\partial u_1}{\partial z}-\\
&-\left(\frac{\Delta_5}{c_{11}}+e_{15}\right)\frac{\partial^2u_2}{\partial\theta\partial z}+\left(\frac{\Delta_3}{c_{11}}\frac{\partial^2}{\partial z^2}-\frac{e_{15}}{r^2}\frac{\partial^2}{\partial\theta^2}\right)u_3-\frac{1}{r}D_1.
\end{aligned}
\tag{3.13}
$$

The boundary conditions for the corresponding system of equations (3.13) are as follows:

$$
\begin{aligned}
&c_{11}\frac{\partial u_1}{\partial r}+c_{12}\frac{1}{r}\left(\frac{\partial u_2}{\partial\theta}+u_1\right)+c_{13}\frac{\partial u_3}{\partial r}+e_{13}\frac{\partial\Phi}{\partial r}=0,\\
&\frac{1}{r}\frac{\partial u_1}{\partial\theta}+\frac{\partial u_2}{\partial r}-\frac{\partial u_2}{r}=0,\ c_{55}\frac{\partial u_1}{\partial z}+c_{55}\frac{\partial u_3}{\partial r}+e_{15}\frac{\partial\Phi}{\partial r}=0,\ \Phi=0,
\end{aligned}
\tag{3.14}
$$

where

$$
\begin{aligned}
&\Delta=c_{55}\varepsilon_{11}+e_{15}^2,\ \Delta_1=c_{13}\left(c_{11}-c_{12}\right),\ \Delta_2=c_{13}^2-c_{11}c_{33},\ \Delta_3=c_{13}e_{13}-c_{11}e_{33},\\
&\Delta_4=c_{11}^2-c_{12}^2,\ \Delta_5=e_{13}\left(c_{11}-c_{12}\right),\ \Delta_6=c_{11}\varepsilon_{33}+e_{13}^2;
\end{aligned}
$$

Radial polarization:

$$
\begin{aligned}
\frac{\partial T_1}{\partial r}&=\frac{1}{r}\left(\frac{\Delta_2}{\Delta}-1\right)T_1-\frac{1}{r}\frac{\partial T_5}{\partial\theta}-\frac{\partial T_5}{\partial z}-\left(\frac{\Delta_2}{r^2\Delta}-\rho\frac{\partial^2}{\partial t^2}\right)u_1+\\
&+\frac{1}{r^2}\frac{\Delta_2}{\Delta}\frac{\partial u_2}{\partial\theta}-\frac{\Delta_3}{r\Delta}\frac{\partial u_3}{\partial z}-\frac{\Delta_4}{r\Delta}D_1,\\
\frac{\partial T_6}{\partial r}&=\frac{\Delta_2}{r\Delta}\frac{\partial T_1}{\partial\theta}-\frac{2}{r}T_5+\frac{\Delta_6}{r^2\Delta}\frac{\partial u_1}{\partial\theta}-\left(\frac{\Delta_6}{r^2\Delta}\frac{\partial^2}{\partial\theta^2}+\frac{\partial^2}{\partial z^2}-\rho\frac{\partial^2}{\partial t^2}\right)u_2+
\end{aligned}
$$

$$
\begin{aligned}
& + \frac{\Delta_5}{r\Delta}\frac{\partial^2 u_3}{\partial\theta\partial z} + \frac{\Delta_1}{r\Delta}\frac{\partial D_1}{\partial\theta}, \\
\frac{\partial T_5}{\partial r} &= -\frac{\Delta_2}{\Delta}\frac{\partial T_1}{\partial z} - \frac{1}{r}T_5 + \frac{\Delta_5}{r\Delta}\frac{\partial u_1}{\partial z} + \left(\frac{\Delta_5}{\Delta} + c_{66}\right)\frac{\partial^2 u_2}{\partial\theta\partial z} - \\
& - \left(\frac{c_{66}}{r^2}\frac{\partial^2}{\partial\theta^2} + \frac{\Delta_6}{\Delta}\frac{\partial^2}{\partial z^2} - \rho\frac{\partial^2}{\partial t^2}\right)u_3 - \frac{\Delta_7}{\Delta}\frac{\partial D_1}{\partial z}, \\
\frac{\partial \Phi}{\partial r} &= \frac{e_{33}}{\Delta}T_1 + \frac{\Delta_1}{r\Delta}u_1 - \frac{\Delta_1}{r\Delta}\frac{\partial u_2}{\partial\theta} + \frac{\Delta_1}{\Delta}\frac{\partial u_3}{\partial z} - \frac{c_{33}}{\Delta}D_1, \\
\frac{\partial u_1}{\partial r} &= \frac{\varepsilon_{33}}{\Delta}T_1 - \frac{\Delta_2}{r\Delta}u_1 + \frac{\Delta_2}{r\Delta}\frac{\partial u_2}{\partial\theta} - \frac{\Delta_2}{\Delta}\frac{\partial u_3}{\partial z} + \frac{e_{33}}{\Delta}D_1, \\
\frac{\partial u_2}{\partial r} &= \frac{1}{c_{55}}T_6 + \frac{e_{51}}{rc_{55}}\frac{\partial\Phi}{\partial\theta} + \frac{1}{r}\frac{\partial u_1}{\partial\theta} + \frac{1}{r}u_2, \quad \frac{\partial u_3}{\partial r} = \frac{1}{c_{55}}T_5 + \frac{e_{51}}{c_{55}}\frac{\partial\Phi}{\partial z} + \frac{\partial u_1}{\partial z}, \\
\frac{\partial D_1}{\partial r} &= -\frac{e_{51}}{c_{55}}\frac{\partial T_6}{\partial\theta} - \frac{e_{51}}{c_{55}}\frac{\partial T_5}{\partial z} + \left(\frac{1}{r^2}\frac{\partial^2}{\partial\theta^2} + \frac{\partial^2}{\partial z^2}\right)\frac{\Delta_7}{c_{55}}\Phi - \frac{1}{r}D_1.
\end{aligned}
\tag{3.15}
$$

The boundary conditions for the corresponding system of equations (3.15) are as follows:

$$
\begin{aligned}
& c_{33}\frac{\partial u_1}{\partial r} + c_{13}\frac{1}{r}\left(\frac{\partial u_3}{\partial\theta} + u_1\right) + c_{13}\frac{\partial u_3}{\partial r} + e_{33}\frac{\partial\Phi}{\partial r} = 0, \\
& c_{55}\left(\frac{1}{r}\frac{\partial u_1}{\partial\theta} + \frac{\partial u_2}{\partial r} - \frac{\partial u_2}{r}\right) + e_{15}\frac{1}{r}\frac{\partial\Phi}{\partial\theta} = 0, \\
& c_{55}\frac{\partial u_1}{\partial z} + c_{55}\frac{\partial u_3}{\partial r} + e_{15}\frac{\partial\Phi}{\partial z} = 0, \ \Phi = 0,
\end{aligned}
\tag{3.16}
$$

where

$$
\begin{aligned}
& \Delta = c_{33}\varepsilon_{33} + e_{33}^2, \ \Delta_1 = c_{33}e_{13} - c_{13}e_{33}, \ \Delta_2 = c_{13}\varepsilon_{33} + e_{13}e_{33}, \\
& \Delta_3 = (c_{13} - c_{33})\,\Delta_2 + (e_{33} - e_{13})\,\Delta_1 + (c_{13} - c_{11})\,\Delta, \\
& \Delta_4 = (c_{13} - c_{33})\,\Delta_2 + (e_{33} - e_{13})\,\Delta_1 + (c_{13} - c_{12})\,\Delta, \\
& \Delta_5 = c_{12}\Delta + e_{13}\Delta_1 - c_{13}\Delta_2, \ \Delta_6 = c_{11}\Delta + e_{13}\Delta_1 - c_{13}\Delta_2, \ \Delta_7 = c_{55}\varepsilon_{11} + e_{15}^2;
\end{aligned}
\tag{3.17}
$$

Circumferential polarization:

$$
\begin{aligned}
\frac{\partial T_1}{\partial r} &= \frac{1}{r}\left(\frac{c_{13}}{c_{11}} - 1\right)T_1 + \frac{1}{r}\frac{\partial T_6}{\partial\theta} - \frac{\partial T_5}{\partial z} - \frac{\Delta_1}{r^2 c_{11}}\frac{\partial\Phi}{\partial\theta} - \\
& - \left(\frac{\Delta_2}{r^2 c_{11}} - \rho\frac{\partial^2}{\partial t^2}\right)u_1 - \frac{\Delta_2}{r^2 c_{11}}\frac{\partial u_2}{\partial\theta} - \frac{\Delta_4}{r\Delta}\frac{\partial u_3}{\partial z}, \\
\frac{\partial T_6}{\partial r} &= \frac{c_{13}}{rc_{11}}\frac{\partial T_1}{\partial\theta} - \frac{2}{r}T_6 - \left(e_{15}\frac{\partial^2}{\partial z^2} - \frac{\Delta_1}{r^2 c_{11}}\frac{\partial^2}{\partial\theta^2}\right)\Phi + \frac{\Delta_2}{r^2 c_{11}}\frac{\partial u_1}{\partial\theta} - \\
& - \left(\frac{\Delta_2}{r^2 c_{11}}\frac{\partial^2}{\partial\theta^2} + c_{55}\frac{\partial^2}{\partial z^2} + \rho\frac{\partial^2}{\partial t^2}\right)u_2 - \left(\frac{\Delta_3}{c_{11}} + c_{55}\right)\frac{1}{r}\frac{\partial^2 u_3}{\partial\theta\partial z},
\end{aligned}
$$

$$
\begin{aligned}
\frac{\partial T_5}{\partial r} &= \frac{c_{12}}{c_{11}}\frac{\partial T_1}{\partial z} - \frac{1}{r}T_5 + \left(\frac{\Delta_6}{c_{11}} + e_{15}\right)\frac{\partial^2 \Phi}{\partial \theta \partial z} + \frac{\Delta_3}{rc_{11}}\frac{\partial u_1}{\partial z} - \\
&- \left(\frac{\Delta_3}{c_{11}} + c_{55}\right)\frac{\partial^2 u_2}{\partial \theta \partial z} - \left(\frac{c_{55}}{r^2}\frac{\partial^2}{\partial \theta^2} + \frac{\Delta_7}{c_{11}}\frac{\partial^2}{\partial z^2} + \rho\frac{\partial^2}{\partial t^2}\right)u_1, \qquad (3.18)\\
\frac{\partial \Phi}{\partial r} &= \frac{e_{15}}{\Delta}T_6 - \frac{c_{55}}{\Delta}D_1,\ \frac{\partial u_1}{\partial r} = \frac{1}{c_{11}}T_1 + \frac{e_{13}}{rc_{11}}\frac{\partial \Phi}{\partial \theta} - \frac{c_{13}}{rc_{11}}u_1 + \frac{c_{13}}{rc_{11}}\frac{\partial u_2}{\partial \theta} + \frac{c_{12}}{c_{11}}\frac{\partial u_3}{\partial z},\\
\frac{\partial u_2}{\partial r} &= \frac{\varepsilon_{11}}{\Delta}T_6 + \frac{1}{r}\frac{\partial u_1}{\partial \theta} + \frac{1}{r}u_2 + \frac{e_{51}}{\Delta}D_1,\ \frac{\partial u_3}{\partial r} = \frac{1}{c_{66}}T_5 - \frac{\partial u_1}{\partial z},\\
\frac{\partial D_1}{\partial r} &= -\frac{e_{15}}{c_{11}}\frac{1}{r}\frac{\partial T_1}{\partial \theta} + \left(\frac{\Delta_8}{c_{11}}\frac{1}{r^2}\frac{\partial^2}{\partial \theta^2} + \varepsilon_{11}\frac{\partial^2}{\partial z^2}\right)\Phi + \frac{\Delta_1}{r^2 c_{11}}\frac{\partial u_1}{\partial \theta} + \\
&+ \left(\frac{\Delta_1}{r^2}\frac{\partial^2}{\partial \theta^2} - e_{15}\frac{\partial^2}{\partial z^2}\right)u_2 - \left(\frac{\Delta_6}{c_{11}} + e_{15}\right)\frac{\partial^2 u_3}{\partial \theta \partial z} - \frac{1}{r}D_1.
\end{aligned}
$$

The boundary conditions for the corresponding system of equations (3.18) are as follows:

$$
\begin{aligned}
&c_{11}\frac{\partial u_1}{\partial r} + c_{13}\frac{1}{r}\left(\frac{\partial u_2}{\partial \theta} + u_1\right) + c_{12}\frac{\partial u_2}{\partial r} + e_{13}\frac{1}{r}\frac{\partial \Phi}{\partial \theta} = 0,\\
&c_{55}\left(\frac{1}{r}\frac{\partial u_1}{\partial \theta} + \frac{\partial u_2}{\partial r} - \frac{\partial u_2}{r}\right) + e_{15}\frac{\partial \Phi}{\partial r} = 0,\ \frac{\partial u_1}{\partial z} + \frac{\partial u_3}{\partial r} = 0;\ \Phi = 0, \qquad (3.19)
\end{aligned}
$$

where

$$
\begin{aligned}
&\Delta = c_{55}\varepsilon_{11} + e_{15}^2,\ \Delta_1 = c_{13}e_{13} - c_{11}e_{33},\ \Delta_2 = c_{11}c_{33} - c_{13}^2,\\
&\Delta_3 = (c_{11} - c_{12})\,c_{13},\ \Delta_4 = (c_{11} - c_{13})\,e_{13},\ \Delta_5 = c_{11}e_{33} - c_{13}e_{13},\\
&\Delta_6 = (c_{11} - c_{12})\,e_{13},\ \Delta_7 = c_{11}^2 - c_{12}^2,\ \Delta_8 = c_{11}\varepsilon_{33} + e_{13}^2.
\end{aligned}
$$

3.2.2 *Solution Method of the Problem*

To solve the problem, we will employ an efficient analytical–numerical approach. In the first step, we apply the method of separating variables. Let us represent the components of the stress tensors and the vectors of displacements, electric-flux density, and electrostatic potential as standing circumferential waves and traveling axial waves. As a result, the original three-dimensional problem of electroelasticity for partial differential equations is reduced to a boundary-value eigenvalue problem for ordinary differential equations

axial polarization:

$$
\begin{aligned}
T_1(r,\theta,z,t) &= \lambda T_1(r)\sin m\theta \sin(kz - \omega t),\\
T_6(r,\theta,z,t) &= \lambda T_6(r)\cos m\theta \sin(kz - \omega t),\\
T_5(r,\theta,z,t) &= \lambda T_5(r)\sin m\theta \cos(kz - \omega t),
\end{aligned}
$$

$$\begin{aligned}
\Phi\left(r,\theta,z,t\right) &= h\sqrt{\frac{\lambda}{\varepsilon_0}}\Phi\left(r\right)\sin m\theta\cos\left(kz-\omega t\right),\\
u_1\left(r,\theta,z,t\right) &= hu_1\left(r\right)\sin m\theta\sin\left(kz-\omega t\right),\\
u_2\left(r,\theta,z,t\right) &= hu_2\left(r\right)\cos m\theta\sin\left(kz-\omega t\right),\\
u_3\left(r,\theta,z,t\right) &= hu_3\left(r\right)\sin m\theta\cos\left(kz-\omega t\right),\\
D_1\left(r,\theta,z,t\right) &= \sqrt{\varepsilon_0\lambda}D_1\left(r\right)\sin m\theta\cos\left(kz-\omega t\right);
\end{aligned} \tag{3.20}$$

radial polarization:

$$\begin{aligned}
T_1\left(r,\theta,z,t\right) &= \lambda T_1\left(r\right)\sin m\theta\cos\left(kz-\omega t\right),\\
T_6\left(r,\theta,z,t\right) &= \lambda T_6\left(r\right)\cos m\theta\cos\left(kz-\omega t\right),\\
T_5\left(r,\theta,z,t\right) &= \lambda T_5\left(r\right)\sin m\theta\sin\left(kz-\omega t\right),\\
\Phi\left(r,\theta,z,t\right) &= h\sqrt{\frac{\lambda}{\varepsilon_0}}\Phi\left(r\right)\cos m\theta\cos\left(kz-\omega t\right),\\
u_1\left(r,\theta,z,t\right) &= hu_1\left(r\right)\cos m\theta\cos\left(kz-\omega t\right),\\
u_2\left(r,\theta,z,t\right) &= hu_2\left(r\right)\sin m\theta\cos\left(kz-\omega t\right),\\
u_3\left(r,\theta,z,t\right) &= hu_3\left(r\right)\cos m\theta\sin\left(kz-\omega t\right),\\
D_1\left(r,\theta,z,t\right) &= \sqrt{\varepsilon_0\lambda}D_1\left(r\right)\cos m\theta\cos\left(kz-\omega t\right);
\end{aligned} \tag{3.21}$$

circumferential polarization:

$$\begin{aligned}
T_1\left(r,\theta,z,t\right) &= \lambda T_1\left(r\right)\cos m\theta\cos\left(kz-\omega t\right),\\
T_6\left(r,\theta,z,t\right) &= \lambda T_6\left(r\right)\sin m\theta\cos\left(kz-\omega t\right),\\
T_5\left(r,\theta,z,t\right) &= \lambda T_5\left(r\right)\cos m\theta\cos\left(kz-\omega t\right),\\
\Phi\left(r,\theta,z,t\right) &= h\sqrt{\frac{\lambda}{\varepsilon_0}}\Phi\left(r\right)\sin m\theta\cos\left(kz-\omega t\right),\\
u_1\left(r,\theta,z,t\right) &= hu_1\left(r\right)\cos m\theta\cos\left(kz-\omega t\right),\\
u_2\left(r,\theta,z,t\right) &= hu_2\left(r\right)\sin m\theta\cos\left(kz-\omega t\right),\\
u_3\left(r,\theta,z,t\right) &= hu_3\left(r\right)\cos m\theta\sin\left(kz-\omega t\right),\\
D_1\left(r,\theta,z,t\right) &= \sqrt{\varepsilon_0\lambda}D_1\left(r\right)\sin m\theta\cos\left(kz-\omega t\right).
\end{aligned} \tag{3.22}$$

After the transformations we get at a boundary-value eigenvalue problem for the systems of ordinary differential equations,

$$\frac{\mathrm{d}\mathbf{R}}{\mathrm{d}r} = \mathbf{A}(x,\Omega)\mathbf{R}. \tag{3.23}$$

By applying the symbols $\lambda_1 = -\dfrac{\tilde{x}^2\Delta_4}{c_{11}}$, $\lambda_2 = \dfrac{m^2\tilde{x}^2\Delta_4}{c_{11}} + k^2c_{55}$, $\lambda_3 = m^2\tilde{x}^2c_{55} - \dfrac{k^2\Delta_2}{c_{11}}$ the matrix **A** for the case of the axial polarization is given by

$$\begin{Vmatrix}
\tilde{x}\left(\dfrac{c_{12}}{c_{11}}-1\right) & m\tilde{x} & k & \dfrac{k\tilde{x}\Delta_5}{c_{11}} & \lambda_1-\Omega^2 & -\dfrac{\Delta_4}{c_{11}}m\tilde{x}^2 & -\dfrac{k\tilde{x}\Delta_1}{c_{11}} & 0 \\
-\dfrac{m\tilde{x}c_{12}}{c_{11}} & -2\tilde{x} & 0 & \left(\dfrac{\Delta_5}{c_{11}}+\tilde{e}_{15}\right)mk\tilde{x} & \dfrac{m\tilde{x}^2\Delta_2}{c_{11}} & \lambda_2-\Omega^2 & mk\tilde{x}\left(\dfrac{\Delta_1}{c_{11}}+c_{55}\right) & 0 \\
-\dfrac{kc_{13}}{c_{11}} & 0 & -\tilde{x} & m^2\tilde{x}^2\tilde{e}_{15}-k^2\dfrac{\Delta_3}{c_{11}} & -\dfrac{k\tilde{x}\Delta_1}{c_{11}} & mk\tilde{x}\left(\dfrac{\Delta_1}{c_{11}}+c_{55}\right) & \lambda_3-\Omega^2 & 0 \\
\dfrac{\tilde{e}_{15}}{\Delta} & 0 & 0 & 0 & 0 & 0 & 0 & -\dfrac{\tilde{c}_{55}}{\Delta} \\
\dfrac{1}{\tilde{c}_{11}} & 0 & 0 & 0 & -\dfrac{\tilde{x}\tilde{c}_{12}}{\tilde{c}_{11}} & \dfrac{m\tilde{x}\tilde{c}_{12}}{\tilde{c}_{11}} & \dfrac{k\tilde{c}_{13}}{\tilde{c}_{11}} & 0 \\
0 & \dfrac{1}{c_{66}} & 0 & 0 & -m\tilde{x} & \tilde{x} & 0 & 0 \\
0 & 0 & \dfrac{\tilde{\varepsilon}_{11}}{\Delta} & 0 & -k & 0 & 0 & \dfrac{\tilde{e}_{15}}{\Delta} \\
-\dfrac{k\tilde{e}_{13}}{c_{11}} & 0 & 0 & -m^2\tilde{x}^2\tilde{\varepsilon}_{11}-\dfrac{k^2\Delta_6}{\tilde{c}_{11}} & -\dfrac{k\tilde{x}\Delta_5}{c_{11}} & \left(\tilde{e}_{15}+\dfrac{\Delta_5}{\tilde{c}_{11}}\right)mk\tilde{x} & m^2\tilde{x}^2\tilde{e}_{15}-\dfrac{k^2\Delta_3}{\tilde{c}_{11}} & -\tilde{x}
\end{Vmatrix} \tag{3.24}$$

By applying the symbols $\lambda_1 = -\dfrac{\tilde{x}^2\Delta_3}{\Delta}$, $\lambda_2 = \dfrac{m^2\tilde{x}^2\Delta_6}{\Delta} + k^2c_{66}$, $\lambda_3 = m^2\tilde{x}^2c_{66} + \dfrac{k^2\Delta_6}{\Delta}$ the matrix **A** for the case of the radial polarization is given by

$$\begin{Vmatrix}
\tilde{x}\left(\dfrac{\Delta_2}{\Delta}-1\right) & m\tilde{x} & k & 0 & \lambda_1-\Omega^2 & \dfrac{m\tilde{x}^2\Delta_3}{\Delta} & \dfrac{k\tilde{x}\Delta_4}{\Delta} & -\dfrac{\tilde{x}\Delta_1}{\Delta} \\
-\dfrac{m\tilde{x}\Delta_2}{\Delta} & -2\tilde{x} & 0 & 0 & -\dfrac{m\tilde{x}^2\Delta_6}{\Delta} & \lambda_2-\Omega^2 & mk\tilde{x}\left(\dfrac{\Delta_5}{\Delta}+c_{66}\right) & \dfrac{m\tilde{x}\Delta_1}{\Delta} \\
-\dfrac{k\Delta_2}{\Delta} & 0 & -\tilde{x} & 0 & -\dfrac{k\tilde{x}\Delta_5}{\Delta} & mk\tilde{x}\left(\dfrac{\Delta_5}{\Delta}+c_{66}\right) & \lambda_3-\Omega^2 & \dfrac{k\Delta_1}{\Delta} \\
\dfrac{\tilde{e}_{33}}{\Delta} & 0 & 0 & 0 & \dfrac{\tilde{x}\Delta_1}{\Delta} & -\dfrac{m\tilde{x}\Delta_1}{\Delta} & -\dfrac{k\Delta_1}{\Delta} & -\dfrac{\tilde{c}_{33}}{\Delta} \\
\dfrac{\tilde{\varepsilon}_{33}}{\Delta} & 0 & 0 & 0 & -\dfrac{\tilde{x}\Delta_2}{\Delta} & \dfrac{m\tilde{x}\Delta_2}{\Delta} & \dfrac{k\Delta_2}{\Delta} & \dfrac{\tilde{e}_{33}}{\Delta} \\
0 & \dfrac{1}{c_{55}} & 0 & -\dfrac{m\tilde{x}e_{15}}{c_{55}} & -m\tilde{x} & \tilde{x} & 0 & 0 \\
0 & 0 & \dfrac{1}{\tilde{c}_{55}} & -\dfrac{k\tilde{e}_{51}}{\tilde{c}_{55}} & -k & 0 & 0 & 0 \\
0 & \dfrac{m\tilde{x}\tilde{e}_{15}}{\tilde{c}_{55}} & \dfrac{k\tilde{e}_{15}}{\tilde{c}_{55}} & -\left(m^2\tilde{x}^2+k^2\right)\dfrac{\Delta_7}{\tilde{c}_{55}} & 0 & 0 & 0 & -\tilde{x}
\end{Vmatrix} \tag{3.25}$$

By applying the symbols $\lambda_1 = \dfrac{\tilde{x}^2\Delta_2}{\tilde{c}_{11}}$, $\lambda_2 = \dfrac{m^2\tilde{x}^2\Delta_2}{\tilde{c}_{11}} + k^2\tilde{c}_{55}$, $\lambda_3 = m^2\tilde{x}^2\tilde{c}_{55} + \dfrac{k^2\Delta_7}{\tilde{c}_{11}}$ the matrix **A** for the case the circumferential polarization is given by

$$\left\| \begin{matrix} \tilde{x}\left(\frac{\tilde{c}_{13}}{\tilde{c}_{11}}-1\right) & m\tilde{x} & k & -\frac{m\tilde{x}^2\Delta_1}{\tilde{c}_{11}} & \lambda_1-\Omega^2 & -\frac{m\tilde{x}^2\Delta_2}{\tilde{c}_{11}} & -\frac{k\tilde{x}\Delta_3}{\tilde{c}_{11}} & 0 \\ -\frac{m\tilde{x}\tilde{c}_{13}}{\tilde{c}_{11}} & -2\tilde{x} & 0 & k^2\tilde{e}_{15}-\frac{m^2\tilde{x}^2\Delta_1}{\tilde{c}_{11}} & -\frac{m\tilde{x}^2\Delta_2}{\tilde{c}_{11}} & \lambda_2-\Omega^2 & \left(\frac{\Delta_3}{\tilde{c}_{11}}+\tilde{c}_{55}\right)mk\tilde{x} & 0 \\ -\frac{k\tilde{c}_{12}}{\tilde{c}_{11}} & 0 & -\tilde{x} & -\left(\frac{\Delta_6}{\tilde{c}_{11}}+\tilde{e}_{15}\right)mk\tilde{x} & -\frac{k\tilde{x}\Delta_3}{\tilde{c}_{11}} & \left(\frac{\Delta_3}{\tilde{c}_{11}}+\tilde{c}_{55}\right)mk\tilde{x} & \lambda_3-\Omega^2 & 0 \\ 0 & \frac{\tilde{e}_{15}}{\Delta} & 0 & 0 & 0 & 0 & 0 & -\frac{\tilde{c}_{55}}{\Delta} \\ \frac{1}{\tilde{c}_{11}} & 0 & 0 & \frac{m\tilde{x}\tilde{e}_{15}}{\tilde{c}_{11}} & -\frac{\tilde{x}c_{13}}{\tilde{c}_{11}} & \frac{m\tilde{x}\tilde{c}_{13}}{\tilde{c}_{11}} & -\frac{kc_{12}}{\tilde{c}_{11}} & 0 \\ 0 & \frac{\tilde{\varepsilon}_{11}}{\Delta} & 0 & 0 & -m\tilde{x} & \tilde{x} & 0 & \frac{\tilde{e}_{15}}{\Delta} \\ 0 & 0 & \frac{1}{\tilde{c}_{66}} & 0 & -k & 0 & 0 & 0 \\ \frac{m\tilde{x}\tilde{e}_{13}}{\tilde{c}_{11}} & 0 & 0 & -\left(\frac{m^2\tilde{x}^2\Delta_8}{\tilde{c}_{11}}+k^2\tilde{\varepsilon}_{11}\right) & -\frac{m\tilde{x}^2\Delta_1}{\tilde{c}_{11}} & k^2\tilde{e}_{15}-\frac{m^2\tilde{x}^2\Delta_1}{\tilde{c}_{11}} & -\left(\frac{\Delta_6}{\tilde{c}_{11}}+\tilde{e}_{15}\right)mk\tilde{x} & -\tilde{x} \end{matrix} \right\| \tag{3.26}$$

The boundary homogeneous conditions have the form

$$B_1\mathbf{R}(-1)=0,\ B_2\mathbf{R}(1)=0 \tag{3.27}$$

where the matrices B_1andB_2 are given by

$$B_1=\left\| \begin{matrix} 1 & 0 & 0 & 0 \\ 0 & 1 & 0 & 0 \\ 0 & 0 & 1 & 0 \\ 0 & 0 & 0 & -1 \end{matrix} \right\|,\quad B_2=\left\| \begin{matrix} 1 & 0 & 0 & 0 \\ 0 & 1 & 0 & 0 \\ 0 & 0 & 1 & 0 \\ 0 & 0 & 0 & 1 \end{matrix} \right\|.$$

The boundary-value problem (3.23)–(3.27) is solved by using the stable numerical discrete-orthogonalization method in combination with the incremental search method.

3.2.3 *Numerical Results and Analysis Propagation Characteristics of Electric Elastic Waves in a Laminated Piezoceramic Cylinder with Layers Polarized in the Axial Direction*

We will consider the propagation of acoustoelectric waves in a three-layer hollow cylinder. The cylinder has three layers. The face layers have thickness $h/2$ each, while the thickness of the middle layer is h. The face layers are made of steel with the following characteristics:

$$E=21\cdot 10^{10}\mathrm{N/m^2},\ \nu=0.28,$$

The middle layer is made of PZT-4 piezoceramics with the following characteristics:

$$c_{11} = 13.9 \cdot 10^{10}\ \mathrm{N/m^2},\ c_{12} = 7.43 \cdot 10^{10}\ \mathrm{N/m^2},\ c_{13} = 7.78 \cdot 10^{10}\ \mathrm{N/m^2},$$
$$c_{33} = 11.5 \cdot 10^{10}\ \mathrm{N/m^2},\ c_{55} = 2.56 \cdot 10^{10}\ \mathrm{N/m^2},\ e_{13} = -5.2\,\mathrm{C/m^2},$$
$$e_{15} = 12.7\ \mathrm{C/m^2},\ e_{33} = 15.1\ \mathrm{C/m^2},\ \varepsilon_{11}/\varepsilon_0 = 730,$$
$$\varepsilon_{33}/\varepsilon_0 = 635,\ \rho_p = 7.5 \cdot 10^3\ \mathrm{kg/m^2}.$$

If $\varepsilon = 0$ and $k = 0$ we arrive at the problem of the vibrations of a plane layer. For example, the frequencies of a single-layer metal cylinder are defined by the formulas

$$U(n) = \frac{\pi}{2} n \sqrt{\frac{(1-\nu)\,E}{(1+\nu)\,(1-2\nu)\,\rho_M}} = 0;\ \ 2.905;\ \ 5.81;\ \ldots\ ;\ n = 0,\ 1,\ 2\ \ldots\ ;$$

$$V(n) = W(n) = \frac{\pi}{2} n \sqrt{\frac{E}{2(1+\nu)\rho_M}} = 0;\ \ 1.606;\ \ 3.211;\ \ldots\ ;\ n = 0,\ 1,\ 2\ \ldots\ .$$

For a single-layer PZT-4 cylinder, we have

$$U(n) = \frac{\pi}{2} n \sqrt{c_{11}/\rho_n} = 0;\ 2.138;\ 4.277;\ \ldots; n = 0, 1, 2, \ldots;$$
$$W(2n) = \pi n \sqrt{(c_{55} + e_{15}^2/\varepsilon_{11})/\rho_n} = 0; 1.913; 3.826; \ldots; n = 0, 1, 2, \ldots;$$
$$V(n) = \frac{\pi}{2} n \sqrt{c_{66}/\rho_n} = 0;\ 1.003;\ 2.007;\ \ldots\ ;\ n = 0,\ 1,\ 2\ \ldots\ ; \tag{3.28}$$
$$W(2n+1) = \lambda_n \sqrt{c_{55} + \frac{e_{15}^2}{\varepsilon_{11}}} = 0.925;\ 2.859; \ldots,$$

where λ are the roots of the transcendent equation $\lambda \cos \lambda - \dfrac{e_{15}^2}{e_{15}^2 + c_{55}\varepsilon_{11}} \sin \lambda = 0.$

Figure 3.1 shows the first six frequencies as functions of the dimensionless wave number $\zeta = kh/\pi$ for $m = 1$ and $\varepsilon = 0.25$.The solid lines represent the layered cylinder, while the dashed lines represent the homogeneous PZT-4 cylinder of the same geometry. The curves are denoted as in [9]. As can be seen, the metal layers make the material stiffer, i.e., increase the natural frequencies.

Figure 3.2 shows the first six frequencies as functions of the dimensionless wave number $\zeta = kh/\pi$. The solid lines correspond to the layered cylinder, while the dashed lines to the solid steel cylinder. As can be seen, the natural frequencies of the layered cylinder are lower than the natural frequencies of the steel cylinder. Hence, the natural frequency of the layered cylinder is between the natural frequency of the single-layer piezoceramic cylinder and the frequency of the single-layer steel cylinder. This is illustrated in Fig. 3.3. The solid lines correspond to the layered

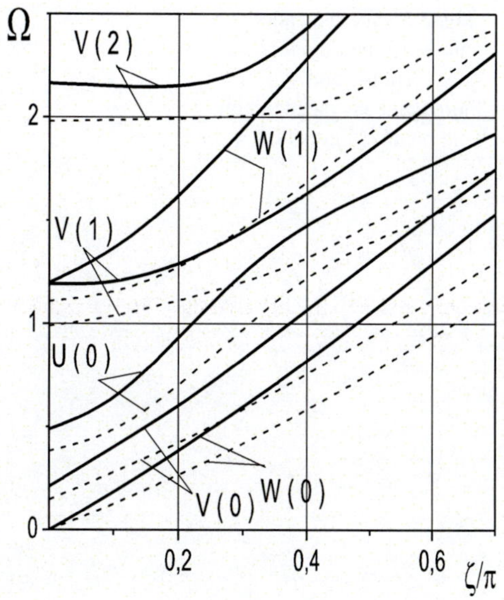

Fig. 3.1 Six dispersion curves for inhomogeneous and homogeneous piezoceramic cylinders ($m = 1$)

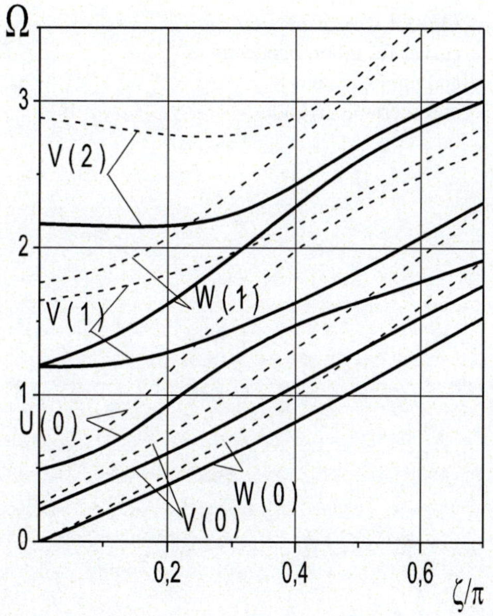

Fig. 3.2 Six dispersion curves for inhomogeneous piezoceramic and homogeneous metal cylinders ($m = 1$)

cylinder, the dashed lines to the piezoceramic cylinder, and the dash-and-dot lines to the steel cylinder. The material and geometry of the cylinder are the same as above.

Figure 3.4 shows the first six frequencies as functions of the dimensionless wave number for $m = 2$. The material and geometry of the cylinder are the same as in Fig. 3.1. The solid line represents the layered cylinder, while the dashed line the PZT-4 cylinder. Figure 3.5 shows the first six frequencies as functions of the dimensionless wave number for $m = 2$. The material and geometry of the cylinder are the same as

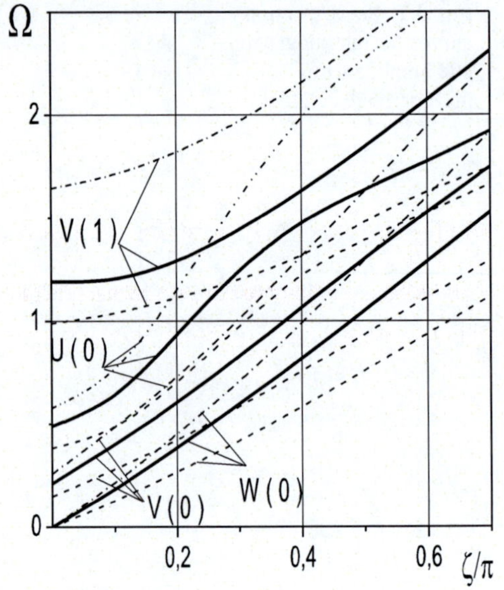

Fig. 3.3 Six dispersion curves for inhomogeneous piezoceramic and homogeneous piezoceramic and metal cylinders ($m = 1$)

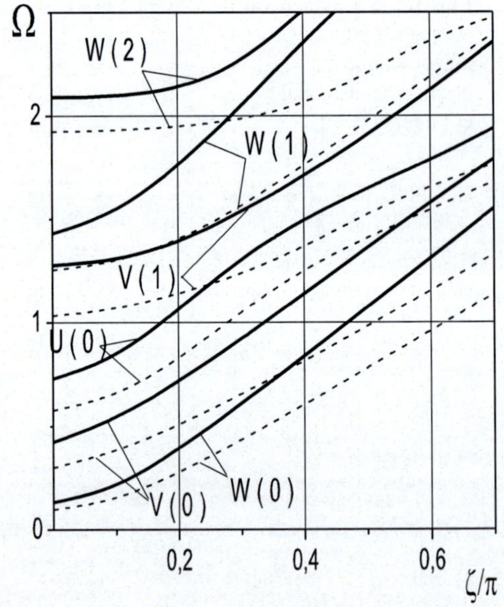

Fig. 3.4 Six dispersion curves for inhomogeneous and homogeneous piezoceramic cylinders ($m = 2$)

in Fig. 3.2. The solid line represents the layered cylinder, and the dashed line the steel cylinder.

The boundary-value problem (3.23), (3.24), (3.25) is known to be mathematically similar to the problem of the free nonaxisymmetric vibrations of a layered cylinder hinged at the ends. Consider a three-layer cylinder with layers similar to the above cylinder (inner radius $R_{\text{in}} = 3$, outer radius $R_{\text{out}} = 5$, length $L = 10$). With such geometric characteristics, $h/R_0 = 0.25$ is equal to the values used in the previous

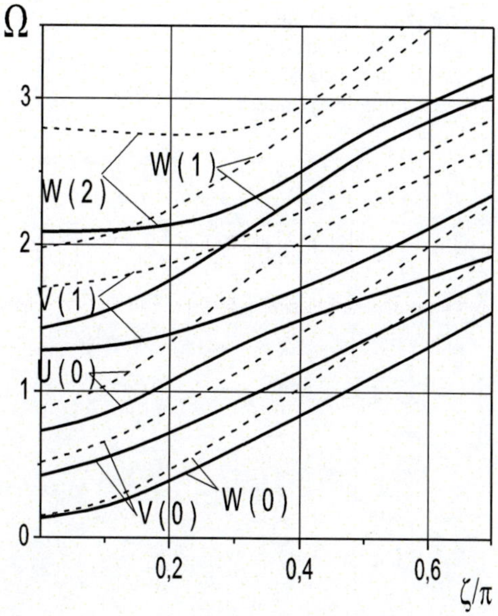

Fig. 3.5 Six dispersion curves for inhomogeneous piezoceramic and homogeneous metal cylinders ($m = 2$)

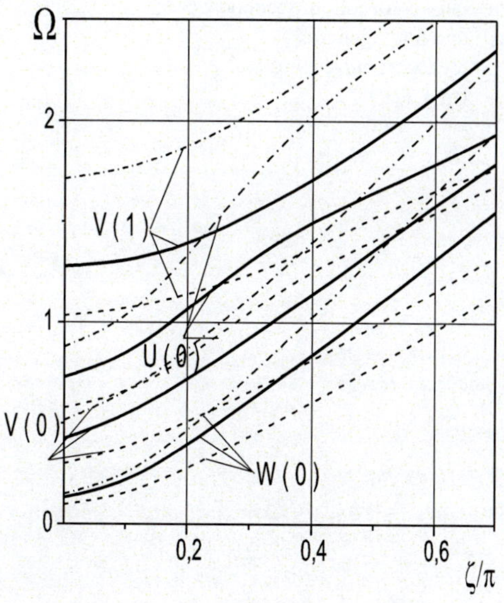

Fig. 3.6 Comparison of the behavior of the dispersion curves presented in Figs. 3.4 and 3.5

problem represented in Figs. 3.1, 3.2, 3.3, 3.4, 3.5, and 3.6. Moreover, it is also necessary to test the values $m = 0, 3, 4, \ldots$.

An analysis of the frequency spectrum shows that the first four values of m ($m = 0, 1, 2, 3$) are sufficient to determine the first five natural frequencies. The results are presented in Figs. 3.7, 3.8, 3.9, and 3.10 for different values of m (the solid lines represent the inhomogeneous cylinder, and the dashed lines represent the homogeneous PZT-4 cylinder).

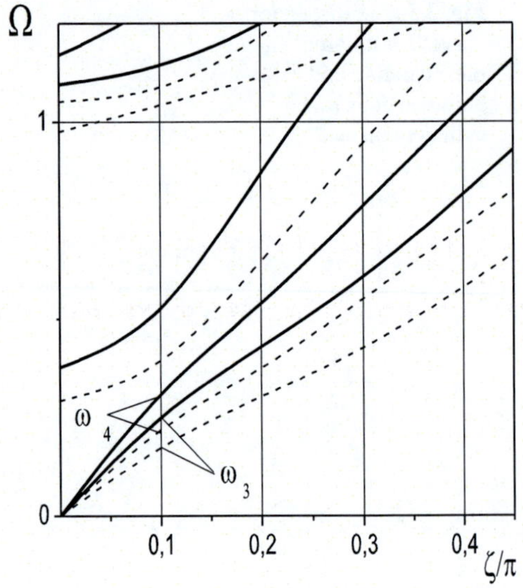

Fig. 3.7 First dispersion curves for inhomogeneous and homogeneous piezoceramic cylinders (m = 0)

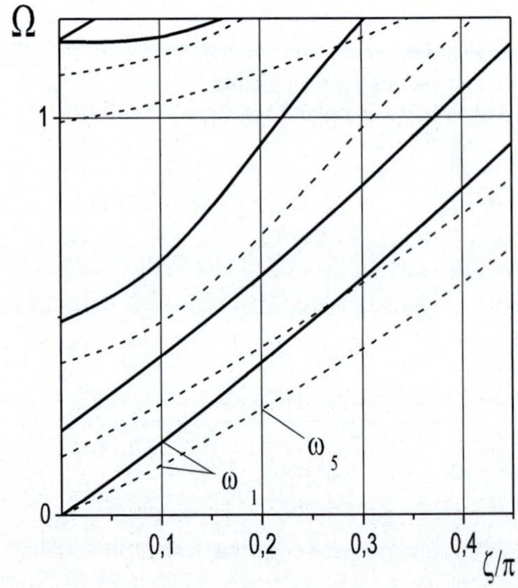

Fig. 3.8 First dispersion curves for inhomogeneous and homogeneous piezoceramic cylinders (m = 1)

Figures 3.7, 3.8, 3.9, and 3.10: First dispersion curves for inhomogeneous and homogeneous piezoceramic cylinders, four values of m ($m = 0, 1, 2, 3$).

The frequencies lie at the intersection of the respective dispersion branches and the values$\varsigma = 0,1;\;\; 0,2;\;\; 0,3\; \ldots$, which should be arranged in ascending order. Table 3.1 summarizes the numerical values of the first five frequencies of the three-layer cylinder (case 1) and the homogeneous piezoceramic cylinder (case 2). The table also indicates the number (m) of circumferential half-waves, the number (n) of axial half-waves, and the relative error (Δ) due to ignoring the inhomogeneity of the cylinder.

Fig. 3.9 First dispersion curves for inhomogeneous and homogeneous piezoceramic cylinders (m = 2)

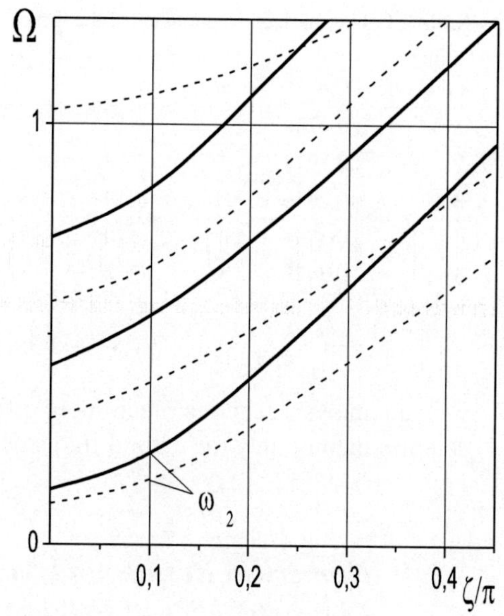

Fig. 3.10 First dispersion curves for inhomogeneous and homogeneous piezoceramic cylinders (m = 3)

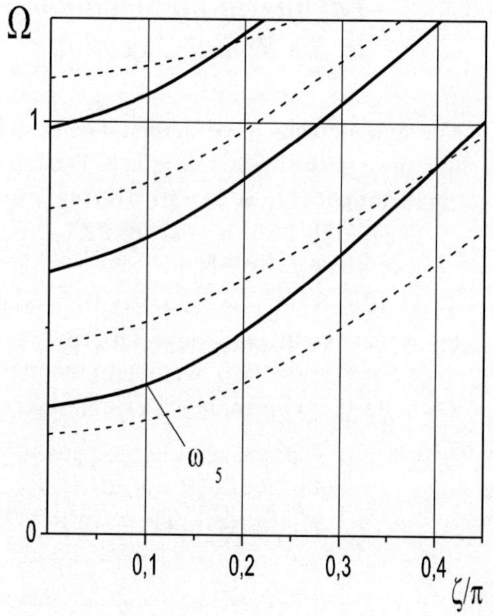

As can be seen from the table, the first natural frequency is not a natural frequency of axisymmetric vibrations. Only the third and fourth natural frequencies are frequencies of axisymmetric vibrations. The third frequency is a frequency of longitudinal vibrations, and the fourth frequency is a frequency of torsional vibrations. The natural frequencies of the solid PZT-4 cylinder calculated above are in agreement with the data obtained with the approach developed in [5]. Naturally, performing

Table 3.1 Comparison frequencies of the three-layer cylinder and the homogeneous piezoceramic cylinder

Frequency	Case 1	m	n	Case 2	m	n	Δ, %
1	0.1784	1	1	0.1235	1	1	30.8
2	0.1973	2	1	0.1422	2	1	27.9
3	0.2656	0	1	0.1835	0	1	30.9
4	0.3294	0	1	0.2347	0	1	28.7
5	0.3477	3	1	0.2665	1	2	23.4

such an analysis requires much more effort. This disadvantage is compensated by determining not only the natural frequencies, but also the vibration modes.

3.2.4 *Numerical Results and Analysis Propagation Characteristics of Electric Elastic Waves in a Laminated Piezoceramic Cylinder with Layer Polarized in the Radial Direction*

The cylinder has three layers. The face layers have thickness $h/2$ each, while the thickness of the middle layer is h. The face layers are made of steel with the following characteristics: $E = 21 \cdot 10^{10}\ \mathrm{N/m^2}$, $\nu = 0.28$, $\rho_M = 7.85 \cdot 10^3\ \mathrm{kg/m^2}$.

The middle layer is made of PZT-4 piezoceramics with the following characteristics: $c_{11} = 13.9 \cdot 10^{10}\ \mathrm{N/m^2}, c_{12} = 7.43 \cdot 10^{10}\ \mathrm{N/m^2}, c_{13} = 7.78 \cdot 10^{10}\ \mathrm{N/m^2}, c_{33} = 11.5 \cdot 10^{10}\ \mathrm{N/m^2}$, $c_{55} = 2.56 \cdot 10^{10}\ \mathrm{N/m^2}$, $e_{33} = 15.1\ \mathrm{C/m^2}$, $e_{13} = -5.2\ \mathrm{C/m^2}$, $e_{15} = 12.7\ \mathrm{C/m^2}$, $\varepsilon_{11}/\varepsilon_0 = 730$, $\varepsilon_{33}/\varepsilon_0 = 635$, $\rho_n = 7.5 \cdot 10^3\ \mathrm{kg/m^2}$.

If $\varepsilon = 0$ and $k = 0$, we obtain the problem of the vibrations of a plane layer. For example, the frequencies of a single-layer metal cylinder are defined by the formulas

$$U(n) = \frac{\pi}{2} n \sqrt{\frac{(1-\nu)\,E}{(1+\nu)\,(1-2\nu)\,\rho_<}} = 0;\ 2.905;\ 5.81; \ldots;\ n = 0, 1, 2, \ldots;$$

$$V(n) = W(n) = \frac{\pi}{2} n \sqrt{\frac{E}{2\,(1+\nu)\,\rho_<}} = 0;\ 1.606;\ 3.211; \ldots\ n = 0, 1, 2, \ldots.$$

For a single-layer PZT-4 cylinder, we have

$$U(2n) = \pi n \sqrt{\frac{1}{\rho_n}\left(c_{11} + \frac{e_{11}^2}{\varepsilon_{11}}\right)} = 0;\ 4.325;\ 8.649; \ldots; n = 0, 1, 2, \ldots;$$

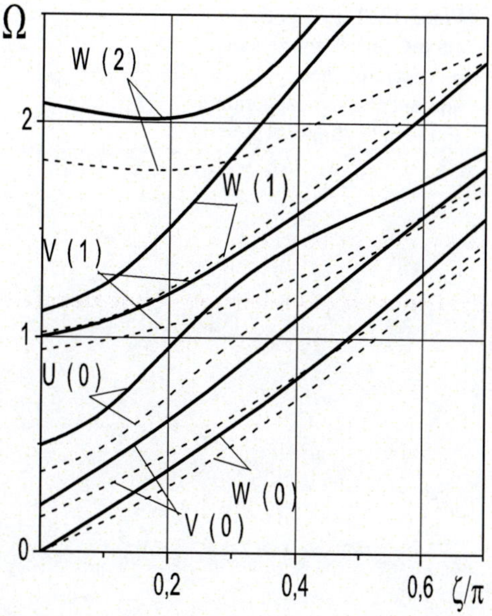

Fig. 3.11 Six dispersion curves for inhomogeneous and homogeneous piezoceramic cylinders ($m = 1$)

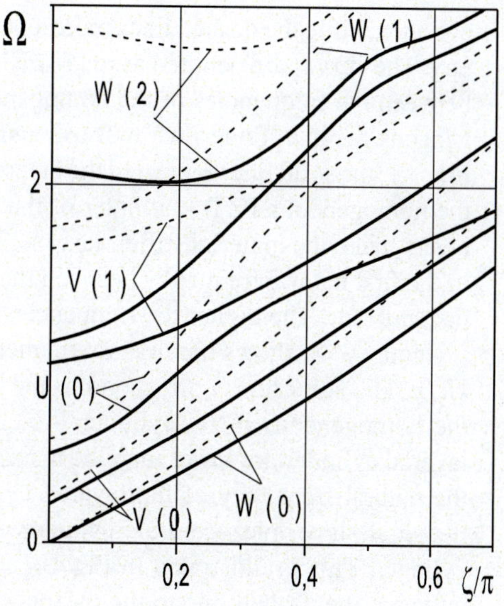

Fig. 3.12 Six dispersion curves for inhomogeneous piezoceramic and homogeneous metal cylinders ($m = 1$)

$$U(2n+1) = \lambda_n \sqrt{\frac{1}{\rho_n}\left(c_{11} + \frac{e_{11}^2}{\varepsilon_{11}}\right)} = 1.995;\ 6.729;\ \ldots;$$

$$V(n) = W(n) = \frac{\pi}{2} n \sqrt{\frac{c_{55}}{\rho_n}} = 0;\ 0.918;\ 1.835;\ \ldots;\ n = 0,\ 1,\ 2\ldots.$$

Figure 3.11 shows the first six frequencies as functions of the dimensionless wave number $\zeta = kh/\pi$ for $m = 1$ and $\varepsilon = 0, 25$. The solid lines represent the layered

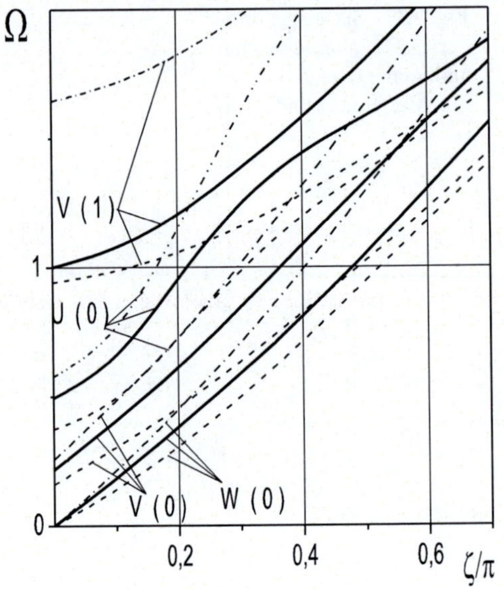

Fig. 3.13 Six dispersion curves for inhomogeneous piezoceramic and homogeneous piezoceramic and metal cylinders ($m = 1$)

cylinder, while the dashed lines represent the solid PZT-4 cylinder of the same geometry. The curves are denoted as in [14]. Since the frequency of the layered cylinder is between the frequencies of the homogeneous piezoceramic cylinder and the homogeneous metal cylinder, we will use a similar notation for the layered cylinder. In Fig. 3.1, the solid lines represent the layered cylinder, while the dashed lines represent the homogeneous PZT-4 cylinder of the same geometry. As can be seen, the metal layers make the material stiffer, i.e., increase the natural frequencies. The difference in the first natural frequency between the layered and the homogeneous cylinders is insignificant. The higher the frequencies, the greater the difference.

Figure 3.12 shows the first six frequencies as functions of the wave number $\zeta = kh/\pi$.The solid lines correspond to the layered cylinder, while the dashed lines to the homogeneous metal cylinder. As can be seen, the natural frequencies of the layered cylinder are lower than the natural frequencies of the metal cylinder. Hence, the natural frequency of the layered cylinder is between the natural frequency of the single-layer piezoceramic cylinder and the frequency of the single-layer metal cylinder. This is illustrated in Fig. 3.13. The solid line corresponds to the layered cylinder, the dashed line to the piezoceramic cylinder, and the dash-and-dot line to the metal cylinder. The material and geometry of the cylinder are the same as above.

Figure 3.14 shows the first six frequencies as functions of the dimensionless wave number for $m = 2$. The material and geometry of the cylinder are the same as in Fig. 3.11. The solid line represents the layered cylinder, and the dashed line the PZT-4 cylinder. Figure 3.15 shows the first six frequencies as functions of the dimensionless wave number for $m = 2$. The material and geometry of the cylinder are the same as in Fig. 3.12. The solid line represents the layered cylinder and the dashed line the steel cylinder.

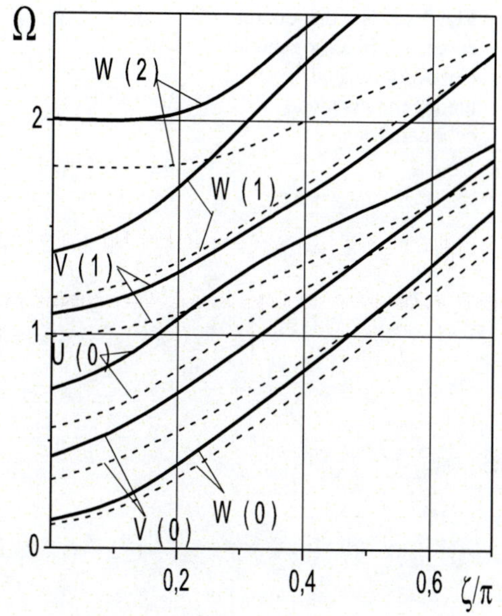

Fig. 3.14 Six dispersion curves for inhomogeneous and homogeneous piezoceramic cylinders ($m = 2$)

The boundary-value problem (3.23), (3.24), (3.26) is known to be mathematically similar to the problem of free nonaxisymmetric vibrations of a layered cylinder hinged at the ends. Consider a three-layer cylinder with layers similar to the above cylinder (inner radius $R_{\text{in}} = 3$, outer radius $R_{\text{out}} = 5$, length $L = 10$). With such geometric characteristics, $\varepsilon = h/R_0 = 0.25$ is equal to the values used in the previous problem represented in Figs. 3.11, 3.12, 3.13, 3.14, 3.15 and 3.16. Moreover, it is also necessary to test the values $m = 0$, $m = 1$, $m = 2, \ldots$.

An analysis of the frequency spectrum shows that the first four values of m ($m = 0, 1, 2, 3$) are sufficient to determine the first five natural frequencies: The results are presented in Figs. 3.17, 3.18, 3.19, and 3.20 for different values of m(the solid lines represent the inhomogeneous cylinder, and the dashed lines represent the solid PZT-4 cylinder).

Figures 3.17, 3.18, 3.19 and 3.20: First dispersion curves for inhomogeneous and homogeneous piezoceramic cylinders for values of m ($m = 0, 1, 2, 3$).

It can be seen that the frequencies lie at the intersection of the respective dispersion branches and the values $\zeta = 0.1; 0.2; 0.3, \ldots$, which should be arranged in ascending order. Table 3.2 summarizes the numerical values of the first five frequencies of the three-layer cylinder (case 1) and the solid piezoceramic cylinder (case 2). The table also indicates the number (m) of circumferential half-waves, the number k) of axial half-waves, and the relative error (Δ) due to ignoring the inhomogeneity of the cylinder.

As can be seen from the table, the first natural frequency is not a natural frequency of axisymmetric vibrations. Only the third and fourth natural frequencies are frequencies of axisymmetric vibrations. The third frequency is a frequency of longitudinal vibrations, and the fourth frequency is a frequency of torsional vibrations. The natu-

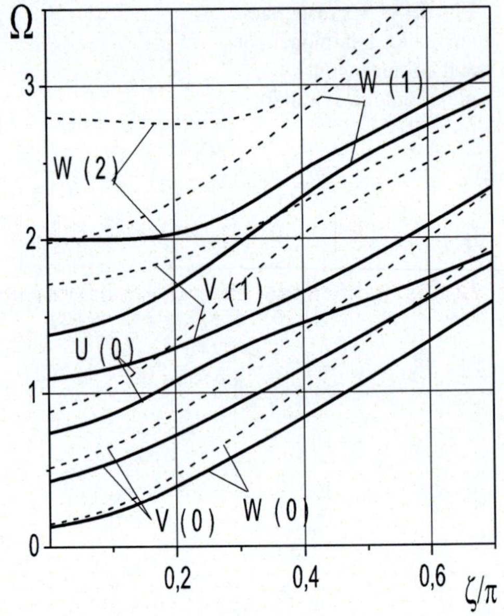

Fig. 3.15 Six dispersion curves for inhomogeneous piezoceramic and homogeneous metal cylinders ($m = 2$)

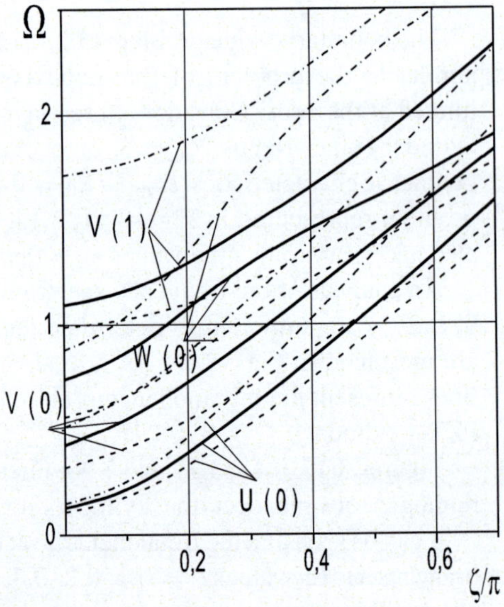

Fig. 3.16 Comparison of the behavior of the dispersion curves presented in Figs. 3.14 and 3.15

Fig. 3.17 First dispersion curves for inhomogeneous and homogeneous piezoceramic cylinders (m = 0)

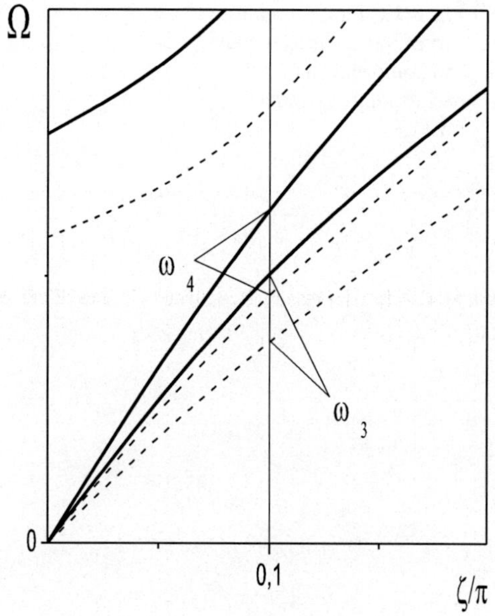

Fig. 3.18 First dispersion curves for inhomogeneous and homogeneous piezoceramic cylinders (m = 1)

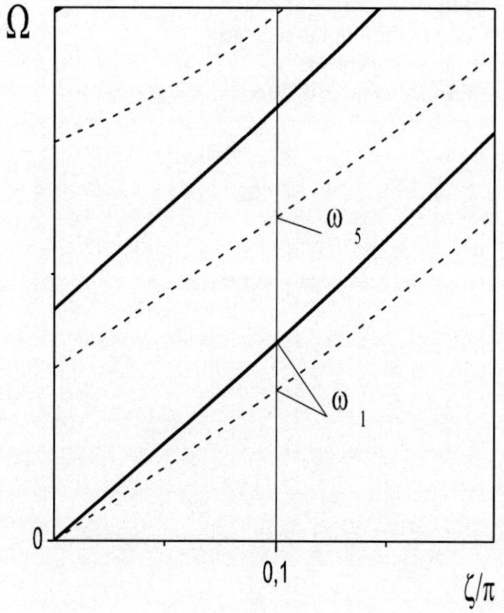

Fig. 3.19 First dispersion curves for inhomogeneous and homogeneous piezoceramic cylinders (m = 2)

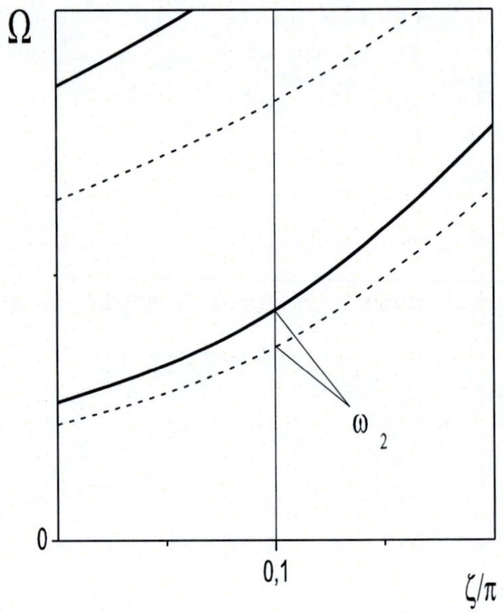

Fig. 3.20 First dispersion curves for inhomogeneous and homogeneous piezoceramic cylinders (m = 3)

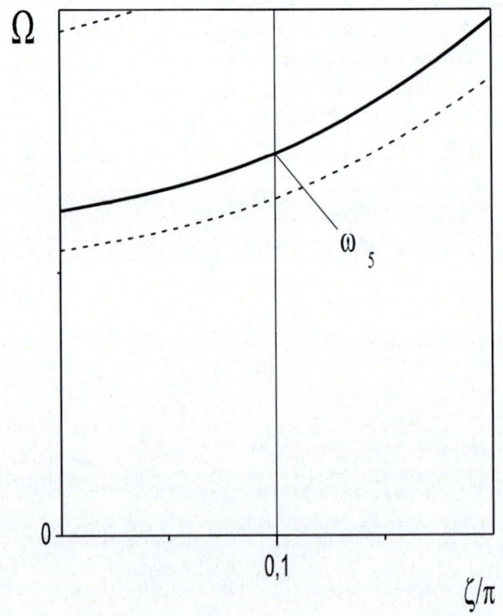

ral frequencies of a solid PZT-4 cylinder calculated above are in agreement with the data obtained with the approach developed in [14]. It is natural that the efforts taken in both cases are absolutely different. Here, however, not only the natural frequencies have been determined, but also the vibration modes have been identified.

The problem of the propagation of nonaxisymmetric waves in a hollow layered cylinder with piezoceramic and metal layers has been formulated and solved. The case

Table 3.2 Comparison frequencies of the three-layer cylinder and the homogeneous piezoceramic cylinder

No. ω_i	Case 1	m	k	Case 2	m	k	Δ, %
1	0.1824	1	1	0.1368	1	1	25
2	0.1965	2	1	0.1645	2	1	16.3
3	0.2713	0	1	0.2007	0	1	26
4	0.3332	0	1	0.2487	0	1	25.4
5	0.3467	3	1	03023	1	1	12.8

where the lateral surfaces are free from mechanical loads and are short-circuited has been examined. To solve the problem, an efficient numerical analytic approach has been used. The mathematically equivalent problem of the free vibrations of a hinged hollow layered cylinder with piezoceramic and steel layers has been solved as well. The numerical analysis performed has revealed that (i) the first natural frequency corresponds to a vibration mode with one axial half-wave and one circumferential half-wave; (ii) the second natural frequency corresponds to a vibration mode with one axial half-wave and one circumferential wave; (iii) the third natural frequency corresponds to an axisymmetric mode with one axial half-wave.

The steel layers substantially increase the natural frequencies. Of the first five natural frequencies, the third one shows the maximum increase (26%).

3.2.5 *Numerical Results and Analysis Propagation Characteristics of Electric Elastic Waves in a Laminated Piezoceramic Cylinder, with Layer Polarized in the Circumferential Direction*

The cylinder has three layers. The face layers have thickness $h/2$ each, while the thickness of the middle layer is h. The face layers are made of steel with the following characteristics: $E = 21 \cdot 10^{10}\ \text{N/m}^2$, $\nu = 0.28$, $\rho_S = 7.85 \cdot 10^3\ \text{kg/m}^2$.The middle layer is made of PZT-4 piezoceramics with the following characteristics:

$$c_{11} = 13.9 \cdot 10^{10}\ \text{N/m}^2, c_{12} = 7.43 \cdot 10^{10}\ \text{N/m}^2, c_{13} = 7.78 \cdot 10^{10}\ \text{N/m}^2,$$
$$c_{33} = 11.5 \cdot 10^{10}\ \text{N/m}^2, c_{55} = 2.56 \cdot 10^{10}\ \text{N/m}^2, e_{13} = -5.2\ \text{C/m}^2,$$
$$e_{15} = 12.7\ \text{C/m}^2, e_{33} = 15.1\ \text{C/m}^2, \varepsilon_{11}/\varepsilon_0 = 730, \varepsilon_{33}/\varepsilon_0 = 635,$$
$$\rho_p = 7.5 \cdot 10^3\ \text{kg/m}^2.$$

The curves are denoted as in [9]. If $\varepsilon = 0$ and $k = 0$, we have the problem of the vibrations of a plane layer. For example, the frequencies of a single-layer metal cylinder are defined by the formulas

$$U(n) = \frac{\pi}{2} n \sqrt{\frac{(1-\nu)\,E}{(1+\nu)\,(1-2\nu)\,\rho_M}} = 0;\ 2.905;\ \ 5.81; \ldots; n = \ 0, 1, 2, ; \ldots;$$

$$V(n) = W(n) = \frac{\pi}{2} n \sqrt{\frac{E}{2\,(1+\nu)\,\rho_M}} = 0;\ 1.606;\ 3.211; ; \ldots; n = 0, 1, 2, \ldots.$$

For a single-layer PZT-4 cylinder, we have

$$U(2n) = \frac{\pi}{2} n \sqrt{\frac{c_{11}}{\rho_n}} = 0;\ 2.138;\ 4.277; \ldots\ n = 0, 1, 2 \ldots;$$

$$V(2n) = \pi n \sqrt{\frac{1}{\rho_n}\left(c_{55} + \frac{e_{15}^2}{\varepsilon_{11}}\right)} = 2.580;\ 5.159; \ldots;$$

$$V(2n-1) = \lambda_n \sqrt{\frac{1}{\rho_n}\left(c_{55} + \frac{e_{15}^2}{\varepsilon_{11}}\right)} = 0.962;\ 3.782; \ldots;$$

$$W(n) = \frac{\pi}{2} n \sqrt{\frac{c_{66}}{\rho_n}} = 0;\ 1.003;\ 2.007; \ldots n = 0, 1, 2 \ldots;$$

here λ_n are the non-zero roots of the transcendent equation $\lambda \cos\lambda - \dfrac{e_{15}^2}{e_{15}^2 + c_{55}\varepsilon_{11}}$ $\sin\lambda = 0$.

Figure 3.21 shows the first six frequencies as functions of the dimensionless wave number $\zeta = kh/\pi$ for $m = 1$ and $\varepsilon = 0.25$. The curves are denoted as in [13]. The solid lines represent the layered cylinder, while the dashed lines represent the solid PZT-4 cylinder of the same geometry. As can be seen, the metal layers make the material stiffer, i.e., increase the natural frequencies. Figure 3.22 shows the first six frequencies as functions of the dimensionless wave number $\zeta = kh/\pi$. The solid lines correspond to the layered cylinder, while the dashed lines to the homogeneous metal cylinder. As can be seen, the natural frequencies of the layered cylinder are lower than the natural frequencies of the metal cylinder. Hence, the natural frequency of the layered ball is between the natural frequency of the single-layer piezoceramic cylinder and the frequency of the single-layer steel cylinder. This is illustrated in Fig. 3.23. The solid lines correspond to the layered cylinder, the dashed lines to the piezoceramic cylinder, and the dash-and-dot lines to the steel cylinder. The material and geometry of the cylinder are the same as above.

Figure 3.24 shows the first six frequencies as functions of the dimensionless wave number for $m = 2$.The material and the geometry of the cylinder are the same as in Fig. 3.21. The solid line represents the layered cylinder, while the dashed line the PZT-4.

Figure 3.25 shows the first six frequencies as functions of the dimensionless wave number for $m = 2$. The material and geometry of the cylinder are the same as in Fig. 3.22. The solid line represents the layered cylinder, while the dashed line the

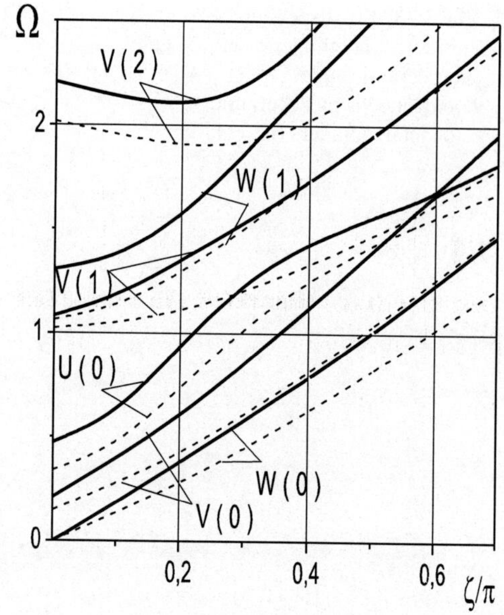

Fig. 3.21 Six dispersion curves for inhomogeneous and homogeneous piezoceramic cylinders ($m = 1$)

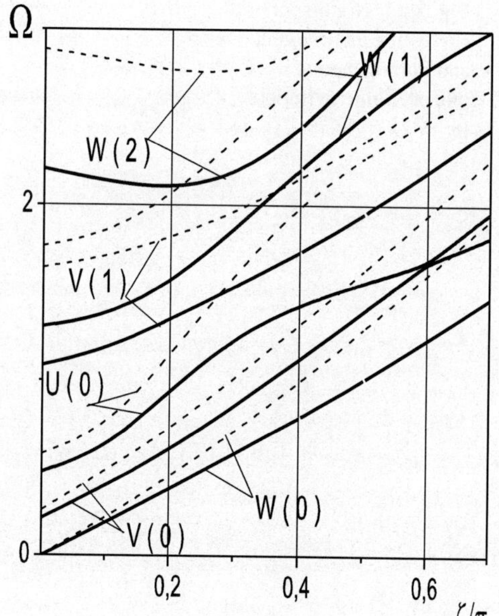

Fig. 3.22 Six dispersion curves for inhomogeneous piezoceramic and homogeneous metal cylinders ($m = 1$)

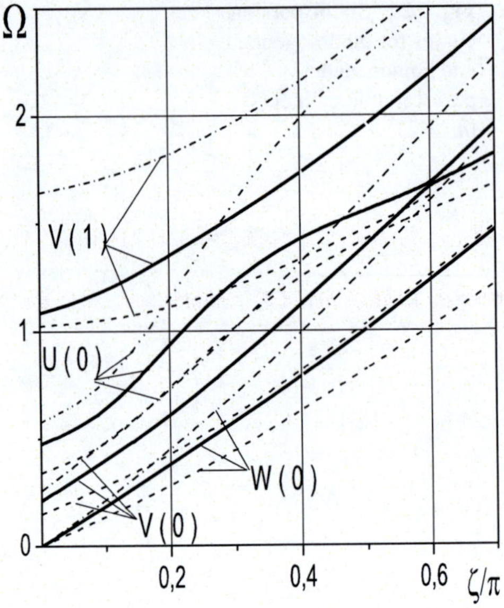

Fig. 3.23 Six dispersion curves for inhomogeneous piezoceramic and homogeneous piezoceramic and metal cylinders ($m = 1$)

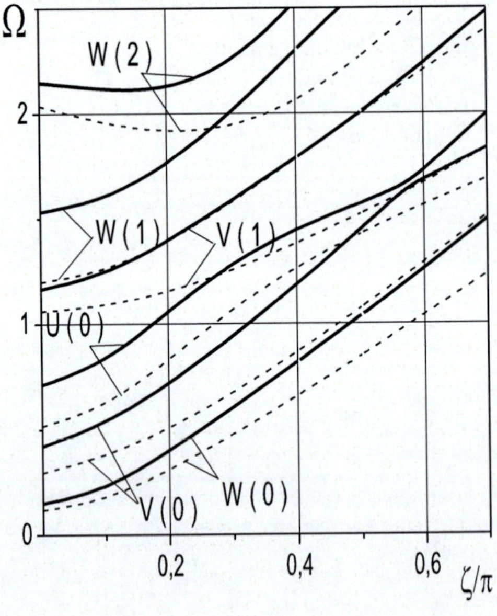

Fig. 3.24 Six dispersion curves for inhomogeneous and homogeneous piezoceramic cylinders ($m = 2$)

steel cylinder. The curves for the first four frequencies in Figs. 3.24 and 3.25 are superimposed in Fig. 3.26.

Figures 3.17, 3.18, 3.19 and 3.20: First dispersion curves for inhomogeneous and homogeneous piezoceramic cylinders for values of m ($m = 0, 1, 2, 3$).

An analysis of the frequency spectrum shows that the first four values of m ($m = 0, 1, 2, 3$) are sufficient to determine the first five natural frequencies. The results are presented in Figs. 3.27, 3.28, 3.29, and 3.30 for the different values of m (the solid

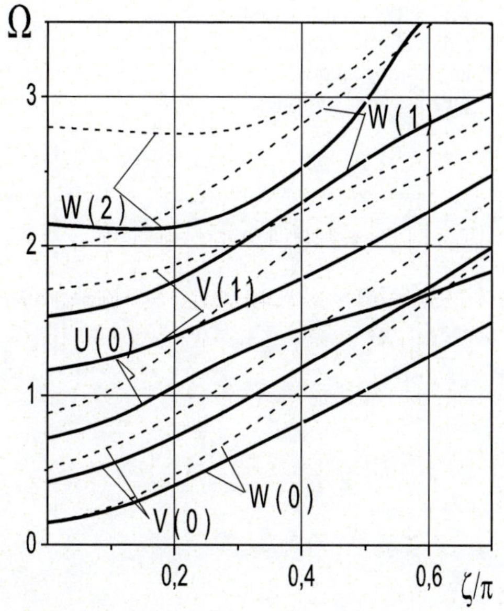

Fig. 3.25 Six dispersion curves for inhomogeneous piezoceramic and homogeneous metal cylinders

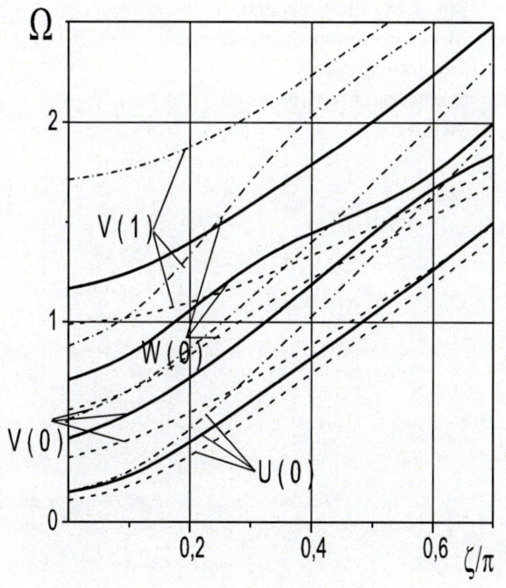

Fig. 3.26 Comparison of the behavior of the dispersion curves presented in Figs. 3.24 and 3.25

lines represent the inhomogeneous cylinder, and the dashed lines represent the solid PZT-4 cylinder). The corresponding frequencies are located at the intersection of the corresponding dispersion branches and the values $\zeta = 0.1; 0.2;\ 0.3; \ldots$. Table 3.3 shows the numerical values of the first five frequencies. As for other polarization directions, the error in neglecting the heterogeneity factor will be higher than in neglecting the associated electric field.

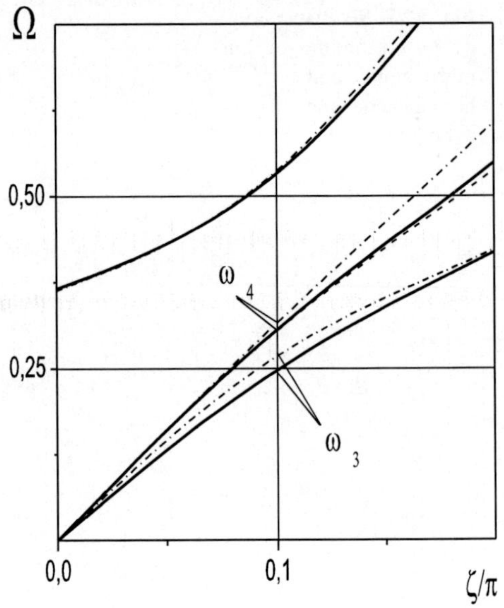

Fig. 3.27 First dispersion curves for inhomogeneous and homogeneous piezoceramic cylinders (m = 0)

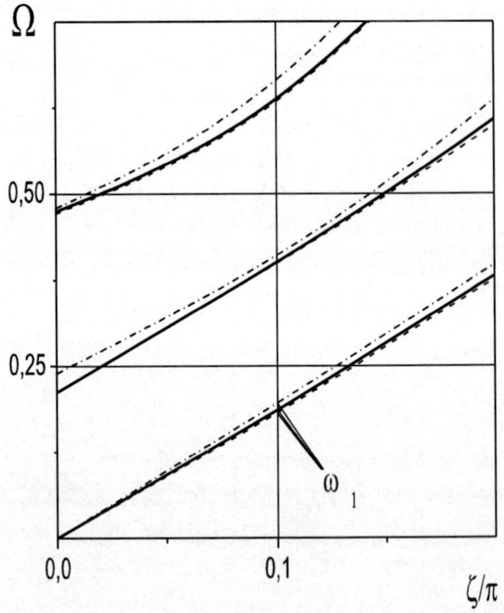

Fig. 3.28 First dispersion curves for inhomogeneous and homogeneous piezoceramic cylinders (m = 1)

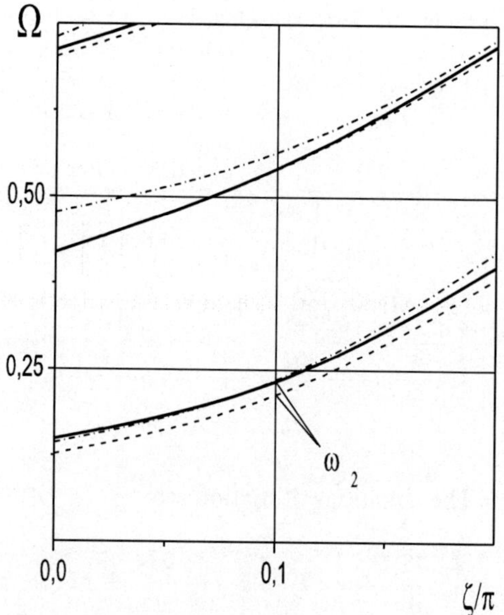

Fig. 3.29 First dispersion curves for inhomogeneous and homogeneous piezoceramic cylinders (m = 2)

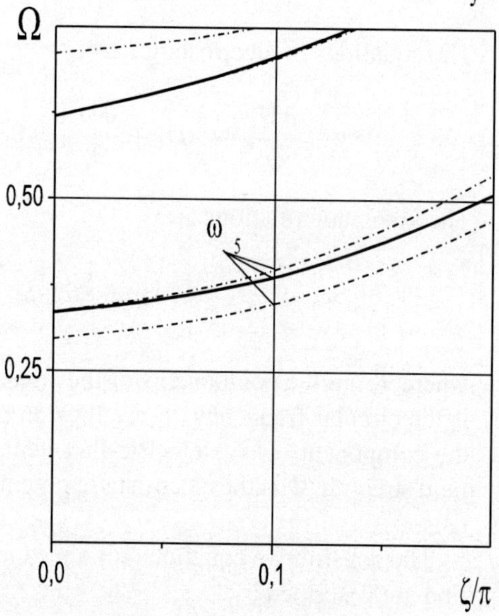

Fig. 3.30 First dispersion curves for inhomogeneous and homogeneous piezoceramic cylinders (m = 3)

3.3 Axisymmetric Problem on Propagation of Forced Acoustoelectric Waves in Inhomogeneous Cylinder Made of Functionally Gradient Piezoceramics

3.3.1 Problem Statement. Basic Equations for Inhomogeneous Cylinders

The axisymmetric longitudinal equations of wave motion in a cylindrical coordinate system (r, θ, z) are described as follows:

Table 3.3 Comparison frequencies of the three-layer cylinder based on different models

No of frequency	Layer cylinder	Number of half waves in axial direction	Number of half-waves in circular direction	The theory of effective moduli	Relative error (%)	Without piezoeffect	Relative error (%)
1	0.1834	1	1	0.1953	6.5	0.1787	2.6
2	0.2124	1	2	0.1931	9.1	0.2120	0.2
3	0.2667	1	0	0.2988	12.0	0.2655	0.4
4	0.3242	1	0	0.3230	0.3	0.3242	0
5	0.3807	1	3	0.3336	12.4	0.3726	2.1

The equations of motion are:

$$\frac{\partial T_1}{\partial r} + \frac{1}{r}(T_1 - T_2) + \frac{\partial T_5}{\partial z} + \rho\omega^2 u_1 = 0; \ \frac{\partial T_5}{\partial r} + \frac{1}{r}T_5 + \frac{\partial T_3}{\partial z} + \rho\omega^2 u_3 = 0; \tag{3.29}$$

The equations of electrostatics are:

$$\frac{\partial D_1}{\partial r} + \frac{1}{r}D_1 + \frac{\partial D_3}{\partial z} = 0, \ E_1 = -\frac{\partial \Phi}{\partial r}, \ E_3 = -\frac{\partial \Phi}{\partial z}; \tag{3.30}$$

The kinematic relations are:

$$S_1 = \frac{\partial u_1}{\partial r}, \ S_2 = \frac{1}{r}u_1, \ S_3 = \frac{\partial u_3}{\partial u_z}, \ S_5 = \frac{\partial u_3}{\partial r} + \frac{\partial u_1}{\partial z}, \tag{3.31}$$

where T_i are the components of the stress tensor, ρ is the density of the material, ω is the circular frequency, u_i are the components of the displacement vector, D_i are the components of the electric-flux density, E_i are the components of the electric field strength, Φ is the electrostatic potential, and S_i are the components of the strain tensor.

The constitutive equations for a piezoelectric material polarized in a radial direction are described by:

$$\begin{aligned} T_1 &= c_{33}S_1 + c_{13}S_2 + c_{13}S_3 - e_{33}E_1, \\ T_2 &= c_{13}S_1 + c_{11}S_2 + c_{12}S_3 - e_{13}E_1, \\ T_3 &= c_{13}S_1 + c_{12}S_2 + c_{11}S_3 - e_{13}E_1, \ T_5 = 2c_{55}S_5 - e_{15}E_3, \\ D_1 &= e_{33}S_1 + e_{13}S_2 + e_{13}S_3 + \varepsilon_{33}E_1; \ D_3 = 2e_{15}S_5 + \varepsilon_{33}E_3. \end{aligned} \tag{3.32}$$

where c_{ij} are the components of the tensor of the elastic moduli, e_{ij} are the components of the piezomodules tensor, ε_{ij} are the components of the permittivity tensor. The above components are functions of the radial coordinate.

Consider a material consisting of two components: steel and piezoceramics. The characteristics of the material vary across the thickness as follows:

$$P(r) = (P_m - P_p)V(r) + P_p, \tag{3.33}$$

where $V(z)$ is the volume fraction of the ceramics, which is described as

$$V(r) = \left(\frac{r - R_0}{2h} + \frac{1}{2}\right)^n. \tag{3.34}$$

The boundary conditions on the lateral surfaces of the cylinder (forr $= R_0 \pm h$) are: the surfaces are free of external forces, $T_1 = T_5 = 0$, and covered with thin electrodes to which a harmonic voltage is applied, $\Phi = \pm V_0 e^{\mathrm{i}(kz-\omega t)}$ (R_0 is the radius of the mid-surface of the cylinder; h is the half-thickness of the cylinder).

The resolving vector of the mixed type is

$$\mathbf{R} = \{T_1,\ T_5,\ \Phi,\ u_1,\ u_3,\ D_1\}^T. \tag{3.35}$$

By resolving the system (3.29)–(3.32) for the vector R and performing some transformations, we obtain

$$\begin{aligned}
\frac{\partial T_1}{\partial r} &= \frac{1}{r}\left(\frac{\Delta_2}{\Delta} - 1\right)T_1 - \frac{\partial T_5}{\partial z} - \left(\frac{\Delta_3}{r^2\Delta} - \rho\frac{\partial^2}{\partial t^2}\right)u_1 - \frac{\Delta_4}{r\Delta}\frac{\partial u_3}{\partial z} - \frac{\Delta_1}{r\Delta}D_1,\\
\frac{\partial T_5}{\partial r} &= \frac{\Delta_2}{\Delta}\frac{\partial T_1}{\partial z} - \frac{1}{r}T_5 + \frac{\Delta_5}{r\Delta}\frac{\partial u_1}{\partial z} - \left(\frac{\Delta_6}{\Delta}\frac{\partial^2}{\partial z^2} - \rho\frac{\partial^2}{\partial t^2}\right)u_3 - \frac{\Delta_1}{\Delta}\frac{\partial D_1}{\partial z},\\
\frac{\partial \Phi}{\partial r} &= \frac{e_{33}}{\Delta}T_1 + \frac{\Delta_1}{r\Delta}u_1 + \frac{\Delta_1}{\Delta}\frac{\partial u_3}{\partial z} - \frac{c_{33}}{\Delta}D_1,\\
\frac{\partial u_1}{\partial r} &= \frac{\varepsilon_{33}}{\Delta}T_1 - \frac{\Delta_2}{r\Delta}u_1 - \frac{\Delta_2}{\Delta}\frac{\partial u_3}{\partial z} + \frac{e_{33}}{\Delta}D_1,\\
\frac{\partial u_3}{\partial r} &= \frac{1}{c_{55}}T_5 - \frac{\partial u_1}{\partial z} - \frac{e_{51}}{rc_{55}}\frac{\partial \Phi}{\partial z},\quad \frac{\partial D_1}{\partial r} = -\frac{e_{51}}{c_{55}}\frac{\partial T_5}{\partial z} + \frac{\Delta_7}{c_{55}}\frac{\partial^2 \Phi}{\partial z^2} - \frac{1}{r}D_1.
\end{aligned} \tag{3.36}$$

3.3.2 Method for Solving Problem and Analysis of Numerical Results

Let us search for the problem solution in the form of waves running in the axial direction

$$\begin{aligned}
&T_1\,(r,z,t) = \mathrm{i}\lambda T_1\,(r)\,e^{\mathrm{i}(kz-\omega t)},\ T_5\,(r,z,t) = \lambda T_5(r)e^{\mathrm{i}(kz-\omega t)},\\
&\Phi\,(r,z,t) = h\sqrt{\frac{\lambda}{\varepsilon_0}}\Phi(r)e^{\mathrm{i}(kz-\omega t)},\ u_1\,(r,z,t) = \mathrm{i}hu_1(r)e^{\mathrm{i}(kz-\omega t)},
\end{aligned} \tag{3.37}$$

$$u_3(r,z,t) = hu_3(r)e^{i(kz-\omega t)},\ D_1(r,z,t) = \sqrt{\varepsilon_0\lambda}D_1(r)e^{i(kz-\omega t)}.$$

With (3.37), the original two-dimensional problem of electroelasticity in partial derivatives can be reduced to a boundary-value problem in ordinary differential equations

$$\frac{d\mathbf{R}}{dx} = \mathbf{A}(x,\ \Omega)\,\mathbf{R} \tag{3.38}$$

with boundary conditions

$$\mathbf{B}_1\mathbf{R}(-1) = \mathbf{C}_1,\ \mathbf{B}_2\mathbf{R}(1) = \mathbf{C}_2, \tag{3.39}$$

where the vector $\mathbf{C}_1^T = \{0, 0, -V_0, 0, 0, 0\}$, vector $\mathbf{C}_2^T = \{0, 0, +V_0, 0, 0, 0\}$.

There the following notation is used:

$$\Omega = \omega h\sqrt{\frac{\rho}{\lambda}},\ \tilde{c}_{ij} = \frac{c_{ij}^0}{\lambda},\ \tilde{e}_{ij} = \frac{e_{ij}^0}{\sqrt{\varepsilon_0\lambda}},\ \tilde{\varepsilon}_{ij} = \frac{\varepsilon_{ij}^0}{\varepsilon_0},\ x = \frac{r-R_0}{h},$$

where ρ is the density of the cylinder material, R_0 is the radius of the mid-surface, ε_0 is the vacuum permittivity, $\lambda = 10^{10}\mathrm{N/m}^2$.

The problem is solved by the stable discrete-orthogonalization method.

The matrix $\mathbf{A}(x,\Omega)$ has the form

$$\left\|\begin{matrix}
\tilde{x}\left(\frac{\Delta_2}{\Delta}-1\right) & k & 0 & -\frac{\tilde{x}^2\Delta_3}{\Delta}-\Omega^2 & \frac{k\tilde{x}\Delta_4}{\Delta} & -\frac{\tilde{x}\Delta_1}{\Delta} \\
-\frac{k\Delta_2}{\Delta} & -\tilde{x} & 0 & -\frac{k\tilde{x}\Delta_5}{\Delta} & \frac{k^2\Delta_6}{\Delta}-\Omega^2 & \frac{k\Delta_1}{\Delta} \\
\frac{\tilde{e}_{33}}{\Delta} & 0 & 0 & \frac{\tilde{x}\Delta_1}{\Delta} & -\frac{k\Delta_1}{\Delta} & -\frac{\tilde{c}_{33}}{\Delta} \\
\frac{\tilde{\varepsilon}_{33}}{\Delta} & 0 & 0 & -\frac{\tilde{x}\Delta_2}{\Delta} & \frac{k\Delta_2}{\Delta} & \frac{\tilde{e}_{33}}{\Delta} \\
0 & \frac{1}{\tilde{c}_{55}} & -\frac{k\tilde{e}_{51}}{\tilde{c}_{55}} & -k & 0 & 0 \\
0 & \frac{k\tilde{e}_{15}}{\tilde{c}_{55}} & -k^2\frac{\Delta_7}{\tilde{c}_{55}} & 0 & 0 & -\tilde{x}
\end{matrix}\right\|$$

The problem (3.10), (3.11) is solved by the stable discrete-orthogonalization method.

Let us consider results of the numerical analysis of the boundary-value problem (3.10), (3.11). The expression (3.6) is the general formula for the physical–mechanical characteristics of a material, P_p and P_m are the appropriate characteristics of the ceramics and metal. The power of the volume fraction of the ceramics in (3.6) can vary within $0 \le n < 1000$. The structure is completely metallic if $n = 0$, the structure is completely metallic; if $n = \infty$ it is piezoceramic (Fig. 3.31).

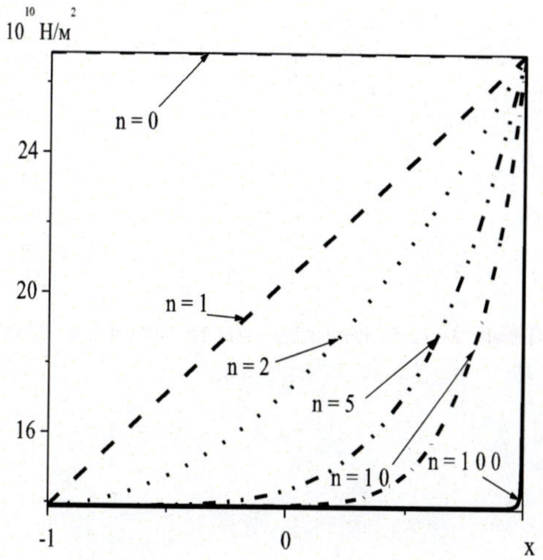

Fig. 3.31 Physical–mechanical characteristics of the inhomogeneous material of cylinder

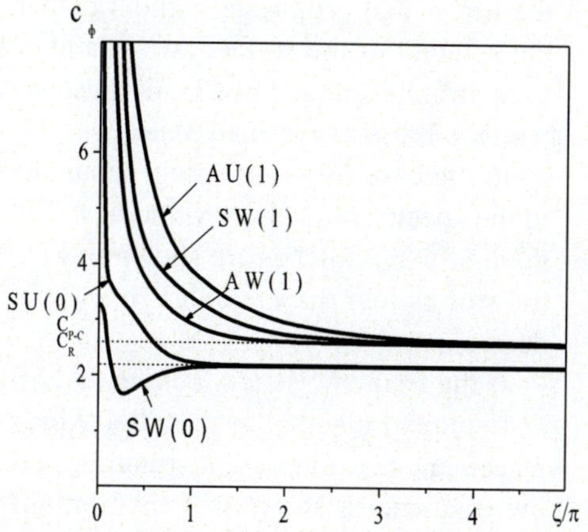

Fig. 3.32 The phase velocities of propagating waves for homogeneous piezoceramic cylinder

As noted in [8], for a homogeneous problem (free motions), the dispersion relations demonstrate qualitatively the piezoceramic distinctions. This is better seen from consideration of the phase velocities of propagating waves.Thus, the first two waves $SW(0)$ and $AU(0)$ in a short-wave range for a homogeneous cylinder made of PZT 4 piezoceramics (Fig. 3.32) transform into the Rayleigh-type surface wave. The velocity of these waves is lower than the least from those of body waves in an infinite space

$$c_R < \min\left(\sqrt{c_{55}/\rho};\ \sqrt{(c_{33} + e_{33}^2\varepsilon_{33})/\rho}\right).$$

A sophisticated treatment of the displacement distribution in these waves is given below. In the figures, the wave notation is the same as in [9]. The notation $SW(0)$ means that a wave is generated ($k = 0$) as symmetric longitudinal vibrations (planar vibrations), the notation $AU(0)$ means that wave is generated as asymmetric radial vibrations. The other branches in the short-wave range rather quickly transform into

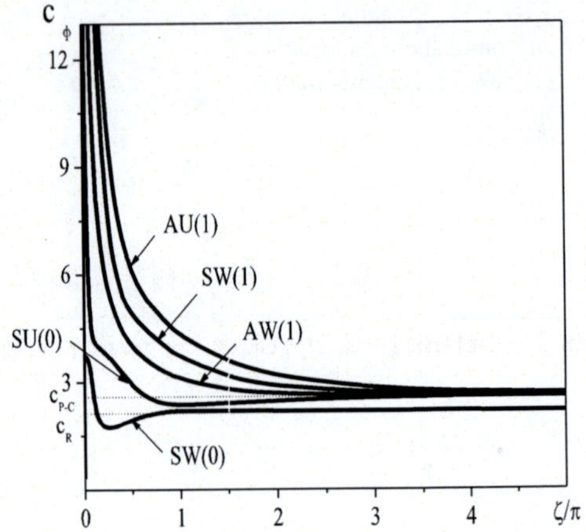

Fig. 3.33 The phase velocities of propagating waves for inhomogeneous cylinder

the waves that propagate without dispersion with constant velocity, which exceeds the velocity of the surface waves and is lower than the velocity of the body waves in an infinite space. Let us call these waves Pochhammer–Cree waves (by analogy with the Lamb waves in a plate).

In the case of an inhomogeneous material, we can see essential transformation in the spectrum of phase velocities. The phase velocities of propagating waves for the first five branches are shown in Fig. 3.33. As can be seen, only the first branch transforms into dispersionless ones, while all the others propagate essentially with dispersion.

If the frequencies are locking ($\zeta = 0$), the purely elastic longitudinal vibrations and coupled electroelastic radial vibrations [9] are held. The analysis of the displacement distributions in running waves in the immediate vicinity of the locking frequencies shows that the motions kept the manner adopted in the notation. Figure 3.34 demonstrates how the displacement amplitudes of the first five branches in the homogeneous cylinder ($n = 1000$) are distributed.

Let us study how the parameter of inhomogeneity affects the distribution of the displacement amplitudes. Indeed, this influence for the first branch $SW(0)$ is so insignificant that it cannot be presented graphically. The displacements in this wave are mainly longitudinal and this is reflected in the notation of this branch. Figure 3.34 demonstrate how the inhomogeneity parameter influences on the displacement amplitudes for these second branch $AU(0)$ ($\zeta = 0, 01\pi$) at different magnitudes of the inhomogeneity factor n. The amplitudes of the radial u_r and the longitudinal u_z displacements are shown by solid and dotted lines, respectively. For these waves, the radial displacements dominate. The nearly linear distribution of displacements across the thickness is typical for the branches $SW(0)$ 0 and $AU(0)$. Figures 3.35, 3.36, 3.37, 3.38, 3.39, and 3.40 show how the inhomogeneity factor n affects the distribution of the displacement amplitudes for the branches $AW(1)$, $SW(1)$ and $AU(1)$, respectively ($\zeta = 0, 01\pi$).

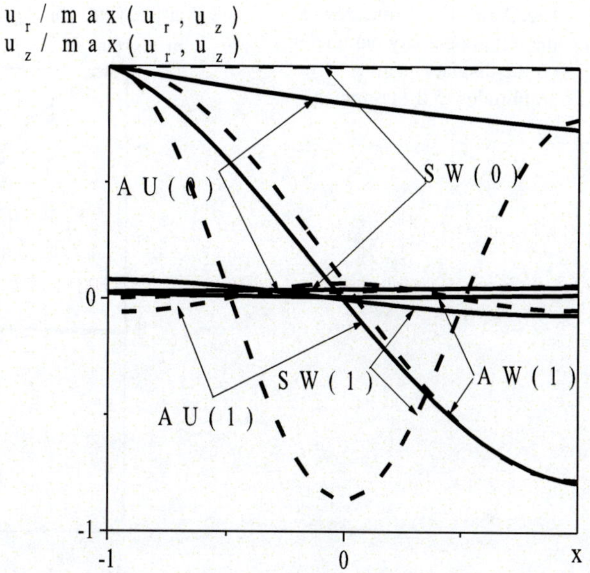

Fig. 3.34 The displacement amplitudes of the first five branches in the homogeneous cylinder

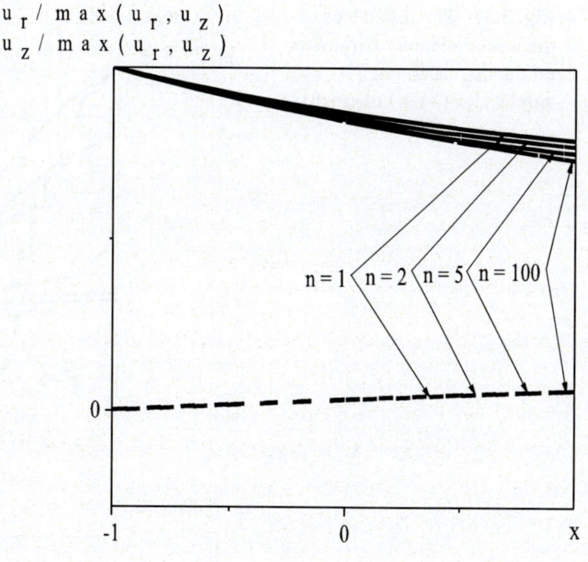

Fig. 3.35 The influences of the inhomogeneity parameter on the displacement amplitudes ($AW(1)$branch)

If the material is homogeneous, the longitudinal displacements for the $AW(1)$ and $SW(1)$ branches dominate. The number of half-waves per unit increase with the frequency. In the case of an inhomogeneous material, as applied to the $AW(1)$ branch (Fig. 3.35), the inhomogeneity parameter hardly influences the manner with which the displacements are distributed. With regard to the $SW(1)$ branch (Fig. 3.36), we should note that variations in the inhomogeneity parameter may result in pronounced changes in the pattern of the displacement distributions. For the high value of the parameter n, i.e., if the cylinder is made, mostly of piezoceramics, the axial dis-

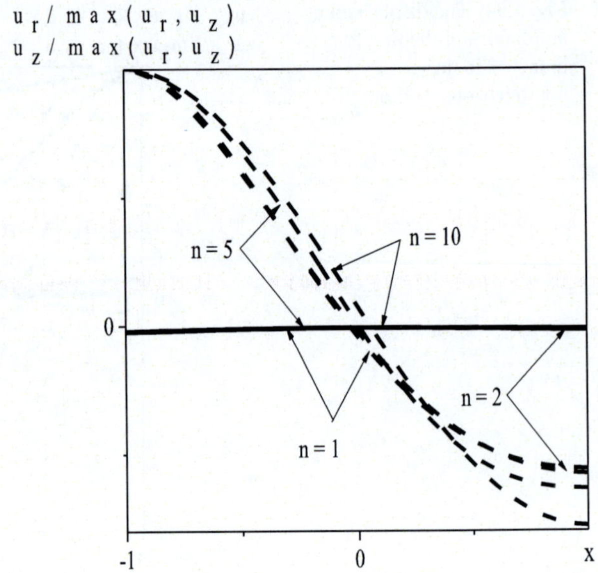

Fig. 3.36 The influences of the inhomogeneity parameter on the displacement amplitudes ($SW(1)$ branch)

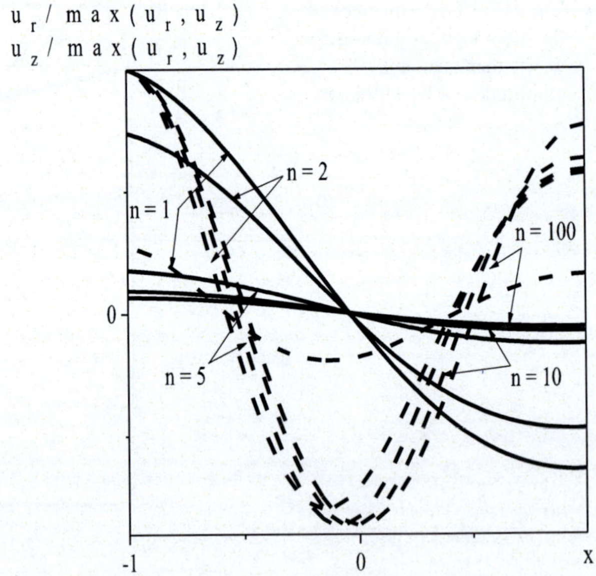

Fig. 3.37 The influences of the inhomogeneity parameter on the displacement amplitudes ($SW(1)$ branch)

placements dominate (the same way as in the case of a homogeneous piezoceramic material). When the parameter n decreases (the volume fraction of the piezoceramics decreases), the displacements become predominantly radial.

Consider how the displacement distributions transform across the cylinder thickness with decreasing wavelength. In the case of a homogeneous material (Fig. 3.39), the first branch $SW(0)$ approaches the Rayleigh-type surface wave propagating over the inside surface of the cylinder. The second $AU(0)$ branch in the short-wave

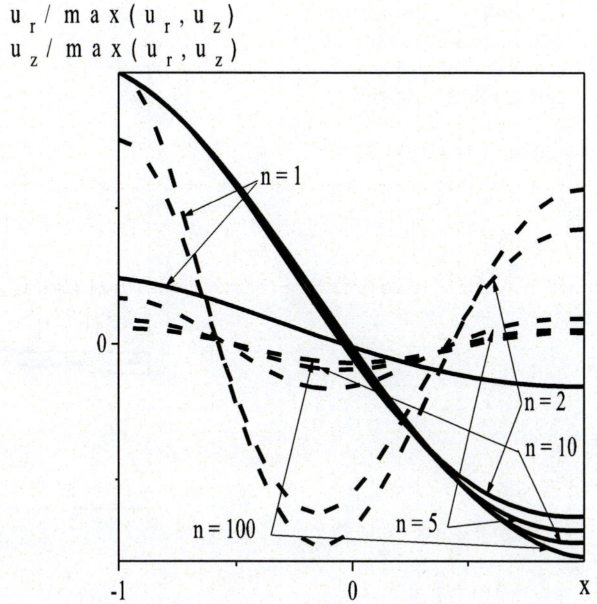

Fig. 3.38 The influences of the inhomogeneity parameter on the displacement amplitudes ($AU(1)$ branch)

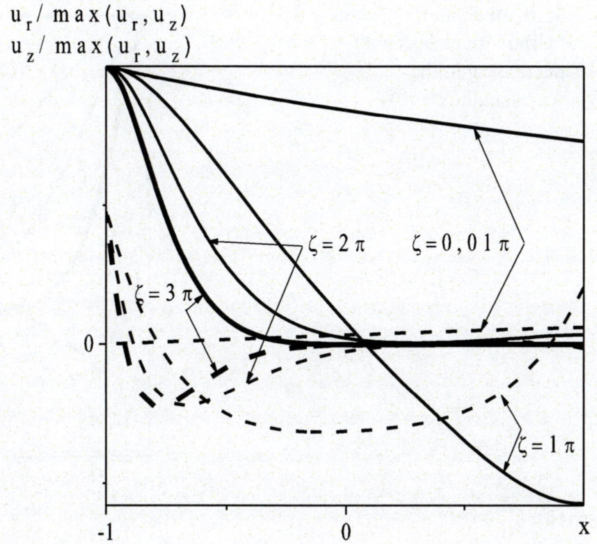

Fig. 3.39 The displacement distributions transform across the homogeneous cylinder thickness with decreasing in the wavelength($SW(0)$)

range approaches the Rayleigh-type surface wave propagating over the inside surface (Fig. 3.40, heavy line).

In the case of an inhomogeneous material of the cylinder ($n = 5$), we can see qualitative distinctions in the pattern of distribution of the displacement amplitudes. As it was noted above, the first $SW(0)$ branch transforms into the Rayleigh-type surface wave as well (Fig. 3.41, heavy line). The second $AU(0)$ branch by this time does not transform into Rayleigh-type surface wave (Fig. 3.42).

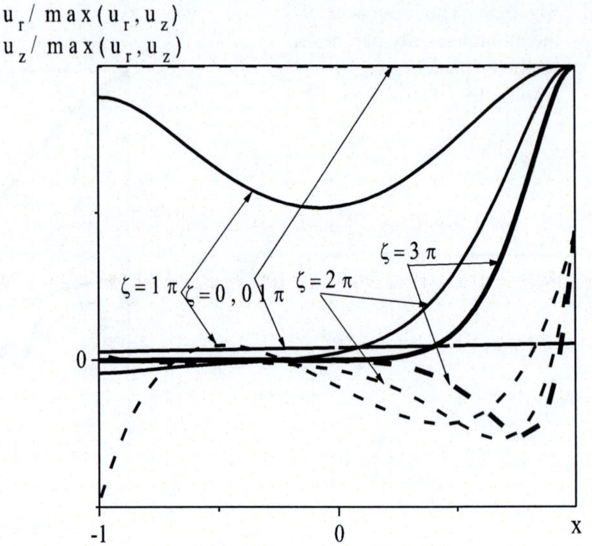

Fig. 3.40 The displacement distributions transform across the homogeneous cylinder thickness with decreasing in the wavelength($AU(0)$)

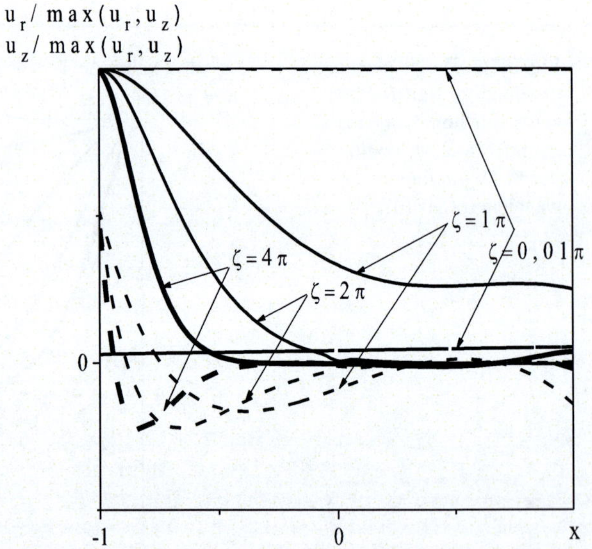

Fig. 3.41 The displacement distributions transform across the inhomogeneous cylinder thickness with decreasing in the wavelength($SW(0)$)

In the case of a homogeneous material, the following $AW(1)$ and $SW(1)$ branches (Figs. 3.43 and 3.44) in the short-wave range approach to the nearly symmetric (with respect to the cylinder mid-surface) or nearly asymmetric distribution of the displacements across the cylinder thickness. The waves $AW(1)$ and $SW(1)$ are generated as longitudinal vibrations, asymmetric, and symmetric, respectively. The displacements become mostly radial with the decreasing wavelength. The radial displacement for the $AW(1)$ wave has one half-wave across the thickness, while the longitudinal displacement has two half-waves. For the $SW(1)$ wave, the radial displacement has two waves, while the longitudinal displacement has three waves.

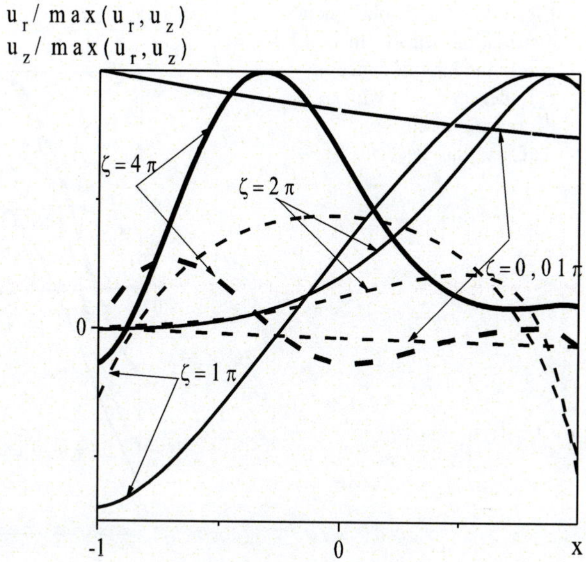

Fig. 3.42 The displacement distributions transform across the homogeneous cylinder thickness with decreasing in the wavelength($AU(0)$)

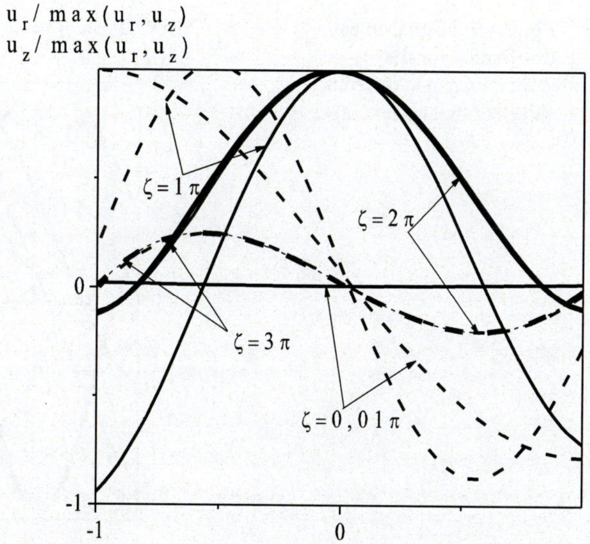

Fig. 3.43 The displacement distributions transform across the inhomogeneous cylinder thickness with decreasing in the wavelength($AW(1)$)

Such tendency is kept for higher branches: the displacements are either nearly symmetric or nearly asymmetric (with the number of half-waves per unit being increased when the serial number of the wave increases by unit).

In the case of an inhomogeneous material, the symmetry of the distribution of displacements about the mid-surface is disturbed. The displacements increase on "softer" fragments of the cylinder and decrease on more "rigid" ones.

Figure 3.45 represents the displacement distribution for the $AW(1)$ branch at different values of the wave number. The displacements for the maximum wave

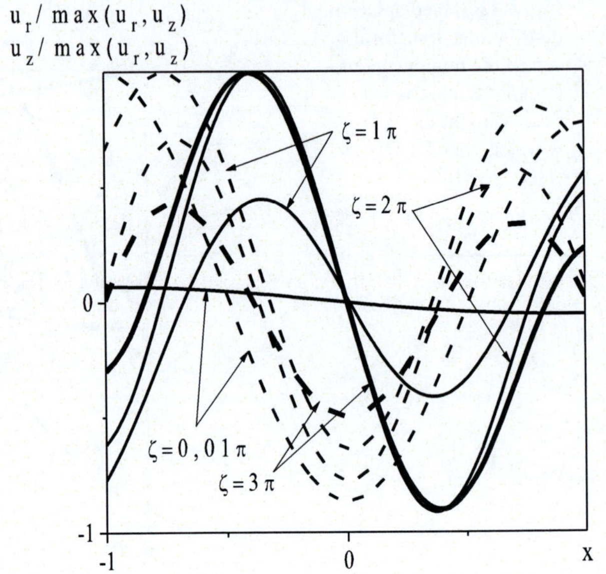

Fig. 3.44 The displacement distributions transform across the inhomogeneous cylinder thickness with decreasing in the wavelength($SW(1)$)

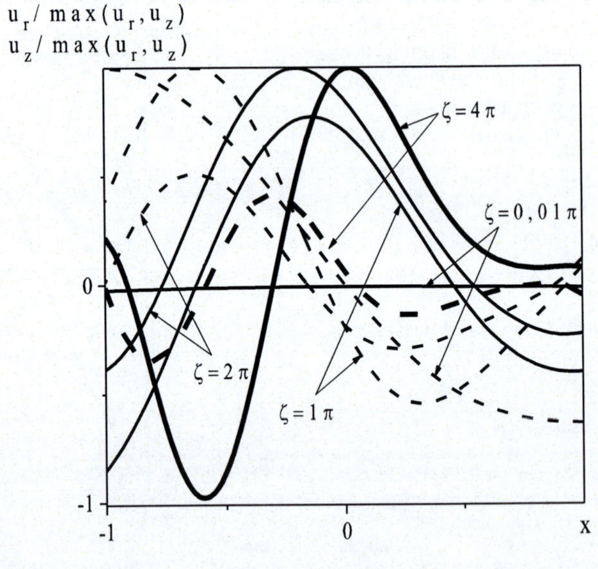

Fig. 3.45 Displacement distribution for the $AW(1)$branch at different values of the wave number

number are designated, as above, by heavy lines. The displacement distribution for the $SW(1)$ branch in the case of different wave numbers is shown in Fig. 3.46

We have established that the inhomogeneity of the cylinder material causes strong distinctions in the pattern of distribution of the displacement amplitudes of running waves. These distinctions are especially strong in the case of the second branch of dispersion relations. Thus, in the case of a homogeneous material, the first branch in the short-wave range transforms into the Rayleigh-type surface wave propagat-

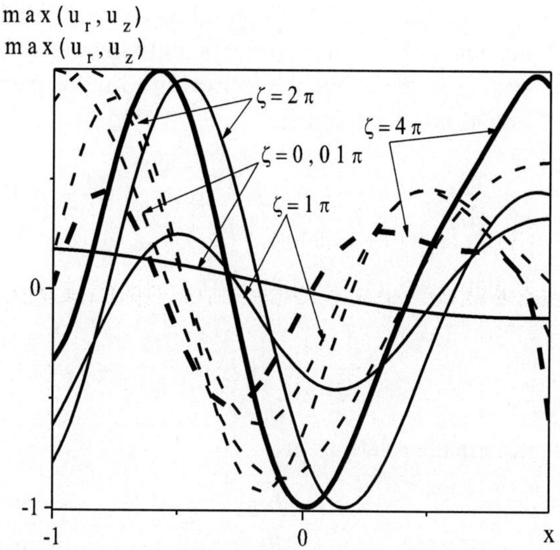

Fig. 3.46 Displacement distribution for the $SW(1)$ branch in the case of different wave numbers

ing over the outside surface of the cylinder. The second branch transforms into the Rayleigh-type surface wave which propagates over the inside surface of the cylinder.

In the case of an inhomogeneous material, only the first branch transforms into the Rayleigh-type surface wave propagating over the surface with minimum moduli values. The second branch does not transform into the surface wave. In the case of the more high branches, the symmetry in the distribution of displacement amplitudes across the thickness and in their shifting to the domain with minimum values of the moduli is broken.

In the case of a homogeneous material, the more high branches in the short-wave range transform into the waves propagating without dispersion. At the same time, the displacement amplitudes are distributed either symmetrically or asymmetrically with respect to the mid-surface when the number of half-waves per unit increases with the wave number. In the case of an inhomogeneous material, the symmetry in the distribution of the displacement amplitudes is broken and motions of the cylinder particles shift to the lower values of the moduli.

3.4 Axisymmetric Problem on Propagation of Forced Acoustoelectric Waves in Layered Hollow Cylinder with Piezoceramic and Metallic Layers

3.4.1 Problem Statement. Basic Equations

We will consider an axisymmetric problem on the propagation of forced acoustoelectric waves in a layered hollow cylinder with piezoceramic and metallic layers.

The axisymmetric longitudinal equations of motion of the ith layer as well as the equations of electrostatics and kinematic relations in the cylindrical coordinate system (r, θ, z) take the following form, respectively [11]:
Equations of motion are

$$\frac{\partial T_1}{\partial r} + \frac{1}{r}(T_1 - T_2) + \frac{\partial T_5}{\partial z} + \rho\omega^2 u_1 = 0, \frac{\partial T_5}{\partial r} + \frac{1}{r}T_5 + \frac{\partial T_3}{\partial z} + \rho\omega^2 u_3 = 0; \tag{3.40}$$

Equations of electrostatics are

$$\frac{\partial D_1}{\partial r} + \frac{1}{r}D_1 + \frac{\partial D_3}{\partial z} = 0, E_1 = -\frac{\partial \Phi}{\partial r}, E_3 = -\frac{\partial \Phi}{\partial z}; \tag{3.41}$$

Kinematic relations are

$$S_1 = \frac{\partial u_1}{\partial r}, S_2 = \frac{1}{r}u_1, S_3 = \frac{\partial u_3}{\partial z}, S_5 = \frac{\partial u_3}{\partial r} + \frac{\partial u_1}{\partial z}, \tag{3.42}$$

where T_i are the components of the stress tensor, ρ is the density of the material, ω is the circular frequency, u_i are the components of the displacement vector, D_i are the components of the electric-flux density, E_i are the components of the electric field strength, Φ is the electrostatic potential, and S_i are the components of the strain tensor.

The constitutive equations for a piezoelectric material polarized in an axial direction are described by

$$\begin{aligned} T_1^i &= c_{11}^i S_1^i + c_{12}^i S_2^i + c_{13}^i S_3^i - e_{13}^i E_3^i; \\ T_2^i &= c_{12}^i S_1^i + c_{11}^i S_2 + c_{13}^i S_3 - e_{13}^i E_3^i; \\ T_3^i &= c_{13}^i S_1^i + c_{13}^i S_2 + c_{33}^i S_3^i - e_{33}^i E_3^i; \\ T_5^i &= 2c_{55}^i S_5^i - e_{15}^i E_1; \; D_1^i = 2e_{15}^i S_5^i + \varepsilon_{11}^i E_1^i; \\ D_3^i &= e_{13}^i S_1^i + e_{13}^i S_2^i + e_{33}^i S_3^i + \varepsilon_{33}^i E_3^i. \end{aligned} \tag{3.43}$$

The constitutive equations for a piezoelectric material polarized in a radial direction are described by

$$\begin{aligned} T_1^i &= c_{33}^i S_1^i + c_{13}^i S_2^i + c_{13}^i S_3^i - e_{33}^i E_1^i, \\ T_2^i &= c_{13}^i S_1^i + c_{11}^i S_2 + c_{12}^i S_3 - e_{31}^i E_1^i; \\ T_3^i &= c_{13}^i S_1^i + c_{12}^i S_2 + c_{11}^i S_3^i - e_{31}^i E_1^i, \\ T_5^i &= 2c_{55}^i S_5^i - e_{51}^i E_3; \; D_3^i = 2e_{51}^i S_5^i + \varepsilon_{33}^i E_3^i, \\ D_1^i &= e_{33}^i S_1^i + e_{31}^i S_2^i + e_{31}^i S_3^i + \varepsilon_{33}^i E_1^i, \end{aligned} \tag{3.44}$$

where c_{ij} are the components of the tensor of the elastic moduli, e_{ij} are the components of the piezomodules tensor, and ε_{ij} are the components of the permittivity tensor.

The constitutive relations for the ith metallic layer are

$$\begin{aligned}
T_1^i &= \frac{(1-\nu^i)\,E^i}{(1+\nu^i)\,(1-2\nu^i)}S_1^i + \frac{\nu^i E^i}{(1+\nu^i)\,(1-2\nu^i)}S_2^i + \frac{\nu^i E^i}{(1+\nu^i)\,(1-2\nu^i)}S_3^i,\\
T_2^i &= \frac{\nu^i E^i}{(1+\nu^i)\,(1-2\nu^i)}S_1^i + \frac{(1-\nu^i)\,E^i}{(1+\nu^i)\,(1-2\nu^i)}S_2 + \frac{\nu^i E^i}{(1+\nu^i)\,(1-2\nu^i)}S_3,\\
T_3^i &= \frac{\nu^i E^i}{(1+\nu^i)\,(1-2\nu^i)}S_1^i + \frac{\nu^i E^i}{(1+\nu^i)\,(1-2\nu^i)}S_2 + \frac{(1-\nu^i)\,E^i}{(1+\nu^i)\,(1-2\nu^i)}S_3^i,\\
T_5^i &= 2\frac{E^i}{2\,(1+\nu^i)}S_5^i,
\end{aligned} \tag{3.45}$$

where ν is Poisson's ratio, E is the Young modulus (hereafter the index i will be omitted).

We prescribe the following boundary conditions on the lateral surfaces of the cylinder (for $r = R_0 \pm h$):

- the surfaces are not covered by electrodes: $D_1|_{r=R_o\pm h} = 0$;
- the outside surface of the cylinder is free of external forces (for $r = R_0 \pm h$): $T_1 = T_5 = 0$;
- the inside surface is under harmonically varying pressure

$$T_1|_{r=R_0-h} = Pe^{i(kz-\omega t)},\;\; T_5|_{r=R_0-h} = 0.$$

Here R_0 is the radius of the mid-surface of the cylinder; h is its half-thickness.

The vector-functions of unknowns in mixed form is

$$\mathbf{R} = \{T_1,\; T_5,\; \Phi,\; u_1,\; u_3,\; D_1\}^T \tag{3.46}$$

By resolving the system (3.40)–(3.43) for $\mathbf{R}$ and performing some transformations, we get

$$\begin{aligned}
\frac{\partial T_1}{\partial r} &= \frac{1}{r}\left(\frac{\Delta_2}{\Delta}-1\right)T_1 - \frac{\partial T_5}{\partial z} - \left(\frac{\Delta_3}{r^2\Delta} - \rho\frac{\partial^2}{\partial t^2}\right)u_1 - \frac{\Delta_4}{r\Delta}\frac{\partial u_3}{\partial z} - \frac{\Delta_1}{r\Delta}D_1;\\
\frac{\partial T_5}{\partial r} &= \frac{\Delta_2}{\Delta}\frac{\partial T_1}{\partial z} - \frac{1}{r}T_5 + \frac{\Delta_5}{r\Delta}\frac{\partial u_1}{\partial z} - \left(\frac{\Delta_6}{\Delta}\frac{\partial^2}{\partial z^2} - \rho\frac{\partial^2}{\partial t^2}\right)u_3 - \frac{\Delta_1}{\Delta}\frac{\partial D_1}{\partial z},\\
\frac{\partial \Phi}{\partial r} &= \frac{e_{33}}{\Delta}T_1 + \frac{\Delta_1}{r\Delta}u_1 + \frac{\Delta_1}{\Delta}\frac{\partial u_3}{\partial z} - \frac{c_{33}}{\Delta}D_1,\\
\frac{\partial u_1}{\partial r} &= \frac{\varepsilon_{33}}{\Delta}T_1 - \frac{\Delta_2}{r\Delta}u_1 - \frac{\Delta_2}{\Delta}\frac{\partial u_3}{\partial z} + \frac{e_{33}}{\Delta}D_1,\\
\frac{\partial u_3}{\partial r} &= \frac{1}{c_{55}}T_5 - \frac{\partial u_1}{\partial z} - \frac{e_{51}}{rc_{55}}\frac{\partial \Phi}{\partial z},\;\; \frac{\partial D_1}{\partial r} = -\frac{e_{51}}{c_{55}}\frac{\partial T_5}{\partial z} + \frac{\Delta_7}{c_{55}}\frac{\partial^2 \Phi}{\partial z^2} - \frac{1}{r}D_1,
\end{aligned} \tag{3.47}$$

where:

$$\Delta = c_{33}\varepsilon_{33} + e_{33}^2,\ \Delta_1 = c_{33}e_{13} - c_{13}e_{33},\ \Delta_2 = c_{13}\varepsilon_{33} + e_{13}e_{33},$$
$$\Delta_3 = (c_{13} - c_{33})\,\Delta_2 + (e_{33} - e_{13})\,\Delta_1 + (c_{13} - c_{11})\,\Delta,$$
$$\Delta_4 = (c_{13} - c_{33})\,\Delta_2 + (e_{33} - e_{13})\,\Delta_1 + (c_{13} - c_{12})\,\Delta,$$
$$\Delta_5 = c_{12}\Delta + e_{13}\Delta_1 - c_{13}\Delta_2,\ \Delta_6 = c_{11}\Delta + e_{13}\Delta_1 - c_{13}\Delta_2,\ \Delta_7 = c_{55}\varepsilon_{11} + e_{15}^2.$$

3.4.2 *Solving the Axisymmetric Boundary-Value Problems and Analysis of Numerical Results*

Let us represent the solution in the form of waves traveling in the axial direction

$$\begin{aligned} &T_1\,(r,z,t) = \mathrm{i}\lambda T_1\,(r)\,e^{\mathrm{i}(kz-\omega t)},\ T_5\,(r,z,t) = \lambda T_5(r)e^{\mathrm{i}(kz-\omega t)},\\ &\Phi\,(r,z,t) = h\sqrt{\frac{\lambda}{\varepsilon_0}}\Phi(r)e^{\mathrm{i}(kz-\omega t)},\ u_1\,(r,z,t) = \mathrm{i}hu_1(r)e^{\mathrm{i}(kz-\omega t)},\\ &u_3\,(r,z,t) = hu_3(r)e^{\mathrm{i}(kz-\omega t)},\ D_1\,(r,z,t) = \sqrt{\varepsilon_0\lambda}D_1(r)e^{\mathrm{i}(kz-\omega t)}. \end{aligned} \tag{3.48}$$

With (3.47), the original two-dimensional electroelastic problem in partial derivatives (3.46) can be reduced to a boundary-value problem for ordinary differential equations [8]:

$$\frac{\mathrm{d}\mathbf{R}}{\mathrm{d}x} = \mathbf{A}\,(x,\Omega)\,\mathbf{R}, \tag{3.49}$$

with boundary conditions

$$\mathbf{B}_1\mathbf{R}\,(-1) = \mathbf{C}_1,\ \mathbf{B}_2\mathbf{R}\,(1) = \mathbf{C}_2, \tag{3.50}$$

where $\mathbf{C}_1^T = \{0, 0, 0, 0, 0, 0\}$ and $\mathbf{C}_2^T = \{P, 0, 0, 0, 0, 0\}$.

Here

$$\Omega = \omega h\sqrt{\frac{\rho_0}{\lambda}},\ \tilde{c}_{ij} = \frac{c_{ij}}{\lambda},\ \tilde{e}_{ij} = \frac{e_{ij}}{\sqrt{\varepsilon_0\lambda}},\ \tilde{\varepsilon}_{ij} = \frac{\varepsilon_{ij}}{\varepsilon_0},\ x = \frac{r - R_0}{h},$$

where ρ is the density of the cylinder material, R_0 is the radius of the mid-surface, and ε_0 is the vacuum permittivity.

To solve problem (3.48), (3.49), we use the stable discrete-orthogonalization method in combination with the step-by-step search method.

Let us consider the results of numerical analysis of problem (3.48), (3.49) for a three-layer cylinder whose face layers are made of PZT ceramic while the core layer is made of steel with ($\nu = 0, 28$, $E = 21$) the curvature parameter $\varepsilon = 0, 25$. The thicknesses of the face and core layers are $h$and$2h$, respectively.

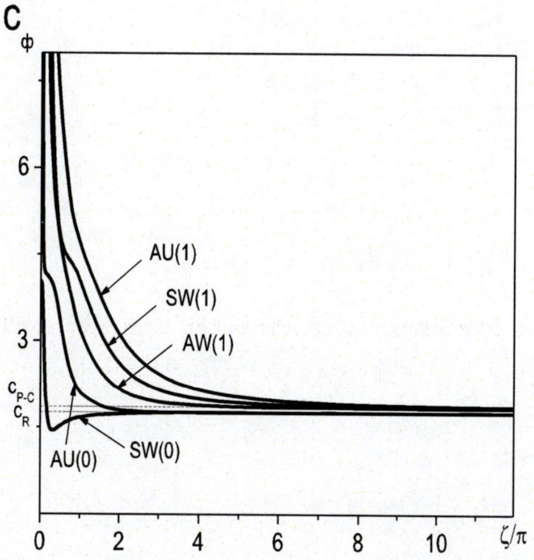

Fig. 3.47 Phase velocities for the first five waves in the case of a homogeneous piezoceramic cylinder

Figure 3.47 shows the phase velocities for the first five waves in the case of a homogeneous cylinder made of PZT piezoceramic with lateral surfaces without electrodes. The notation is the same as in [9]. The notation SW(0) means that the waves that are represented ($k = 0$) are asymmetric longitudinal oscillations (oscillations in expansion/compression wave), while $AU(0)$ are asymmetric (flexural) radial oscillations. It follows from Fig. 3.47 that the first two branches $SW(0)$ and $AU(0)0$ within a short-wave range tend to a similar velocity and propagate practically without dispersion. The kinematic analyses of these branches show that they represent Rayleigh-type waves. The other branches also tend to waves propagating without dispersion and are called Pochhammer–Chree waves, by analogy to the Lamb waves for a plate.

Figure 3.48 shows the phase velocities for the first five waves in the layered cylinder. As can be seen, the first two waves transform into a Rayleigh-type wave. The higher branches in the short-wave range pairwise converge, as can be seen for the $AW(1)$ and$SW(1)$ branches.

Let us analyze how the distribution of displacements varies in the layered cylinder across the thickness depending on the wavelength. Figure 3.49 shows the distribution of the displacement amplitudes over the thickness $-1.0 \leq h \leq 1.0$ in the first wave $SW(0)$ for different values of the wavelength. The solid line corresponds to the amplitudes of the radial displacements u_r, and the dashed line to the longitudinal displacements u_z. The distribution of the amplitudes of the displacements in the short-wave domain is shown by heavy lines. In the case of long waves, the displacement amplitudes are distributed nearly linearly. The displacement waves are concentrated with decreasing length near the outside lateral surface. In this case, the displacements are distributed as in Rayleigh waves.

Figure 3.50 shows the distribution of the displacement amplitudes across the thickness in the second wave $AU(0)$. Note that in the case of long waves this distribution is linear. With decreasing wavelength, the displacements are concentrated on

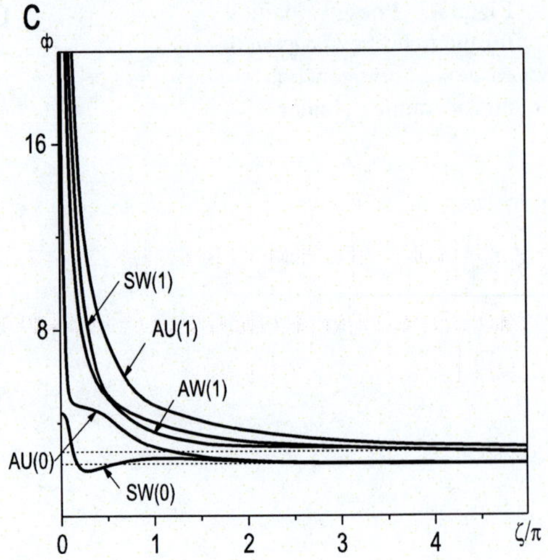

Fig. 3.48 Phase velocities for the first five waves in the case of an inhomogeneous layered cylinder

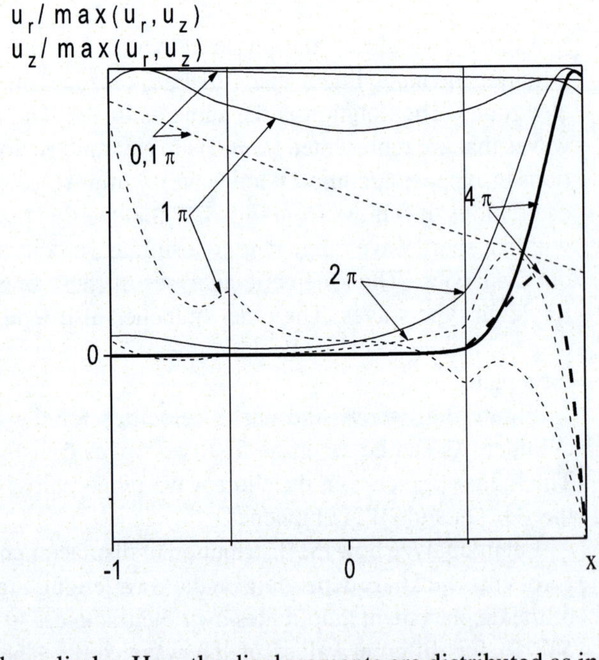

Fig. 3.49 Displacement of wave $SW(0)$ amplitudes across the thickness

the inside lateral surface of the cylinder. Here the displacements are distributed as in Rayleigh waves.

Figure 3.51 shows the distribution of the displacement amplitudes across the thickness in the third wave $AW(1)$. As in the case of long waves, the displacement amplitudes are distributed across the thickness nonlinearly. With the decrease in the wavelength, the displacements are concentrated in the softer piezoceramic layers, whereas the steel layer remains undeformed. The displacements peak in the piezoceramic face layer.

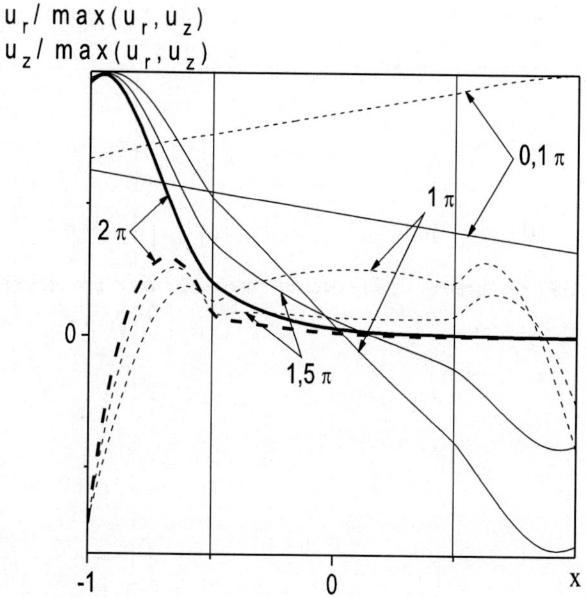

Fig. 3.50 Displacement of wave $AU(0)$ amplitudes across the thickness

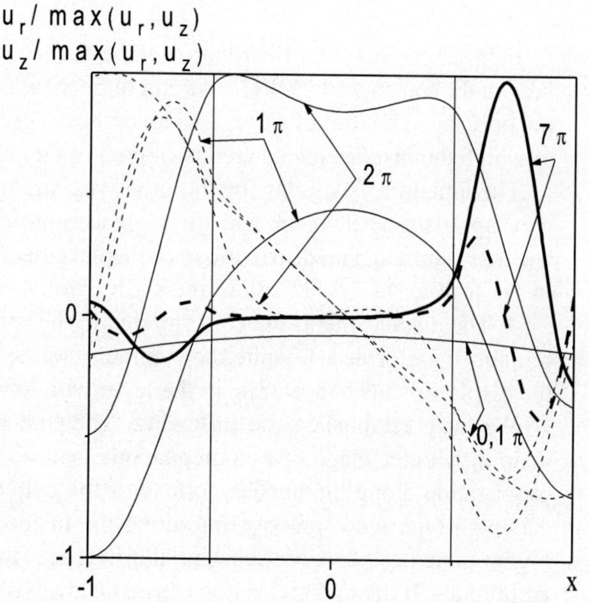

Fig. 3.51 Displacement of wave $AW(1)$ amplitudes across the thickness

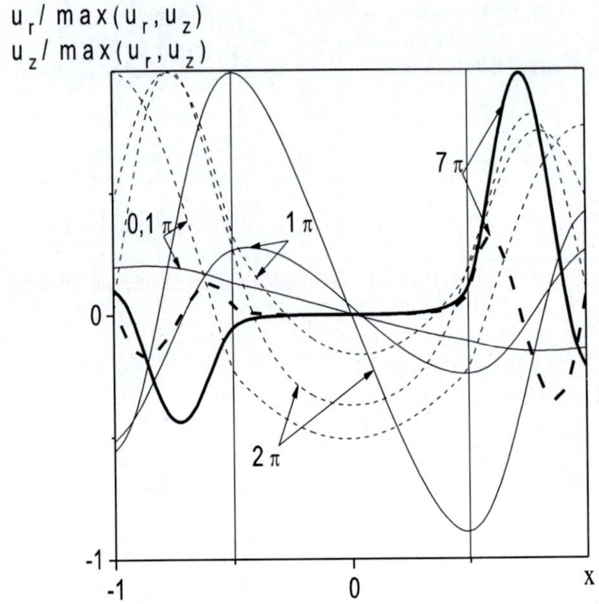

Fig. 3.52 Displacement of wave $SW(1)$ amplitudes across the thickness

Figure 3.52 shows the distribution of the displacement amplitudes across the thickness in the fourth wave $SW(1)$. As can be seen, with the decrease in the wavelength, not only the velocities of waves in both branches become similar, but also the manner in which the displacements are distributed in them.

The situation is similar for the other two branches. In the short-wave domain, they approach each other and the displacement distribution across the thickness becomes similar. The distributions of displacement amplitudes across the thickness in the fifth wave $AU(1)$ and in the sixth wave $AW(2)$are shown in Figs. 3.53 and 3.54. The displacements are concentrated in the softer layer, whereas the steel layer remains undeformed. Despite the symmetry of the structure about the mid-surface, the displacements concentrate in the layer with lower curvature.

We have established the following. The first branch in the short-wave range, as in a cylinder made of a homogeneous material, tends to a Rayleigh-type wave propagating along the outside surface of the cylinder. The second wave tends to a Rayleigh-type wave propagating along the inside surface of the cylinder. For the higher branches, there are substantial differences in the distribution of displacement amplitudes. If the material is homogeneous, with the decrease in the wavelength, the displacement amplitudes distribute either symmetrically or asymmetrically about the cylinder mid-surface. If the material is inhomogeneous, with decrease in the wavelength, the dispersion branches approach each other pairwise and the distribution of displacement across the thickness becomes pairwise similar. The displacements concentrate in the softer piezoceramic layer with the lower curvature, whereas the stiffer steel layer remains undeformed.

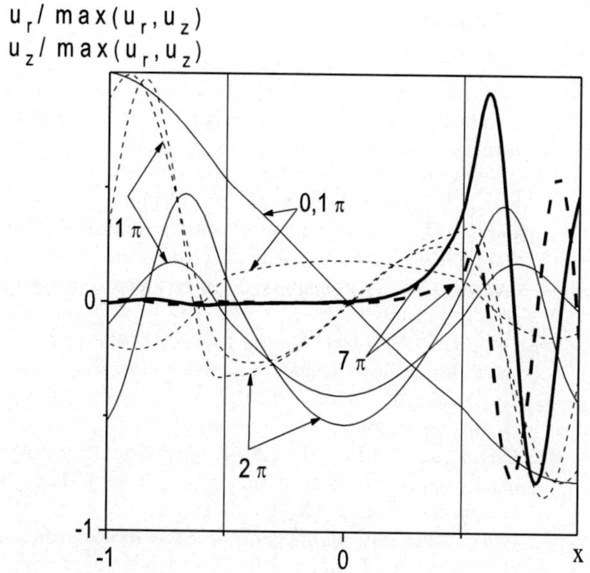

Fig. 3.53 Displacement of $AU(1)$ wave amplitudes across the thickness

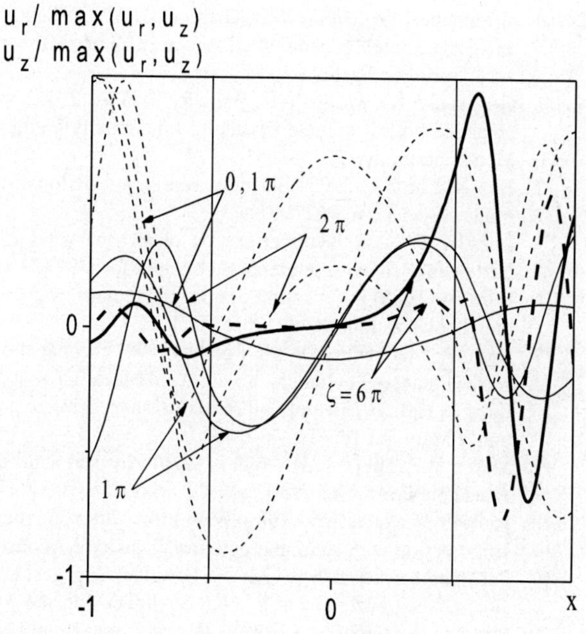

Fig. 3.54 Displacement of $AW(2)$wave amplitudes across the thickness

References

1. Ambadar A, Ferris CD (1965) Wave propagation in piezoelectric two-layered cylindrical shell with hexagonal symmetry. some application for long bone. J A Soc Am 63(3):781–792
2. Birman, V., Byrd, L.W.: Modeling and analysis of functionally graded materials and structures. ASME Appl. Mech. Rev. **60**, 195–216 (1886); Chree, C.: Longitudinal vibrations of a circular bar. Q. J. Pure Appl. Math. **21**, 287–298 (2007)
3. Chree C (1886) Longitudinal vibrations of a circular bar. Q J Pure Appl Math 21:287–298
4. Dai HI, Hong L, Fu YM, Xiao X (2010) Analytical solution for electromagnetothermoelastic behavior of a functionally graded piezoelectric hollow cylinder. Appl Math Model 34(2):343–357
5. Grigorenko AY, Loza IA, Shul'ga NA (1984) Propagation of nonaxisymmetric acoustoelectric waves in a hollow cylinder. Soviet Appl Mech 20(6):517–521
6. Grigorenko AY, Loza IA (2013) Nonaxisymmetric waves in layered hollow cylinders with radially polarized piezoceramic layers. Int Appl Mech 49(6):641–649
7. Grigorenko AY, Loza IA (2014) Nonaxisymmetric waves in layered hollow cylinders with axially polarized piezoceramic layers. Int Appl Mech 50(2):150–158
8. Grigorenko AY, Loza IA (2017) Axisymmetric acoustoelectric waves in a hollow cylinder made of a continuously inhomogeneous piezoelectric material. Int Appl Mech 53(4):374–380
9. Grigorenko AY, Loza IA (2017) Propagation of axisymmetric electroelastic waves in a hollow layered cylinder under mechanical excitation. Int Appl Mech 53(5):562–567
10. Grigorenko AY, Müller WH, Grigorenko YM, Vlaikov GG (2016) Recent developments in anisotropic heterogeneous shell theory. In: General theory and applications of classical theory, vol I. Springer, Berlin
11. Grigorenko AY, Müller WH, Wille R, Loza IA (2012) Nonaxisymmetric vibrations of radially polarized hollow cylinders made of functionally gradient piezoelectric materials. Continuum Mech Thermodyn 24(4–6):515–524
12. Han X, Liu GR (2003) Elastic waves in a functionally graded piezoelectric cylinder. Smart Mater Struct 12(6):962–971
13. Loza IA (1984) Axisymmetric acoustoelectrical wave propagation in a hollow circularly polarized cylindrical waveguide. Soviet Appl Mech 20(12):1103–1106
14. Loza IA (1985) Propagation of nonaxisymmetric waves in hollow piezoceramic cylinder with radial polarization. Soviet Appl Mech 21(1):22–27
15. Loza IA, Medvedev KV, Shul'ga NA (1987) Propagation of nonaxisymmetric acoustoelectric waves in layered cylinders. Soviet Appl Mech 23(8):703–706
16. Loza IA (2012) Torsional vibrations of piezoceramic hollow cylinders with circular polarization. J Math Sci 180(2):146–152
17. Mason WP, Thurston RN (eds) (1976) Physical acoustics: principles and methods, vol 1–7. Academic Press, New York
18. Paul HS, Venkatesan M (1989) Wave propagation in a piezoelectric ceramic cylinder of arbitrary cross section with a circular cylindrical cavity. J Acoust Soc Am 85(1):87–96
19. Pochhammer L (1876) Ueber die Fortpflanzungsgeschwindigkeiten kleiner Schwingungen in einem unbegrenzten isotropen Kreiscylinder. J Reine Angew Math 81:324–336
20. Puzyrev V, Storozhev VI (2011) Wave propagation in axially polarized piezoelectric hollow cylinder of sector cross section. J Sound Vib 330:4508–4518
21. Saravanos DA, Heyliger PR (1999) Mechanics and computational models for laminated piezoelectric beams, plates, and shells. ASME Appl Mech Rev 52:305–320
22. Storozhev VI (1998) Dispersion relations for normal waves in a cylinder made of orthorhombic monocrystal with pyro- and piezoelectric properties. J Math Sc (New-York, Plenum Publishing Corporation) 92(5):4196–4198
23. Shatalov MY, Every AG, Yenwong-Faia AS (2009) Analysis of non-axisymmetric wave propagation in a homogeneous piezoelectric solid circular cylinder of transversely isotropic material. Int J Solids Struct 46(3–4):837–850

24. Shul'ga NA (2002) Propagation of harmonic waves in anisotropic piezoelectric cylinders. Homogeneous piezoceramic waveguides. Int Appl Mech 38(8):933–953
25. Shul'ga NA, Grigorenko AYa, Loza IA (1984) Propagation of nonaxisymmetric acoustoelectric waves in a hollow cylinder. Soviet Appl Mech 20(6):517–521
26. Sun CT, Cheng NC (1974) Piezoelectric waves on a layered cylinder. J Appl Phys 45:4288–4294

Chapter 4
Electroelastic Vibrations of Heterogeneous Piezoceramic Hollow Spheres

Abstract The axisymmetric and nonaxisymmetric problems of natural and forced vibrations of a hollow sphere made of a functionally gradient piezoelectric material theory are considered based on 3D electroelasticity. The properties of the material vary along the radial coordinate. The external surface of the sphere is free of tractions and is either insulated or short-circuited by electrodes for the analysis of the natural vibrations of the system. Two cases of forced vibrations are investigated: an electric excitation—when an electrostatic potential with an alternating sign is applied to the external surface of the spheres; and mechanical excitation—when pressure with an alternating sign is applied to its external surface. Separation of variables and series of the components of the mechanical and electric displacements were used. The electric potential and of the stress tensor is expressed in terms of spherical functions. As a result, the initially three-dimensional problem described by the partial differential equations with variable coefficients is reduced to a boundary-value problem for the systems of the ordinary differential equations. A boundary-value eigenvalue problem for the case of natural vibrations is arrived at in the process. It is solved by discrete-orthogonalization methods combined with a step-by-step search method. The influence of the geometric, mechanical, and electric parameters on the frequency spectrum in the case of nonaxisymmetric natural vibrations of an inhomogeneous piezoceramic thick-walled sphere was analyzed. An inhomogeneous boundary-value problem is obtained for the case of forced vibrations. This problem is solved by a stable discrete-orthogonalization method. The influence of the geometric and electric parameters on the kinematic (mechanical displacement and electrostatic potential) and dynamic (mechanical stress and electric displacement) characteristics was analyzed. Different variants of polarized piezoceramic materials are considered. Significant attention is paid to the validation of the reliability of the results obtained by numerical calculations.

Keywords 3D electroelasticity theory · Piezoceramic sphere · Free and forced vibrations · Inhomogeneous material · Numerical method

A. Ya. Grigorenko et al., *Selected Problems in the Elastodynamics of Piezoceramic Bodies*, Advanced Structured Materials 154,
https://doi.org/10.1007/978-3-030-74199-0_4

4.1 Introduction

Spherically shaped active piezoelectric transformers are widely used in various modern acoustoelectric facilities such as hydroacoustic devices [2, 11]. Because of this, the study of the vibrations of spherical-shaped piezoceramic bodies is an electroelasticity problem of current interest both from a fundamental and the applied point of view.

For the first time, the free radial vibrations of a homogeneous isotropic sphere have been studied in [26]. The problem on vibrations of a sphere in general spatial statement was addressed in [20], where the same problem has been solved in spherical coordinates and classification of vibration modes was proposed. The free and forced vibrations of thick-walled spherical shells as well as the surface Rayleigh waves on a spherical surface for the first time were studied in [10, 21, 22] using the spherical-vector method. Nonaxisymmetric vibrations of solid and hollow spheres made of a homogeneous isotropic material are considered in [25, 27], respectively. It should be noted that the anisotropy of the material makes the analysis of those problems significantly more complex. For example, the natural vibrations of a transversely isotropic sphere were studied in [29]. In addition, another factor that complicates solving the problems on vibrations of spherical-shaped bodies is the inhomogeneity of a material. If the material has a laminated structure, the necessity arises in satisfying the boundary conditions not only on the bounding surfaces but on the interfacial ones as well, which leads to an increase in the number of equations of the resolving system [6]. One more factor complicating the investigation is physical fields coupled with mechanical displacement and stress ones [12]. Research of free axisymmetric vibrations of a homogeneous piezoceramic sphere with the application of the various analytical approaches should also be mentioned [5, 23, 24].

In recent years, a new direction in the material sciences which has actively been developed is the so-called functionally gradient piezoelectric materials (FGPM) [1, 4, 7].

The functionally gradient piezoelectric materials become more widely used due to combining the advantages of bimorphs and materials without interfaces having different thermal expansion coefficients. Trying to allow for continuous variation in material properties, we arrive at the situation where the material modules become functions of one coordinate [8, 9, 13]. This is a serious obstacle for the employment of many numerical methods. Such a problem can be solved with the widely used numerical–analytical approach proposed by the authors of the present book. The approach uses the discrete-orthogonalization method which earlier has been effectively applied in solving the wide class of the problems of elasticity and shells theory of electroelasticity [14, 17–19]. This approach to studying the dynamic behavior of piezoceramic layered and continuously inhomogeneous bodies of cylindrical and spherical shapes was applied [15, 16].

The present chapter considers the problems of the natural and the forced vibrations of the different inhomogeneous structures of the piezoceramic hollow sphere. The

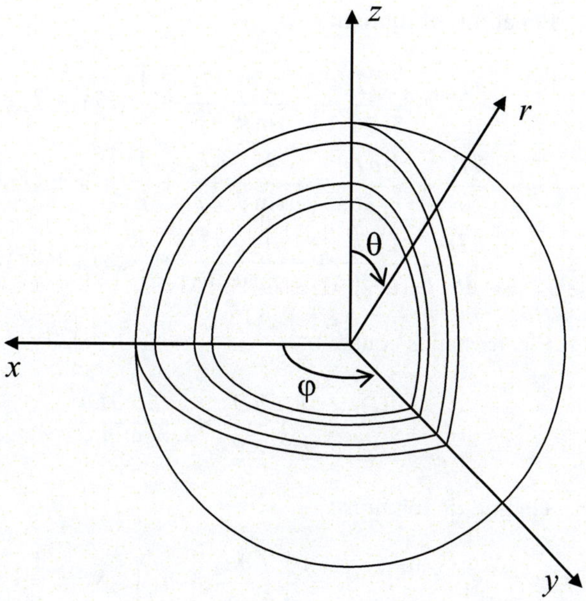

Fig. 4.1 Spherical coordinate system

investigation of this problem based on a solution of coupled 3D electroelasticity and on the effective analytical–numerical discrete–continual approach is proposed.

4.2 Statement of the Problem—Basic Relations

We consider the free vibrations of the piezoceramic sphere with a different kind of structural inhomogeneity of the material (layered inhomogeneity and continuous inhomogeneity) within the framework of the spatial theory of electroelasticity and use a spherical coordinate system. The relationship between Cartesian (x, y, z) and spherical (r, θ, φ) coordinate systems is presented as (Fig. 4.1)

$$x = r\sin\theta\cos\varphi,\; y = r\sin\theta\sin\varphi,\; z = r\cos\theta \quad \Leftrightarrow$$

$$r = \sqrt{x^2 + y^2 + z^2},\; \theta = \mathrm{arctg}\sqrt{\frac{x^2 + y^2}{z^2}},\quad \varphi = \mathrm{arctg}\frac{y}{x}.$$

For the main relations describing the free vibrations of a nonhomogeneous spherical hollow body made from a piezoceramic material, the details were presented in the first chapter (1.144), (1.146), and (1.147):

Equations of motion:

$$\frac{\partial T_1}{\partial r}+\frac{1}{r}\frac{\partial T_6}{\partial \theta}+\frac{1}{r\sin\theta}\frac{\partial T_5}{\partial \varphi}+\frac{1}{r}(2T_1-T_2-T_3+T_6\text{ctg}\theta)=\rho\frac{\partial^2 u_1}{\partial t^2},$$
$$\frac{\partial T_6}{\partial r}+\frac{1}{r}\frac{\partial T_2}{\partial \theta}+\frac{1}{r\sin\theta}\frac{\partial T_4}{\partial \varphi}+\frac{1}{r}(3T_6+(T_2-T_3)\,\text{ctg}\theta)=\rho\frac{\partial^2 u_2}{\partial t^2},$$
$$\frac{\partial T_5}{\partial r}+\frac{1}{r}\frac{\partial T_4}{\partial \theta}+\frac{1}{r\sin\theta}\frac{\partial T_3}{\partial \varphi}+\frac{1}{r}(3T_6+2T_4\text{ctg}\theta)=\rho\frac{\partial^2 u_3}{\partial t^2}; \tag{4.1}$$

Electrostatics equations:

$$\frac{\partial D_1}{\partial r}+\frac{1}{r}\left(\frac{\partial D_2}{\partial \theta}+\frac{1}{\sin\theta}\frac{\partial D_3}{\partial \varphi}+2D_1+D_2\text{ctg}\theta\right)=0; \tag{4.2}$$

Geometric relations:

$$S_1=\frac{\partial u_r}{\partial r},\ S_2=\frac{1}{r}\left(\frac{\partial u_2}{\partial \theta}+u_1\right),\ S_3=\frac{1}{r}\left(\frac{1}{\sin\theta}\frac{\partial u_3}{\partial \varphi}+u_1+u_2\text{ctg}\theta\right),$$
$$2S_4=\frac{1}{r}\left(\frac{\partial u_3}{\partial \theta}+\frac{1}{\sin\theta}\frac{\partial u_2}{\partial \varphi}-2u_3\text{ctg}\theta\right),$$
$$2S_5=\frac{\partial u_3}{\partial r}+\frac{1}{r}\left(\frac{1}{\sin\theta}\frac{\partial u_1}{\partial \varphi}-2u_3\right),\ 2S_6=\frac{\partial u_2}{\partial r}+\frac{1}{r}\left(\frac{\partial u_1}{\partial \theta}-2u_2\right), \tag{4.3}$$

where $\mathbf{T}=\begin{pmatrix} T_1 & T_6 & T_5\\ T_6 & T_2 & T_4\\ T_5 & T_4 & T_3\end{pmatrix}$ is the mechanical stress tensor, $\mathbf{S}=\begin{pmatrix} S_1 & S_6 & S_5\\ S_6 & S_2 & S_4\\ S_5 & S_4 & S_3\end{pmatrix}$ is the strain tensor, u_1, u_2, u_3 are the components of the displacement vector, $\mathbf{E}$ is the electric field strength vector, $\mathbf{D}$ is the electric-flux density vector, and Φ is the electrostatic potential.

The vibrations of the layered inhomogeneous and continuous inhomogeneous piezoceramic bodies of spherical forms in this chapter are studied. Let us present the relations by using Voigt symbols for the components of the tensors of mechanical stresses and deformations, and digital indices for the displacement vectors $\mathbf{u}$ movements, the electric field $\mathbf{E}$ strength, and the electric induction $\mathbf{D}$ ($r\to 1$, $\theta\to 2$, $\varphi\to 3$).

In the case of layered inhomogeneous of the sphere material, when its material consists of piezoceramic and metallic layers for each i layer, we will have the following relations:

Equations of motion for the ith layer:

$$\frac{\partial T_1^i}{\partial r} + \frac{1}{r}\frac{\partial T_6^i}{\partial \theta} + \frac{1}{r\sin\theta}\frac{\partial T_5^i}{\partial \varphi} + \frac{1}{r}\left(2T_1^i - T_2^i - T_3^i + T_6^i \text{ctg}\theta\right) = \rho\frac{\partial^2 u_1^i}{\partial t^2},$$
$$\frac{\partial T_6^i}{\partial r} + \frac{1}{r}\frac{\partial T_2^i}{\partial \theta} + \frac{1}{r\sin\theta}\frac{\partial T_4^i}{\partial \varphi} + \frac{1}{r}\left(3T_6^i + \left(T_2^i - T_3^i\right)\text{ctg}\theta\right) = \rho\frac{\partial^2 u_2^i}{\partial t^2},$$
$$\frac{\partial T_5^i}{\partial r} + \frac{1}{r}\frac{\partial T_4^i}{\partial \theta} + \frac{1}{r\sin\theta}\frac{\partial T_3^i}{\partial \varphi} + \frac{1}{r}\left(3T_6^i + 2T_4^i \text{ctg}\theta\right) = \rho\frac{\partial^2 u_3^i}{\partial t^2}; \tag{4.4}$$

Electrostatics equations for the ith layer:

$$E_1^i = -\frac{\partial \Phi}{\partial r},\ E_2^i = -\frac{1}{r}\frac{\partial \Phi}{\partial \theta},\ E_3^i = -\frac{1}{r\sin\theta}\frac{\partial \Phi}{\partial \varphi},$$
$$\frac{\partial D_1^i}{\partial r} + \frac{2}{r}D_1^i + \frac{1}{r}\left(\frac{\partial D_2^i}{\partial \theta} + \text{ctg}\theta D_2^i\right) + \frac{1}{r\sin\theta}\frac{\partial D_3^i}{\partial \varphi} = 0; \tag{4.5}$$

Geometric relations (Cauchy relations) for ith layer:

$$S_1^i = \frac{\partial u_1^i}{\partial r},\ S_2^i = \frac{1}{r}\frac{\partial u_2^i}{\partial \theta} + \frac{u_1^i}{r},\ S_3^i = \frac{1}{r}\left(u_2^i \text{ctg}\theta + u_1^i\right),$$
$$2S_4^i = \frac{1}{r\sin\theta}\frac{\partial u_2^i}{\partial \varphi} + \frac{1}{r}\frac{\partial u_3^i}{\partial \varphi} - \frac{u_3^i}{r}\text{ctg}\theta, \tag{4.6}$$
$$2S_5^i = \frac{1}{r\sin\theta}\frac{\partial u_1^i}{\partial \varphi} + \frac{\partial u_3^i}{\partial r} - \frac{u_3^i}{r},\ 2S_6^i = \frac{1}{r}\left(\frac{\partial u_1^i}{\partial \theta} - u_2^i\right) + \frac{\partial u_2^i}{\partial r}.$$

The constitutive equations for a piezoelectric material polarized for the ith layer in a radial direction are

$$T_2^i = c_{11}^i S_2^i + c_{12}^i S_3^i + c_{13}^i S_1^i - e_{31}^i E_1^i,\ T_3^i = c_{12}^i S_2^i + c_{11}^i S_3^i + c_{13}^i S_1^i - e_{31}^i E_1^i,$$
$$T_1^i = c_{13}^i S_2^i + c_{13}^i S_3^i + c_{33}^i S_1^i - e_{33}^i E_1^i,\ T_6^i = 2c_{44}^i S_6^i - e_{15}^i E_2^i,$$
$$T_5^i = 2c_{44}^i S_5^i - e_{15}^i E_3^i,\ T_4^i = 2c_{66}^i S_4^i,\ D_1^i = e_{31}^i S_2^i + e_{31}^i S_3^i + e_{33}^i S_1^i + \varepsilon_{33}^i E_1^i, \tag{4.7}$$
$$D_2^i = 2e_{15}^i S_6^i + \varepsilon_{11}^i E_2^i,\ D_3^i = 2e_{15}^i S_5^i + \varepsilon_{11}^i E_3^i;$$

The constitutive equations for the metal material for the ith layer are given by

$$T_1^i = \frac{\left(1-\nu^i\right)E^i}{\left(1+\nu^i\right)\left(1-2\nu^i\right)}S_1^i + \frac{\nu^i E^i}{\left(1+\nu^i\right)\left(1-2\nu^i\right)}S_2^i + \frac{\nu^i E^i}{\left(1+\nu^i\right)\left(1-2\nu^i\right)}S_3^i,$$
$$T_2^i = \frac{\nu^i E^i}{\left(1+\nu^i\right)\left(1-2\nu^i\right)}S_1^i + \frac{\left(1-\nu^i\right)E^i}{\left(1+\nu^i\right)\left(1-2\nu^i\right)}S_2^i + \frac{\nu^i E^i}{\left(1+\nu^i\right)\left(1-2\nu^i\right)}S_3^i,$$
$$T_3^i = \frac{\nu^i E^i}{\left(1+\nu^i\right)\left(1-2\nu^i\right)}S_1^i + \frac{\nu^i E^i}{\left(1+\nu^i\right)\left(1-2\nu^i\right)}S_2^i + \frac{\left(1-\nu^i\right)E^i}{\left(1+\nu^i\right)\left(1-2\nu^i\right)}S_3^i, \tag{4.8}$$

$$T_6^i = 2\frac{E^i}{2\left(1+\nu^i\right)}S_6^i,\ T_5^i = 2\frac{E^i}{2\left(1+\nu^i\right)}S_5^i,\ T_4^i = 2\frac{E^i}{2\left(1+\nu^i\right)}S_4^i,$$

where E^i is Young's modulus for the ith layer, and ν^i is Poisson's ratio for the ith layer.

For the case of a continuously inhomogeneous piezoelectric material, the physical and mechanical characteristics in the relationships (4.7) are not constants, but functions of one of the coordinates. The constitutive equations for radially polarized piezoceramic of continuous inhomogeneity medium in the direction of the radial coordinate r take the form (1.149) [3, 6]:

$$\begin{aligned}
&T_2 = c_{11}(r)\,S_2 + c_{12}(r)\,S_3 + c_{13}(r)\,S_1 - e_{31}(r)\,E_1,\\
&T_3 = c_{12}(r)\,S_2 + c_{11}(r)\,S_3 + c_{13}(r)\,S_1 - e_{31}(r)\,E_1,\\
&T_1 = c_{13}(r)\,S_2 + c_{13}(r)\,S_3 + c_{33}(r)\,S_1 - e_{33}(r)\,E_1,\\
&T_6 = 2c_{44}(r)\,S_6 - e_{15}(r)\,E_2,\ T_5 = 2c_{44}(r)\,S_5 - e_{15}(r)\,E_3,\ T_4 = 2c_{66}(r)\,S_4,\\
&D_2 = 2e_{15}(r)\,S_6 + \varepsilon_{11}(r)\,E_2,\ D_3 = 2e_{15}(r)\,S_5 + \varepsilon_{11}(r)\,E_3,\\
&D_1 = e_{31}(r)\,S_2 + e_{31}(r)\,S_3 + e_{33}(r)\,S_1 + \varepsilon_{33}(r)\,E_1.
\end{aligned} \tag{4.9}$$

These relations have five independent elastic moduli ($c_{11}, c_{12}, c_{13}, c_{33}, c_{44}$), defined at constant (zero) electric field, three piezomoduli (e_{31}, e_{15}, e_{33}), and two electrical permeabilities (ε_{11}, ε_{33}), defined at constant (zero) deformation.

Assuming that all points of the spherical bodies execute harmonic vibrations with an angular frequency ω, and taking into account the following representations (hereafter the superscript "^" will be omitted):

$$\begin{aligned}
&T_j(r,\theta,\varphi,t) = \hat{T}_j(r,\theta,\varphi)e^{\mathrm{i}\omega t},\ \ j = 1, 2, \ldots 6\\
&D_m(r,\theta,\varphi,t) = \hat{D}_m(r,\theta,\varphi)e^{\mathrm{i}\omega t},\ u_m(r,\theta,\varphi,t) = \hat{u}_m(r,\theta,\varphi)e^{\mathrm{i}\omega t},\ m = 1, 2, 3,\\
&\Phi(r,\theta,\varphi,t) = \hat{\Phi}(r,\theta,\varphi)e^{\mathrm{i}\omega t}.
\end{aligned}$$

4.3 Axisymmetric Free Vibrations of Nonhomogeneous Layered Piezoceramic Hollow Spheres

4.3.1 Resolving System

It is known that a system of the partial differential equations describing the free nonaxisymmetric vibrations of the sphere on the spatial theory of the elasticity and the electroelasticity can be reduced to two independent systems of equations that describe the vibrations of the first and second classes, respectively.

Note that the vibrations of the first class do not depend on the influence of the electric field, whereas vibrations of the second class do, which are coupled electric elasticities. In addition, the system of equations that describes second class vibrations completely coincides with the system of equations for axisymmetric electroelastic vibrations of a piezoceramic sphere. For the axisymmetric case, the systems (4.4)–(4.9) have the form $\left(\frac{\partial u_1}{\partial\varphi}=\frac{\partial u_2}{\partial\varphi}=0;\ u_3=0\right)$:

Equations of motion for the ith layer:

$$\frac{\partial T_1^i}{\partial r}+\frac{1}{r}\frac{\partial T_6^i}{\partial\theta}+\frac{1}{r}\left(2T_1^i-T_2^i+T_6^i\text{ctg}\theta\right)+\rho\omega^2u_1^i=0,$$

$$\frac{\partial T_6^i}{\partial r}+\frac{1}{r}\frac{\partial T_2^i}{\partial\theta}+\frac{1}{r}\left(3T_6^i+T_2^i\text{ctg}\theta\right)+\rho\omega^2u_2^i=0; \tag{4.10}$$

Electrostatics equations for the ith layer:

$$\frac{\partial D_1^i}{\partial r}+\frac{1}{r}\frac{\partial D_1^i}{\partial\theta}+\frac{1}{r}\left(2D_1^i+D_2^i\text{ctg}\theta\right)=0,$$

$$E_1^i=-\frac{\partial\Phi^i}{\partial r},\ E_2^i=-\frac{1}{r}\frac{\partial\Phi^i}{\partial\theta}; \tag{4.11}$$

Geometric relations (Cauchy relations) for the ith layer:

$$S_1^i=\frac{\partial u_1^i}{\partial r},\ S_2^i=\frac{1}{r}\frac{\partial u_2^i}{\partial\theta}+\frac{u_1^i}{r},$$

$$S_3^i=\frac{1}{r}\left(u_2^i\text{ctg}\theta+u_1^i\right),\ S_6^i=\frac{\partial u_1^i}{\partial\theta}+\frac{\partial u_2^i}{\partial r}-\frac{u_2^i}{r}; \tag{4.12}$$

The constitutive equations for a piezoelectric material polarized for the ith layer in a radial direction are

$$\begin{aligned}&T_1^i=c_{33}^iS_1^i+c_{13}^iS_2^i+c_{13}^iS_3^i-e_{33}^iE_1^i,\ T_2^i=c_{13}^iS_1^i+c_{11}^iS_2^i+c_{12}^iS_3^i-e_{13}^iE_1^i,\\&T_3^i=c_{13}^iS_1^i+c_{12}^iS_{12}^i+c_{11}^iS_3^i-e_{13}^iE_1^i,\ T_6^i=2c_{55}^iS_6^i-e_{15}^iE_2^i,\\&D_2^i=2e_{15}^iS_6^i+\varepsilon_{11}^iE_2^i,\ D_1^i=e_{31}^iS_2^i+e_{31}^iS_3^i+e_{33}^iS_1^i+\varepsilon_{33}^iE_1^i;\end{aligned} \tag{4.13}$$

The constitutive equations for the metal material for the ith layer read

$$\begin{aligned}T_1^i&=\frac{\left(1-\nu^i\right)E^i}{\left(1+\nu^i\right)\left(1-2\nu^i\right)}S_1^i+\frac{\nu^iE^i}{\left(1+\nu^i\right)\left(1-2\nu^i\right)}S_2^i+\frac{\nu^iE^i}{\left(1+\nu^i\right)\left(1-2\nu^i\right)}S_3^i,\\T_2^i&=\frac{\nu^iE^i}{\left(1+\nu^i\right)\left(1-2\nu^i\right)}S_1^i+\frac{\left(1-\nu^i\right)E^i}{\left(1+\nu^i\right)\left(1-2\nu^i\right)}S_2^i+\frac{\nu^iE^i}{\left(1+\nu^i\right)\left(1-2\nu^i\right)}S_3^i,\\T_3^i&=\frac{\nu^iE^i}{\left(1+\nu^i\right)\left(1-2\nu^i\right)}S_1^i+\frac{\nu^iE^i}{\left(1+\nu^i\right)\left(1-2\nu^i\right)}S_2^i+\frac{\left(1-\nu^i\right)E^i}{\left(1+\nu^i\right)\left(1-2\nu^i\right)}S_3^i,\\T_6^i&=2\frac{E^i}{2\left(1+\nu^i\right)}S_6^i,\ T_5^i=2\frac{E^i}{2\left(1+\nu^i\right)}S_5^i,\ T_4^i=2\frac{E^i}{2\left(1+\nu^i\right)}S_4^i.\end{aligned} \tag{4.14}$$

The boundary conditions on the lateral surfaces of the piezoceramic sphere are free of the external forces:

$$T_1(R_0 \pm h, \theta) = 0, T_6(R_0 \pm h, \theta) = 0. \tag{4.15}$$

Let us consider the two variants of electrical boundary conditions on the lateral surfaces:

(i) continuity of the normal component of the electric displacement to the outside vacuum (no electrodes):

$$D_1(R_0 \pm h, \theta) = 0, \tag{4.16}$$

(ii) a grounded conducting surface (sphere coated with thin short-circuited electrodes):

$$\Phi(R_0 \pm h, \theta) = 0, \tag{4.17}$$

where R_0 denotes the radius of the sphere mid-surface, and h is the half thickness of the sphere.

In the case of the piecewise inhomogeneity (layer structure) of the sphere at the interface of the layers, let us formulate mechanical and electrical conditions, respectively.

The conditions of the rigid mechanical contact without slipping and tearing on the surface of the discontinuity of properties of the sphere are considered. Therefore, we have conditions of continuity of the components of the vector displacements and corresponding stress tensor components:

$$T_1^i = T_1^{i+1}, T_6^i = T_6^{i+1}, u_1^i = u_1^{i+1}, u_2^i = u_2^{i+1}. \tag{4.18}$$

The conditions for the electrical values on the contact surface (interface) of the layers depend on the material type.

If the ith and $(i + 1)$th layers are piezoelectric or dielectric materials, then the conditions (4.9) must be supplemented by the condition of continuity of the normal component vector of electrical induction:

$$D_1^i = D_1^{i+1}. \tag{4.19}$$

If the ith and $(i + 1)$th layers are conductors, we have no additional boundary conditions (we apply the boundary conditions (4.18)).

If the ith layers are piezoelectric and $(i + 1)$th are conductor layers, we get the next boundary conditions (1.178) and (1.179):

$$\Phi|_S = \text{const.}, \tag{4.20}$$

$$\left.\frac{\partial \Phi}{\partial n}\right|_S = -\frac{\rho_S}{\varepsilon_1}. \tag{4.21}$$

In the relationship (4.20), we choose zero value for the electrostatic potential and assume there is no electric field on the surface. Then

$$\Phi^i = 0. \tag{4.22}$$

Hereafter the superscript "*i*" will be omitted.
The resolving vector-function of mixed type is

$$R = \| T_1, T_6, \Phi, u_1, u_2, D_1 \|^T . \tag{4.23}$$

After setting the same (identical) transformations systems (4.10)–(4.13), we get

$$\begin{aligned}
\frac{\partial T_1}{\partial r} &= \frac{2}{r}(\delta_1 - 1)\, T_1 + \frac{\Delta_\theta}{r} T_6 - \left(\frac{2\delta_3}{r^2} + \rho\omega^2\right) u_1 + \frac{\Delta_\theta \delta_3}{r^2} u_2 + \frac{2\delta_2}{r} D_r, \\
\frac{\partial T_6}{\partial r} &= \frac{\delta_1}{r} T_1 - \frac{3}{r} T_6 - \frac{\delta_3}{r^2} u_1 + \left(\frac{\Delta_\theta \delta_4 - 2c_{66}}{r^2} - \rho\omega^2\right) u_2, \\
\frac{\partial \Phi}{\partial r} &= \frac{e_{33}}{\delta} T_1 + \frac{\Delta_\theta \delta_2}{r} u_2 - \frac{c_{33}}{\delta} D_r, \quad \frac{\partial u_1}{\partial r} = \frac{\varepsilon_{33}}{\delta} T_1 - \frac{2\delta_1}{r} u_1 + \frac{\Delta_\theta \delta_1}{r} u_2 + \frac{e_{33}}{\delta} D_r, \\
\frac{\partial u_2}{\partial r} &= \frac{1}{c_{55}} T_6 + \frac{e_{15}}{r c_{55}} \Phi + \frac{1}{r} u_1 + \frac{1}{r} u_2, \quad \frac{\partial D_r}{\partial r} = \frac{\Delta_\theta e_{15}}{r^2 c_{55}} T_1 + \frac{\Delta_\theta \delta_5}{r^2} T_6 - \frac{2}{r} D_r,
\end{aligned} \tag{4.24}$$

where

$$\Delta_\theta * = \frac{\partial *}{\partial \theta} + *\mathrm{ctg}\theta,\ \delta = c_{33}\varepsilon_{33} + e_{33}^2,\ \delta_1 = \frac{c_{13}\varepsilon_{33} + e_{13}e_{33}}{\delta},$$

$$\delta_2 = \frac{c_{13}e_{33} - c_{33}e_{13}}{\delta},\ \delta_3 = 2\left(c_{13}\delta_1 + e_{13}\delta_2\right) - \left(c_{11} + c_{12}\right),$$

$$\delta_4 = \frac{\delta_3}{2} - c_{66},\ \delta_5 = \frac{c_{55}\varepsilon_{11} + e_{15}^2}{c_{55}}.$$

4.3.2 Solution Method for the Problem

The sphere is a closed body. Therefore, let us search for solutions of the components of the stress tensor, the displacement and the electrical induction vectors, and the electrostatic potential in the form:

$$T_1(r,\theta) = \lambda \sum_{k=0}^{\infty} T_1^n(r) P_n(\cos\theta),\, T_6(r,\theta) = \lambda \sum_{k=0}^{\infty} T_6^n(r) \frac{dP_n(\cos\theta)}{d\theta},$$

$$\Phi(r,\theta) = h\sqrt{\frac{\lambda}{\varepsilon_0}} \sum_{k=0}^{\infty} \Phi^n(r) P_n(\cos\theta),\, u_1(r,\theta) = h \sum_{k=0}^{\infty} u_1^n(r) P_n(\cos\theta),$$

$$u_2(r,\theta) = h \sum_{k=0}^{\infty} u_2^n(r) \frac{dP_n(\cos\theta)}{d\theta},\, D_1(r,\theta) = \sqrt{\varepsilon_0\lambda} \sum_{k=0}^{\infty} D_1^n(r) P_n(\cos\theta), \quad (4.25)$$

where $P_n(x)$ are the Legendre polynomials. Hereafter the superscript "n" will be omitted.

Using form (4.25) and considering that the Legendre polynomials satisfy the ordinary differential equation:

$$\frac{dP_n}{d\theta} + P_n \text{ctg}\theta + n(n+1) P_n = 0 \quad (4.26)$$

reduces the system (4.24) in partial derivatives with the corresponding boundary conditions (4.15), (4.22) to a boundary problem by eigenvalues for systems of ordinary differential equations:

$$\frac{dR}{dx} = A(x,\Omega)R,\, BR(-1) = \|0,0,0\|^T,\, BR(+1) = \|0,0,0\|^T, \quad (4.27)$$

where the matrix A has the form:

$$A = \begin{Vmatrix} 2\tilde{\varepsilon}(\delta_1 - 1) & -l\tilde{\varepsilon} & 0 & -2\delta_3\tilde{\varepsilon}^2 - \Omega^2 & -l\delta_3\tilde{\varepsilon}^2 & 2\delta_2\tilde{\varepsilon} \\ \delta_1\tilde{\varepsilon} & -3\tilde{\varepsilon} & 0 & -\delta_3\tilde{\varepsilon}^2 & -(l\delta_4 + 2c_{66})\tilde{\varepsilon}^2 - \Omega^2 & 0 \\ \frac{e_{33}}{\delta} & 0 & 0 & 0 & -l\delta_2\tilde{\varepsilon} & -\frac{c_{33}}{\delta} \\ \frac{\varepsilon_{33}}{\delta} & 0 & 0 & -2\delta_1\tilde{\varepsilon} & -l\delta_1\tilde{\varepsilon} & \frac{e_{33}}{\delta} \\ 0 & \frac{1}{c_{55}} & \frac{e_{15}\tilde{\varepsilon}}{c_{55}} & \tilde{\varepsilon} & \tilde{\varepsilon} & 0 \\ 0 & 0 & -\frac{le_{15}\tilde{\varepsilon}^2}{c_{55}} & -l\delta_5\tilde{\varepsilon}^2 & 0 & -2\tilde{\varepsilon} \end{Vmatrix} \quad (4.28)$$

where

$$l = n(n+1),\, \Omega = \omega h\sqrt{\frac{\rho}{\lambda}};\, \tilde{c}_{ij} = \frac{c_{ij}}{\lambda},\, \tilde{e}_{ij} = \frac{e_{ij}}{\sqrt{\varepsilon_0\lambda}},$$

$$\tilde{\varepsilon}_{ij} = \frac{\varepsilon_{ij}}{\varepsilon_0},\, x = \frac{r - R_0}{h},\, \varepsilon = \frac{h}{R_0},\, \tilde{\varepsilon} = \frac{\varepsilon}{1 + \varepsilon x}.$$

ρ is the density of the material of the sphere, R_0 is the radius of the mid-surface of the sphere, ε_0 is the permittivity of vacuum, and $\lambda = 10^{10}\text{N/m}^2$.

The matrices B_1 and B_2 have the forms:

$$B_1 = B_2 = \begin{Vmatrix} 1 & 0 & 0 \\ 0 & 1 & 0 \\ 0 & 0 & 1 \end{Vmatrix}.$$

The problem is solved by the stable discrete-orthogonalization method and the search method [16].

4.3.3 Analysis of the Calculation Results for the Free Frequencies of the Axisymmetric Vibrations of the Inhomogeneous Sphere Made of Metal and Piezoceramic Materials

Let us present the results of the numerical analysis.

The sphere has three layers. The face layers have thickness $h/2$ each, while the thickness of the middle layer is h. The face layers are made of steel with the following characteristics:

$$E = 21 \cdot 10^{10}\ \text{N/m}^2,\ \nu = 0.28,\ \rho_S = 7.85 \cdot 10^3\ \text{kg/m}^2.$$

The middle layer is made of PZT-4 piezoceramics with the following characteristics:

$$c_{11}^0 = 13.9 \cdot 10^{10}\ \frac{\text{N}}{\text{m}^2},\ c_{12}^0 = 7.43 \cdot 10^{10}\ \frac{\text{N}}{\text{m}^2},\ c_{13}^0 = 7.78 \cdot 10^{10}\ \frac{\text{N}}{\text{m}^2},$$
$$c_{33}^0 = 11.5 \cdot 10^{10}\ \frac{\text{N}}{\text{m}^2},\ c_{55}^0 = 2.56 \cdot 10^{10}\ \frac{\text{N}}{\text{m}^2},\ e_{13}^0 = -5.2\ \frac{\text{C}}{\text{m}^2},$$
$$e_{15}^0 = 12.7\ \frac{\text{C}}{\text{m}^2},\ e_{33}^0 = 15.1\ \frac{\text{C}}{\text{m}^2},\ \varepsilon_{11}^0/\varepsilon_0 = 730,\ \ \varepsilon_{33}^0/\varepsilon_0 = 635.$$

We have the problem of the free vibrations of the plane (flat, planar) layer at $\varepsilon = 0$. In this case, the system of differential equations that describes the marked vibrations will consist of two homogeneous differential equations.

The first equation describes of the thickness vibrations:

$$\tilde{c}_{33}u_1'' + \Omega^2 u_1 + \tilde{e}_{33}\Phi'' = 0,\ \ \tilde{e}_{33}u_1'' - \varepsilon_{33}\Phi'' = 0, \tag{4.29}$$

with corresponding boundary conditions on layer surfaces:

$$\tilde{c}_{33}u_1' + \tilde{e}_{11}\Phi' = 0,\ \ \Phi = 0. \tag{4.30}$$

The second equation describes the shear vibrations:

$$\tilde{c}_{55}u_2'' + \Omega^2 u_2 = 0, \tag{4.31}$$

with corresponding boundary conditions on layer surfaces:

$$\tilde{c}_{55}u_2' = 0. \tag{4.32}$$

The presented problems have an analytical solution.

The solution of the problems (4.29), (4.30) for metal layers reads

$$U(n) = \frac{\pi}{2}n\sqrt{\frac{(1-\nu)E}{(1+\nu)(1-2\nu)\rho_M}} = 0;\ 2.905;\ 5.81;\ \ldots n = 0, 1, 2\ldots \tag{4.33}$$

The solution of the problems (4.31), (4.32) for the metal plane (flat, planar) layers is

$$W(n) = \frac{\pi}{2}n\sqrt{\frac{E}{2(1+\nu)\rho_M}} = 0;\ 1.606;\ 3.211;\ \ldots n = 0, 1, 2\ldots \tag{4.34}$$

The solution of the problems (4.29), (4.30) for the piezoceramic plane (flat, planar) layer is

$$U(2n) = \pi n\sqrt{\frac{1}{\rho_n}\left(c_{33} + \frac{e_{33}^2}{\varepsilon_{33}}\right)} = 0;\ 3.95;\ 7.901;\ \ldots, n = 0, 1, 2, . \tag{4.35}$$

$$U(2n+1) = \lambda_n\sqrt{\frac{1}{\rho_n}\left(c_{33} + \frac{e_{33}^2}{\varepsilon_{33}}\right)} = 1.995;\ 6.729;\ \ldots \tag{4.36}$$

where λ_n are nonzero roots of the transcendent equation

$$\lambda\cos\lambda + \frac{e_{33}^2}{\sqrt{c_{33} + \dfrac{e_{33}^2}{\varepsilon_{33}}}}\sin\lambda = 0.$$

The solution of the problems (4.31), (4.32) for the plane piezoceramic (flat, planar) layer is

$$W(n) = \frac{\pi}{2}n\sqrt{\frac{c_{55}}{\rho_?}} = 0;\ 0.918;\ 1.835;\ \ldots n = 0, 1, 2\ldots \tag{4.37}$$

We can see by the increase of the geometric parameter ε the change of the corresponding natural frequencies. The identification problem exists only in the case of the first two natural frequencies (these frequencies at the value of the geometric parameter $\varepsilon = 0$ are zero). Using the formula for small values ε $(1 + \varepsilon \approx 1)$:

$$U_0\,(0) \approx \varepsilon\sqrt{c_{11} + c_{13} - \frac{e_{13}^2}{c_{33}}},\ W_0\,(0) \approx \varepsilon\sqrt{c_{66}}. \tag{4.38}$$

Figure 4.2 shows the dependence of the first frequencies of the free vibrations of the layered sphere on the geometric parameter ε at $n = 5$. In Fig. 4.2, solid lines correspond to the frequencies of the vibrations of a laminated sphere: the face layers are made of metal and the middle layer is made of piezoceramics, and dashed lines of the free vibration frequency of the homogeneous piezoceramic sphere of the similar geometry (mechanical and electrical material parameters are presented earlier). The presence of the metal layers leads to an increase in the natural frequencies of the sphere and the spherical structure becomes stiffer. Note a slight difference in the values of the first free vibration frequency for a layered and the homogeneous sphere.

In Fig. 4.3, the solid lines correspond to the frequencies of the vibrations of a laminated sphere whose outer layers are made of metal and the inner layer is made of piezoceramics, and the dashed lines correspond to the free vibration frequency of a homogeneous sphere made of metal materials of similar geometry. In this case, the free vibration frequencies of the laminated sphere are less than the corresponding vibration frequencies of the homogeneous sphere of the metal materials. Therefore,

Fig. 4.2 Comparison of the natural vibration frequencies of homogeneous piezoceramic and inhomogeneous layered sphere

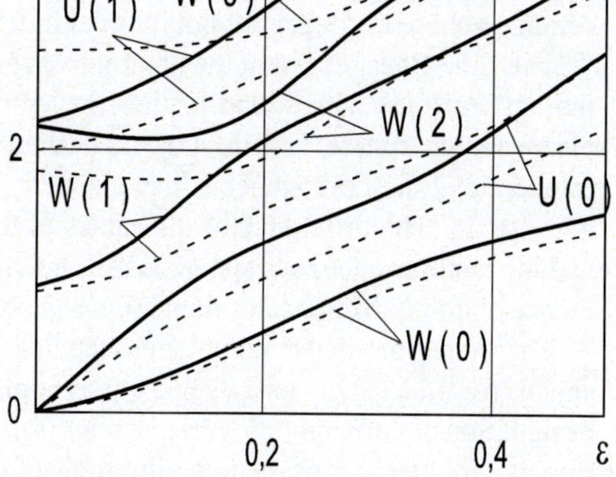

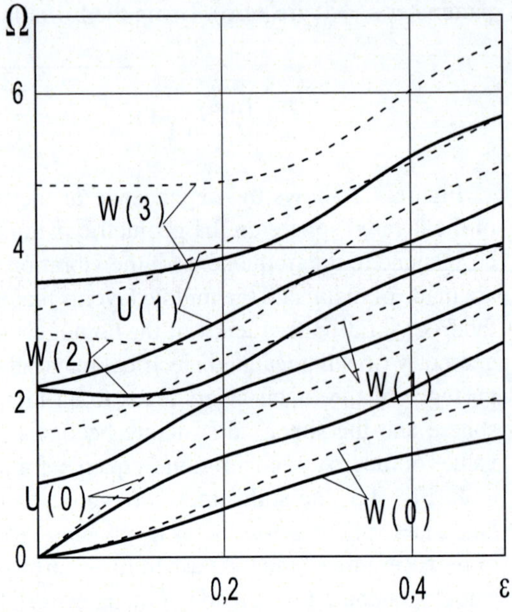

Fig. 4.3 Comparison of the natural vibration frequencies of homogeneous metal and inhomogeneous layered sphere

the free vibration frequencies of the laminated sphere are located in the interval between the free vibration frequencies of the homogeneous of the metal materials and piezoceramic sphere of similar geometry. The behavior of the natural frequencies of the vibrations of a homogeneous and a heterogeneous sphere can be observed in Fig. 4.4. Here, the solid lines correspond to the values of the vibration frequencies of the laminated sphere, the dotted lines correspond to the values of the frequency vibrations of a homogeneous piezoceramic sphere, the stroke dotted lines correspond to the values of frequencies of the vibration of a homogeneous metal sphere.

The relationships of the natural frequencies of layered and homogeneous piezoceramic sphere on Legendre polynomial index n are shown in Figs. 4.5 and 4.6.

The solid lines represent the natural frequencies of the layered sphere. In Fig. 4.5, the dashed lines correspond to the free vibration frequency of the homogeneous piezoceramic sphere. In Fig. 4.6, the dashed lines correspond to the free vibration frequency of the homogeneous metal sphere. The value of the geometric parameter is $\varepsilon = 0.25$. The difference of the values of the first frequency of the homogeneous and the nonhomogeneous sphere is insignificant. However, at higher frequencies, an increase in the difference of their values is observed.

The behavior of the natural frequencies of the layered and the homogeneous sphere for odd values on Legendre polynomial index is illustrated in Fig. 4.7 (the designation of corresponding frequencies is the same as in Fig. 4.4). The values of the first five frequencies of the free vibrations of the homogeneous piezoceramic sphere

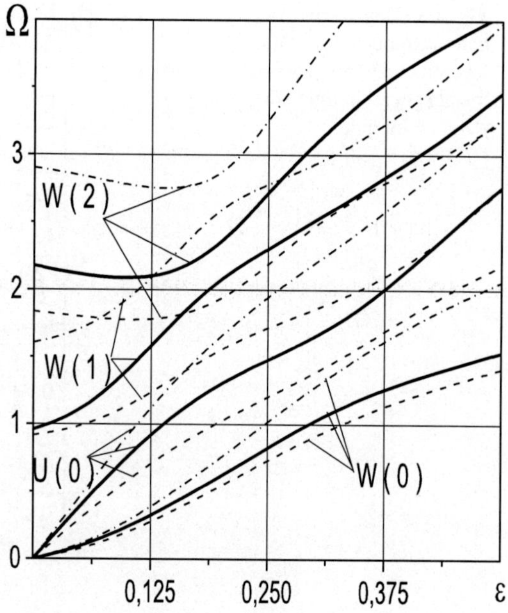

Fig. 4.4 Comparison of the natural frequencies of vibration of laminated and homogeneous piezoceramic and homogeneous metal sphere

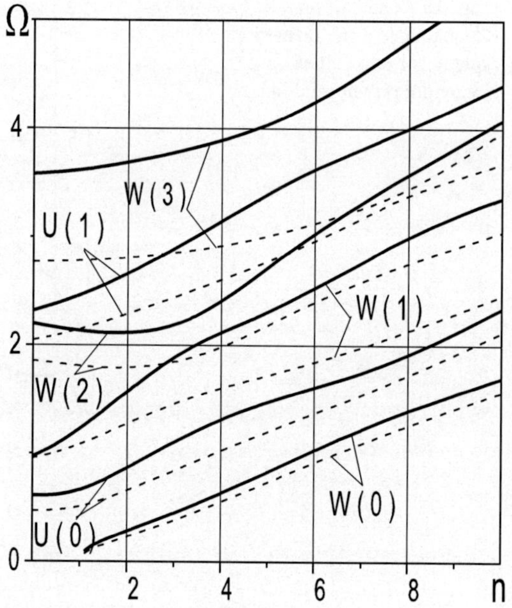

Fig. 4.5 Dependence of natural frequencies of layered and homogeneous piezoceramic sphere on Legendre polynomial index

Fig. 4.6 Dependence of natural frequencies of the layered and homogeneous metal sphere on Legendre polynomial index

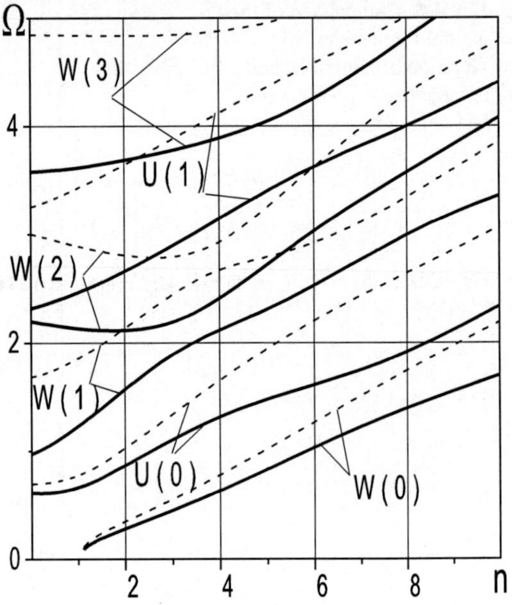

Fig. 4.7 The natural frequencies of the layered sphere for odd values on Legendre polynomial index

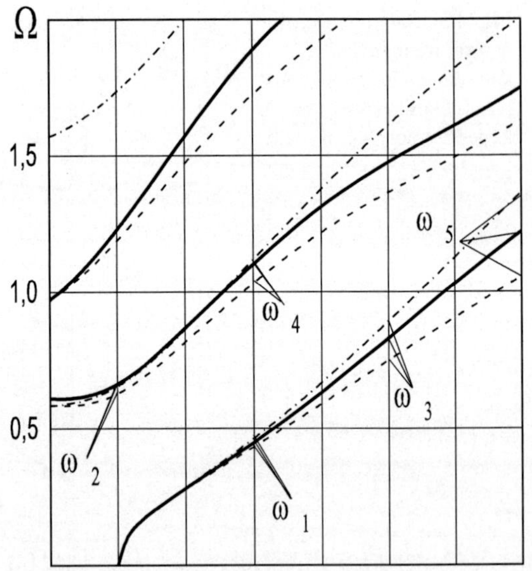

Table 4.1 Comparative analysis of the free vibration frequencies of the homogeneous piezoceramic sphere

No. of frequency	Analytical–numerical method	Power series method	Relative difference, %
1	0.4768	0.4770	0.04
2	0.7200	0.7200	0.00
3	0.9325	0.9320	0.05
4	1.2250	1.2250	0.00
5	1.4420	1.4420	0.00

Table 4.2 Frequencies of axisymmetric free vibrations of a layered piezoceramic sphere, obtained based on different models

No. of frequency	Layered sphere	n	Averaged physical and mechanical modules	Relative error (%)	Without piezoeffect	Relative error (%)
1	0.4455	3	0.4605	3.4	0.4330	2.8
2	0.6642	1	0.6556	1.3	0.6291	5.3
3	0.8314	5	0.8919	7.3	0.7516	9.6
4	1.1080	3	1.1220	1.3	1.039	6.2
5	1.2260	7	1.3660	11.4	1.058	13.7

(material is PZT-4) based on the power series method were obtained in [23, 24]. The results of the calculation of the free frequencies of vibrations of the homogeneous piezoceramic sphere (presented in Fig. 4.7) are compared with the obtained corresponding frequencies obtained in [23] (Table 4.1). The results of the eigenfrequency calculation based on the different approaches practically coincide (almost equal).

We calculate the average values of the three-layer package using

$$\lambda = \frac{\lambda^{\text{steel}} + \lambda^{\text{piezo}}}{2}. \tag{4.39}$$

The values of the first five frequencies of the free vibrations of the inhomogeneous piezoceramic sphere are compared with approximate values of the natural frequencies of the sphere made of a homogeneous material with averaged characteristics calculated by the theory of effective moduli (Table 4.2). Table 4.2 also shows the values of the first five frequencies of the free vibrations of the inhomogeneous sphere neglecting coupling with the electric field. The Legendre polynomial values for each of their frequencies are also given in Table 4.2.

It can be concluded that large values of the Legendre polynomial index show an increase in the difference in the frequency values determined based on different models. When neglecting the piezoelectric effect the free vibration frequency values are more different to the frequency values for the inhomogeneous sphere when compared to the frequencies of the homogeneous sphere based on calculations with the theory of effective moduli.

4.3.4 Analysis of the Calculation Results of the Free Frequencies of the Axisymmetric Vibrations of Continuously Inhomogeneous Piezoceramic Sphere Made of the Piezoceramic Gradient Material

Let us consider the case of a piezoceramic continuously heterogeneous sphere made of the piezoceramic gradient material. The properties of the material are assumed to vary exponentially across the thickness, i.e., $c_{ij} = c_{ij}^{o} e^{sr}$, $e_{ij} = e_{ij}^{o} e^{sr}$, and $\varepsilon_{ij} = \varepsilon_{ij}^{o} e^{sr}$. The parameter s is determined experimentally, and its values are usually in the interval $-2 \leq s \leq 2$. The parameters designated by an index 0 refer to the outside surface of the sphere and have the following values:

$$c_{11}^{0} = 13.9 \cdot 10^{10}\ \frac{\mathrm{N}}{\mathrm{m}^2},\ c_{12}^{0} = 7.43 \cdot 10^{10}\ \frac{\mathrm{N}}{\mathrm{m}^2},\ c_{13}^{0} = 7.78 \cdot 10^{10}\ \frac{\mathrm{N}}{\mathrm{m}^2},$$

$$c_{33}^{0} = 11.5 \cdot 10^{10}\ \frac{\mathrm{N}}{\mathrm{m}^2},\ c_{55}^{0} = 2.56 \cdot 10^{10}\ \frac{\mathrm{N}}{\mathrm{m}^2},\ e_{13}^{0} = -5.2\ \frac{\mathrm{C}}{\mathrm{m}^2},$$

$$e_{15}^{0} = 12.7\ \frac{\mathrm{C}}{\mathrm{m}^2},\ e_{33}^{0} = 15.1\ \frac{\mathrm{C}}{\mathrm{m}^2},\ \varepsilon_{11}^{0}/\varepsilon_0 = 730,\ \varepsilon_{33}^{0}/\varepsilon_0 = 635.$$

The change of the physical and the mechanical characteristics along the thickness of the sphere on the example of the elastic module c_{11} is shown in Fig. 4.8.

The relationship of the first six natural frequencies on the geometric parameter ε is shown in Fig. 4.9. The solid lines represent the natural frequencies of the layered sphere, and the dashed ones refer to the homogeneous piezoceramic sphere of similar geometry and material. The value of the inhomogeneity parameter is $s = 0.5$. The relationships of the first six natural frequencies on the Legendre polynomial index, for a geometrical parameter $\varepsilon = 0.25$, are shown Fig. 4.10.

In Fig. 4.11, the first five natural frequencies of the sphere of the different structures for the odd value of the Legendre polynomial are presented. The solid lines represent the natural frequencies of the layered sphere, the pointed lines correspond to the natural frequencies of the layered sphere without piezoeffect $\left(e_{ij} = 0\right)$, and the dashed lines correspond to the natural frequencies of the homogeneous sphere made of a homogeneous material with averaged characteristics calculated by the theory of effective moduli:

$$\lambda = \frac{\lambda^0}{2h} \int\limits_{R_0-h}^{R_0+h} e^{sr}\, dr = \frac{\lambda^0}{2kh} e^{sR_0-h} \left(e^{2sh} - 1\right). \tag{4.40}$$

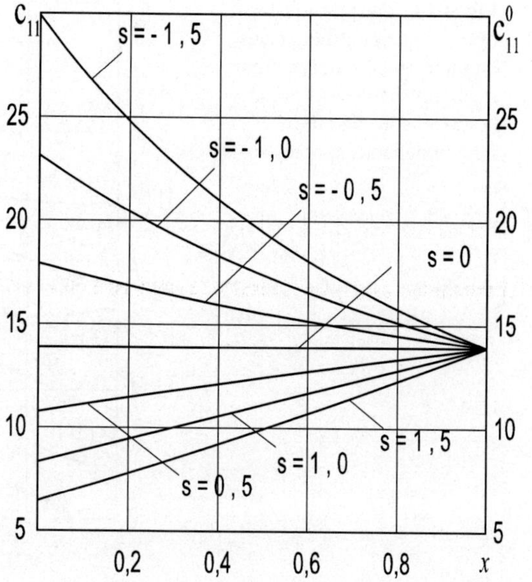

Fig. 4.8 The relationship the value of the elastic module c_{11} on the thickness of the sphere

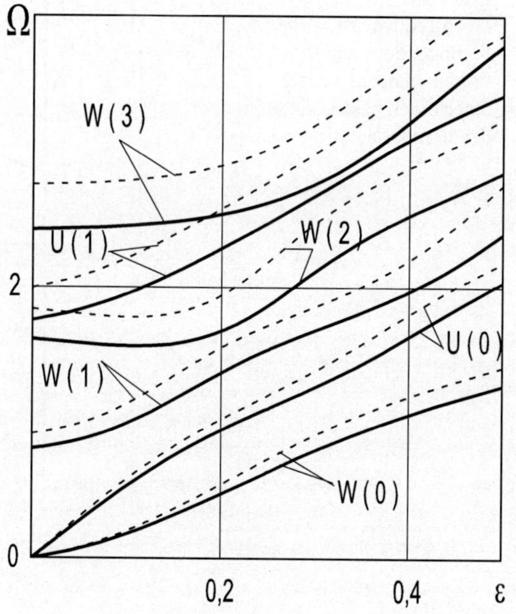

Fig. 4.9 The relationship of the first six natural frequencies on the geometric parameter

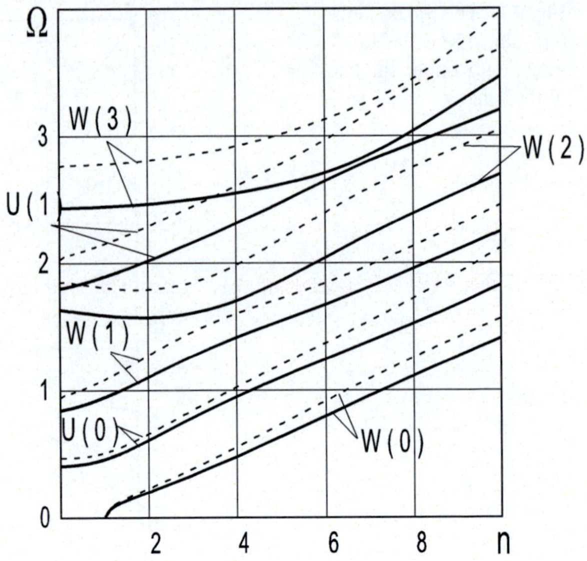

Fig. 4.10 The relationships of the first eigenfrequencies on the Legendre polynomial index for the piezoceramic homogeneous and the inhomogeneous sphere

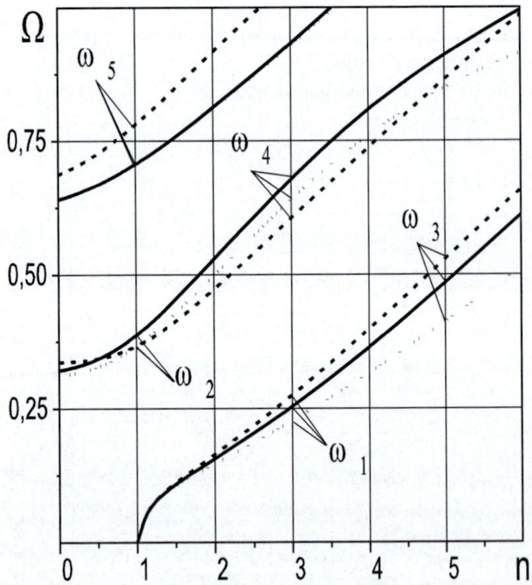

Fig. 4.11 The first five eigenfrequencies of the homogeneous and the inhomogeneous sphere determined based on the different models

The values of the first five frequencies of the free vibrations of the inhomogeneous piezoceramic sphere are compared with approximate values of the natural frequencies of the sphere made of a homogeneous material with averaged characteristics calculated by the theory of effective moduli and with values of the natural frequencies of the inhomogeneous sphere without piezoeffect (Table 4.3). The Legendre polynomial values for each of their frequencies are also given in Table 4.3.

Based on the data in Table 4.3. it can be noted that as in the case of a layered piezoceramic sphere, there is an increase in the difference in the frequency values

Table 4.3 Frequencies of axisymmetric free vibrations of piezoceramic sphere made of FGPM, obtained based on the different models

No. of frequency	Inhomogeneous sphere	n	The theory of effective moduli	Relative error, %	Disregarding piezoeffect	Relative error, %
1	0.2513	3	0.2728	8.6	0.2260	10.1
2	0.3867	1	0.3664	5.2	0.3662	5.3
3	0.4838	5	0.5291	9.4	0.4126	14.7
4	0.6827	3	0.6059	11.2	0.6453	5.5
5	0.7068	1	0.7804	10.4	0.7063	0.1

determined based on different models with large Legendre polynomial values. The data when neglecting the piezoelectric effect are more significantly different from the frequency values for the inhomogeneous sphere in comparison with frequencies of the homogenous sphere defined based on a calculation by the theory of effective moduli.

4.4 Nonaxisymmetric Free Vibrations of Inhomogeneous Piezoceramic Hollow Sphere

4.4.1 Resolving System

Let us present the equations (1.144), (1.146)–(1.148).

The constitutive equations for the piezoceramic sphere are

$$\begin{aligned}
&\Sigma_2 = rT_2 = c_{11}\Xi_2 + c_{12}\Xi_3 + c_{13}\Xi_1 + e_{31}\nabla_2\Phi,\\
&\Sigma_3 = rT_3 = c_{12}\Xi_2 + c_{11}\Xi_3 + c_{13}\Xi_1 + e_{31}\nabla_2\Phi,\\
&\Sigma_1 = rT_1 = c_{13}\Xi_2 + c_{13}\Xi_3 + c_{33}\Xi_1 + e_{33}\nabla_2\Phi,\\
&\Sigma_6 = rT_6 = 2c_{55}\Xi_6 + e_{15}\frac{\partial\Phi}{\partial\theta},\ \Sigma_5 = rT_5 = 2c_{55}\Xi_5 + \frac{e_{15}}{\sin\theta}\frac{\partial\Phi}{\partial\varphi},\\
&\Sigma_4 = rT_4 = 2c_{66}\Xi_4,\ \Delta_1 = rD_1 = e_{31}\Xi_2 + e_{31}\Xi_3 + e_{33}\Xi_2 - \varepsilon_{33}\nabla_2\Phi,\\
&\Delta_2 = rD_2 = 2e_{15}\Xi_6 - \varepsilon_{11}\frac{\partial\Phi}{\partial\theta},\ \Delta_3 = rD_3 = 2e_{15}\Xi_5 - \frac{\varepsilon_{11}}{\sin\theta}\frac{\partial\Phi}{\partial\varphi},
\end{aligned} \tag{4.41}$$

where

$$\nabla_2 = r\partial/\partial r.$$

The tensor Ξ_i is defined as

$$\Xi_1 = rS_1 = \nabla_2 u_1,\ \Xi_2 = rS_2 = \frac{\partial u_2}{\partial \theta} + u_1,\ \Xi_3 = rS_3 = \frac{1}{\sin\theta}\frac{\partial u_3}{\partial \varphi} + u_1 + u_2 \mathrm{ctg}\theta,$$
$$2\Xi_6 = 2rS_6 = \frac{\partial u_1}{\partial \theta} + \nabla_2 u_2 - u_2,\ 2\Xi_5 = 2rS_5 = \frac{1}{\sin\theta}\frac{\partial u_1}{\partial \varphi} + \nabla_2 u_3 - u_3,$$
$$2\Xi_4 = 2rS_4 = \frac{1}{\sin\theta}\frac{\partial u_2}{\partial \varphi} + \frac{\partial u_3}{\partial \theta} - u_3 \mathrm{ctg}\theta. \tag{4.42}$$

Using relations (4.11), the system of the motion equations (4.4) is given in the form:

$$\nabla_2 \Sigma_1 + \csc\theta \frac{\partial \Sigma_5}{\partial \varphi} + \frac{\partial \Sigma_6}{\partial \theta} + \Sigma_1 - \Sigma_2 - \Sigma_3 + \Sigma_6 \mathrm{ctg}\theta = -\rho r^2 \omega^2 u_1,$$
$$\nabla_2 \Sigma_5 + \csc\theta \frac{\partial \Sigma_3}{\partial \varphi} + \frac{\partial \Sigma_4}{\partial \theta} + 2\Sigma_5 + 2\Sigma_4 \mathrm{ctg}\theta = -\rho r^2 \omega^2 u_3,$$
$$\nabla_2 \Sigma_6 + \csc\theta \frac{\partial \Sigma_4}{\partial \varphi} + \frac{\partial \Sigma_2}{\partial \theta} + 2\Sigma_6 + (\Sigma_2 - \Sigma_3)\,\mathrm{ctg}\theta = -\rho r^2 \omega^2 u_2, \tag{4.43}$$

and the electrostatic equations (4.2) read

$$\nabla_2 \Delta_1 + \Delta_1 + \frac{1}{\sin\theta}\frac{\partial}{\partial \theta}(\Delta_2 \sin\theta) + \frac{1}{\sin\theta}\frac{\partial \Delta_3}{\partial \theta} = 0. \tag{4.44}$$

We will now use the method of separation of variables (see [28, 29]). To this end, we a introduce new unknown function of the displacements u_α, u_β, u_γ and stresses Σ_α, Σ_β, respectively. The displacements u_α, u_β, and u_γ are expressed in terms of u_1, u_2, and u_3:

$$u_2 = -\frac{1}{\sin\theta}\frac{\partial u_\alpha}{\partial \varphi} - \frac{\partial u_\beta}{\partial \theta},\ u_3 = \frac{\partial u_\alpha}{\partial \theta} - \frac{1}{\sin\theta}\frac{\partial u_\beta}{\partial \varphi},\ u_1 = u_\gamma. \tag{4.45}$$

From the relations (4.41) and (4.42), we obtain analogous expressions for the stress-related quantities Σ_6 and Σ_5:

$$\Sigma_6 = -\frac{1}{\sin\theta}\frac{\partial \Sigma_\alpha}{\partial \varphi} - \frac{\partial \Sigma_\beta}{\partial \theta},\ \Sigma_5 = \frac{\partial \Sigma_\alpha}{\partial \theta} - \frac{1}{\sin\theta}\frac{\partial \Sigma_\beta}{\partial \varphi}. \tag{4.46}$$

The last two relations (equations) (4.43) from the relations (4.45) and (4.46), can be represented as

$$\frac{\partial}{\partial \theta}\Big[\nabla_2 \Sigma_\beta + 2\Sigma_\beta + c_{11}\nabla_1^2 u_\beta + 2c_{66} u_\beta -$$
$$- (c_{11} + c_{12})\, u_\gamma + r^2 \rho \omega^2 u_\gamma - c_{13}\nabla_2 u_\gamma - e_{31}\nabla_2 \Phi\Big] +$$
$$+ \frac{1}{\sin\theta}\frac{\partial}{\partial \varphi}\left[\nabla_2 \Sigma_\alpha + 2\Sigma_\alpha + c_{66}\left(\nabla_1^2 u_\alpha + 2u_\alpha\right) + r^2 \rho \omega^2 u_\alpha\right] = 0, \tag{4.47}$$

$$\frac{1}{\sin\theta}\frac{\partial}{\partial\varphi}\left[\nabla_2\Sigma_\beta + 2\Sigma_\beta + c_{11}\nabla_1^2 u_\beta + 2c_{66}u_\beta - \right.$$
$$\left. - (c_{11}+c_{12})\,u_\gamma + r^2\rho\omega^2 u_\gamma - c_{13}\nabla_2 u_\gamma - e_{31}\nabla_2\Phi\right] +$$
$$+ \frac{\partial}{\partial\theta}\left[\nabla_2\Sigma_\alpha + 2\Sigma_\alpha + c_{66}\left(\nabla_1^2 u_\alpha + 2u_\alpha\right) + r^2\rho\omega^2 u_\alpha\right] = 0, \quad (4.48)$$

where $\nabla_1^2 = \frac{\partial^2}{\partial\theta^2} + ctg\theta\frac{\partial}{\partial\theta} + \frac{1}{\sin^2\theta}\frac{\partial^2}{\partial\varphi^2}$.

From the relations (4.47) and (4.48), we have

$$\nabla_2\Sigma_\beta + 2\Sigma_\beta + c_{11}\nabla_1^2 u_\beta + 2c_{66}u_\beta - (c_{11}+c_{12})\,u_\gamma +$$
$$+ r^2\rho\omega^2 u_\gamma - c_{13}\nabla_2 u_\gamma - e_{31}\nabla_2\Phi = 0, \quad (4.49)$$

$$\nabla_2\Sigma_\alpha + 2\Sigma_\alpha + c_{66}\left(\nabla_1^2 u_\alpha + 2u_\alpha\right) + r^2\rho\omega^2 u_\alpha = 0. \quad (4.50)$$

The third Eq. (4.43) and Eq. (4.44) can similarly be represented as

$$\nabla_2\Sigma_1 + \Sigma_1 - \nabla_2\Sigma_\beta + (c_{11}+c_{12})\,\nabla_1^2 u_\beta -$$
$$- 2\,(c_{11}+c_{12})\,u_\gamma - 2c_{13}\nabla_2 u_\gamma - 2e_{31}\nabla_2\Phi = 0, \quad (4.51)$$

$$\nabla_2\Delta_1 + \Delta_1 + e_{15}\nabla_1^2 u_\gamma - \varepsilon_{11}\nabla_1^2\Phi + e_{15}\nabla_1^2 u_\beta = 0. \quad (4.52)$$

From the relation (4.41), the fourth and fifth equations using (4.42) read

$$\frac{\partial}{\partial\theta}\left[\Sigma_\beta + c_{55}u_\gamma + e_{15}\Phi - c_{55}\nabla_2 u_\beta + c_{55}u_\beta + r^2\rho\omega^2 u_\beta\right] +$$
$$+ \frac{1}{\sin\theta}\frac{\partial}{\partial\varphi}\left[c_{55}\nabla_2 u_\alpha - c_{55}u_\alpha - \Sigma_\alpha\right] = 0, \quad (4.53)$$

$$\frac{1}{\sin\theta}\frac{\partial}{\partial\varphi}\left[\Sigma_\beta + c_{55}u_\gamma + e_{15}\Phi - c_{55}\nabla_2 u_\beta + c_{55}u_\beta\right] +$$
$$+ \frac{\partial}{\partial\theta}\left[c_{55}\nabla_2 u_\alpha - c_{55}u_\alpha - \Sigma_\alpha\right] = 0; \quad (4.54)$$

after the corresponding transformations (4.53), (4.54), we get

$$\Sigma_\beta + c_{55}u_\gamma + e_{15}\Phi - c_{55}\nabla_2 u_\beta + c_{55}u_\beta + r^2\rho\omega^2 u_\beta = 0, \quad (4.55)$$

$$c_{55}\nabla_2 u_\alpha - c_{55}u_\alpha - \Sigma_\alpha = 0. \quad (4.56)$$

From the relation (4.41), the third and the ninth equations using (4.42) are given by

$$\Sigma_1 = -c_{13}\nabla_1^2 u_\beta + 2c_{13}u_\gamma + c_{33}\nabla_2 u_\gamma + e_{33}\nabla_2\Phi, \tag{4.57}$$

$$\Delta_1 = -e_{31}\nabla_1^2 u_\beta + 2e_{13}u_\gamma + e_{33}\nabla_2 u_\gamma - \varepsilon_{33}\nabla_2\Phi. \tag{4.58}$$

Inspection of the obtained equations shows that the unknown functions u_α and Σ_α are not related to other unknowns. Because of this, all the equations can be divided into two independent sets. From Eqs. (4.50) and (4.56), we have

$$\nabla_2\begin{Bmatrix}\Sigma_\alpha\\ u_\alpha\end{Bmatrix} = \begin{Vmatrix}-2 & -c_{66}\left(\nabla_1^2+2\right)-r^2\rho\omega^2\\ \dfrac{1}{c_{55}} & 1\end{Vmatrix}\begin{Bmatrix}\Sigma_\alpha\\ u_\alpha\end{Bmatrix}. \tag{4.59}$$

From Eqs. (4.49), (4.51), (4.52), (4.55), (4.57), and (4.58), we obtain

$$\nabla_2\left\{\Sigma_1, \Sigma_\beta, \Phi, u_\gamma, u_\beta, \Delta_1\right\}^T = \mathbf{M}\left\{\Sigma_1, \Sigma_\beta, \Phi, u_\gamma, u_\beta, \Delta_1\right\}^T, \tag{4.60}$$

where

$$\mathbf{M} = \begin{Vmatrix}2\delta_1-1 & \nabla_1^2 & 0 & -2\delta_3-r^2\rho\omega^2 & \delta_3\nabla_1^2 & 2\delta_2\\ \delta_1 & -2 & 0 & -\delta_3 & \delta_4\nabla_1^2-2c_{66}-r^2\rho\omega^2 & 0\\ e_{33}/\delta & 0 & 0 & 0 & \delta_2\nabla_1^2 & -c_{33}/\delta\\ \varepsilon_{33}/\delta & 0 & 0 & -2\delta_1 & \delta_1\nabla_1^2 & e_{33}/\delta\\ 0 & 1/c_{55} & e_{15}/c_{55} & 1 & 1 & 0\\ 0 & 0 & e_{15}\nabla_1^2/c_{55} & \delta_5\nabla_1^2 & 0 & -1\end{Vmatrix},$$

and

$$\Delta_1^2 * = \frac{\partial^2 *}{\partial\theta^2} + \mathrm{ctg}\theta\frac{\partial *}{\partial\theta} - \csc^2\theta\frac{\partial^2 *}{\partial\varphi^2}, \delta = c_{33}\varepsilon_{33} + e_{33}^2\delta_1 = c_{13}\varepsilon_{33} + e_{13}e_{33},$$
$$\delta_2 = c_{13}e_{33} - c_{33}e_{13}, \delta_3 = 2\left(c_{13}\delta_1 + e_{13}\delta_2\right) - \left(c_{11} + c_{12}\right);$$
$$\delta_4 = \frac{\delta_3}{2} - c_{66}; \delta_5 = c_{55}\varepsilon_{33} + e_{53}^2.$$

The corresponding boundary conditions will be presented in the next section.

4.4.2 Method of Solution of the Problem

We will search for solutions of the form:

$$\Sigma_1 = \sum_{m=0}^{n}\sum_{n=1}^{\infty} \Sigma_{1n}(r) S_n^m(\theta,\varphi)\,;\ \Sigma_\alpha = \sum_{m=0}^{n}\sum_{n=1}^{\infty} \Sigma_\alpha(r) S_n^m(\theta,\varphi)\,,$$
$$\Sigma_\beta = \sum_{m=0}^{n}\sum_{n=1}^{\infty} \Sigma_\beta(r) S_n^m(\theta,\varphi)\,,\ \Phi = \sum_{m=0}^{n}\sum_{n=1}^{\infty} \Phi_n(r) S_n^m(\theta,\varphi)\,, \tag{4.61}$$
$$u_\alpha = \sum_{m=0}^{n}\sum_{n=1}^{\infty} u_{\alpha n}(r) S_n^m(\theta,\varphi)\,,\ u_\beta = \sum_{m=0}^{n}\sum_{n=1}^{\infty} u_{\beta n}(r) S_n^m(\theta,\varphi)\,,$$
$$u_\gamma = \sum_{m=0}^{n}\sum_{n=1}^{\infty} u_{\gamma n}(r) S_n^m(\theta,\varphi)\,,\ \Delta_1 = \sum_{m=0}^{n}\sum_{n=1}^{\infty} \Delta_{1n}(r) S_n^m(\theta,\varphi)\,,$$

where $S_n^m = P_n^m(\cos\theta)\, e^{im\varphi}$ are spherical harmonic functions, $P_n^m(x)$ are the associated Legendre polynomials, $\lambda = 10^{10}\mathrm{N/m^2}$, and ε_0 is vacuum dielectric constant. The spherical functions are the solution to the following differential equation:

$$\frac{\partial^2 Y_n}{\partial\theta^2} + \mathrm{ctg}\theta\frac{\partial Y_n}{\partial\theta} - \csc^2\theta\frac{\partial^2 Y_n}{\partial\varphi^2} + n(n+1)Y_n = \Delta_1^2 Y_n + n(n+1)Y_n = 0. \tag{4.62}$$

Then the matrix can be represented as

$$\mathbf{M} = \left\| \begin{array}{cccccc} 2\delta_1 - 1 & -l & 0 & -2\delta_3 - r^2\rho\omega^2 & -l\delta_3 & 2\delta_2 \\ \delta_1 & -2 & 0 & -\delta_3 & -l\delta_4 - 2c_{66} - r^2\rho\omega^2 & 0 \\ e_{33}/\delta & 0 & 0 & 0 & -l\delta_2 & -c_{33}/\delta \\ \varepsilon_{33}/\delta & 0 & 0 & -2\delta_1 & -l\delta_1 & e_{33}/\delta \\ 0 & 1/c_{55} & e_{15}/c_{55} & 1 & 1 & 0 \\ 0 & 0 & -le_{15}/c_{55} & -l\delta_5 & 0 & -1 \end{array} \right\|.$$

Using the variables $T_i = \dfrac{\Sigma_i}{r}$, $D_i = \dfrac{\Delta_i}{r}$, we have two independent systems of ordinary differential equations:

$$\frac{\mathrm{d}}{\mathrm{d}r}\mathbf{R}_1 = \mathbf{A}_1\mathbf{R}_1, \tag{4.63}$$

and

$$\frac{\mathrm{d}}{\mathrm{d}r}\mathbf{R}_2 = \mathbf{A}_2\mathbf{R}_2, \tag{4.64}$$

where the matrix $\mathbf{A}_1$ is given by

$$\mathbf{A}_1 = \left\| \begin{array}{cc} -\dfrac{3}{r} & -\dfrac{c_{66}(2-l)}{r^2} - \rho\omega^2 \\ \dfrac{1}{c_{55}} & \dfrac{1}{r} \end{array} \right\|,$$

and the matrix $\mathbf{A}_2$ is given by

$$\mathbf{A}_2 = \left\| \begin{array}{cccccc} \frac{2}{r}(\delta_1 - 1) & -\frac{l}{r} & 0 & -\frac{2\delta_3}{r^2} - \rho\omega^2 & -\frac{l\delta_3}{r^2} & \frac{2\delta_2}{r} \\ \frac{\delta_1}{r} & -\frac{3}{r} & 0 & -\frac{\delta_3}{r^2} & -\frac{l\delta_4 + 2c_{66}}{r^2} - \rho\omega^2 & 0 \\ \frac{e_{33}}{\delta} & 0 & 0 & 0 & -\frac{l\delta_2}{r} & -\frac{c_{33}}{\delta} \\ \frac{\varepsilon_{33}}{\delta} & 0 & 0 & -\frac{2\delta_1}{r} & -\frac{l\delta_1}{r} & \frac{e_{33}}{\delta} \\ 0 & \frac{1}{c_{55}} & \frac{e_{15}}{r c_{55}} & \frac{1}{r} & \frac{1}{r} & 0 \\ 0 & 0 & -\frac{l e_{15}}{r^2 c_{55}} & -\frac{l\delta_5}{r^2} & 0 & -\frac{2}{r} \end{array} \right\| .$$

Let us consider dimensionless values:

$$\begin{aligned} &\tilde{T}_i(r) = \lambda T_i,\ \tilde{\Phi}(r) = h\sqrt{\frac{\lambda}{\varepsilon_0}}\Phi(r), \\ &\tilde{u}_i(r) = h u_i(r),\ \tilde{D}_r(r) = \sqrt{\varepsilon_0\lambda} D_r(r), \end{aligned} \tag{4.65}$$

(hereafter the superscript $\sim$ will be omitted) and

$$l = n(n+1),\ \Omega = \omega h\sqrt{\frac{\rho}{\lambda}},\ \tilde{c}_{ij} = \frac{c_{ij}}{\lambda},\ \tilde{e}_{ij} = \frac{e_{ij}}{\sqrt{\varepsilon_0\lambda}},$$

$$\tilde{\varepsilon}_{ij} = \frac{\varepsilon_{ij}}{\varepsilon_0},\ x = \frac{r - R_0}{h},\ \varepsilon = \frac{h}{R_0},\ \tilde{\varepsilon} = \frac{\varepsilon}{1 + \varepsilon x}.$$

Then the matrix $\mathbf{A}_1$ is given by

$$\mathbf{A}_1 = \left\| \begin{array}{cc} -3\tilde{\varepsilon} & -c_{66}(2 - l)\tilde{\varepsilon}^2 - \Omega^2 \\ \frac{1}{c_{55}} & \tilde{\varepsilon} \end{array} \right\|, \tag{4.66}$$

and the matrix $\mathbf{A}_2$ is given by

$$\mathbf{A}_2 = \begin{Vmatrix} 2\tilde{\varepsilon}\left(\delta_1 - 1\right) & -l\tilde{\varepsilon} & 0 & -2\delta_3\tilde{\varepsilon}^2 - \Omega^2 & -l\delta_3\tilde{\varepsilon}^2 & 2\delta_2\tilde{\varepsilon} \\ \delta_1\tilde{\varepsilon} & -3\tilde{\varepsilon} & 0 & -\delta_3\tilde{\varepsilon}^2 & -\left(l\delta_4 + 2c_{66}\right)\tilde{\varepsilon}^2 - \Omega^2 & 0 \\ \frac{e_{33}}{\delta} & 0 & 0 & 0 & -l\delta_2\tilde{\varepsilon} & -\frac{c_{33}}{\delta} \\ \frac{\varepsilon_{33}}{\delta} & 0 & 0 & -2\delta_1\tilde{\varepsilon} & -l\delta_1\tilde{\varepsilon} & \frac{e_{33}}{\delta} \\ 0 & \frac{1}{c_{55}} & \frac{e_{15}\tilde{\varepsilon}}{c_{55}} & \tilde{\varepsilon} & \tilde{\varepsilon} & 0 \\ 0 & 0 & -\frac{le_{15}\tilde{\varepsilon}^2}{c_{55}} & -l\delta_5\tilde{\varepsilon}^2 & 0 & -2\tilde{\varepsilon} \end{Vmatrix}. \tag{4.67}$$

We obtain two independent systems of differential equations based on the variable separation procedure having the vibrations of the first class with the corresponding matrix (4.66) and the vibrations of the second class with the corresponding matrix (4.67). The vibrations of the first class are elastic (torsional vibrations), and their frequencies can be determined analytically [22]. For the case of a homogeneous hollow sphere, the frequency equation is

$$\frac{2x_1 J_\nu'(x_1) - 3J_\nu(x_1)}{2x_1 Y_\nu'(x_1) - 3Y_\nu(x_1)} = \frac{2x_2 J_\nu'(x_2) - 3J_\nu(x_2)}{2x_2 Y_\nu'(x_2) - 3Y_\nu(x_2)}, \tag{4.68}$$

where $x_1 = \Omega\left(R_0 - h\right)\sqrt{\frac{\rho}{c_{55}}}$, $x_1 = \Omega\left(R_0 + h\right)\sqrt{\frac{\rho}{c_{55}}}$.

The index ν of cylindrical Bessel functions $J_\nu(x_i)$ and $Y_\nu(x_i)$ is determined from the equations

$$2\nu = \sqrt{9 + 8\left(n^2 + n - 2\right)\frac{c_{44}}{c_{55}}}, \quad n = 1, 2, 3, \ldots. \tag{4.69}$$

The analytical solution of the differential equation (4.67) is possible only in the case of a homogeneous sphere. The solution for a layered and continuously nonhomogeneous sphere becomes only possible based on numerical methods.

The second class vibrations completely coincide with the axisymmetric electric elastic vibrations discussed above.

The corresponding boundary conditions have the form:

at the lateral surfaces free of the external forces:

$$T_1(R_0 \pm h, \theta, \varphi) = 0, T_\alpha(R_0 \pm h, \theta, \varphi) = 0, T_\beta(R_0 \pm h, \theta, \varphi) = 0; \tag{4.70}$$

we will consider the two variants of electrical boundary conditions on the lateral surfaces:

(i) continuity of the normal component of the electric displacement to the outside vacuum (no electrodes):

$$D_1(R_0 \pm h, \theta, \varphi) = 0; \tag{4.71}$$

(ii) a grounded conducting surface (sphere coated with thin short-circuited electrodes):

$$\Phi(R_0 \pm h, \theta, \varphi) = 0.$$

On the surface of the discontinuity on properties, we impose conditions of rigid mechanical contact without slipping and tearing off. Therefore, we have condition continuity of the components of the vector displacements and corresponding stress tensor components:

$$\begin{aligned} T_1^i &= T_1^{i+1}, T_\alpha^i = T_\alpha^{i+1}, T_\beta^i = T_\beta^{i+1}, \\ u_\alpha^i &= u_\alpha^{i+1}, u_\beta^i = u_\beta^{i+1}, u_\gamma^i = u_\gamma^{i+1}. \end{aligned} \tag{4.72}$$

The conditions for the electrical values on the contact surface (interface) of the layers depend on the material type.

The conditions for the electrical values on the contact surface (interface) of the layers depend on the material type. If the ith and $(i + 1)$th layers are piezoelectric or dielectric materials, then it is necessary to supplement the condition (4.60) by the condition of continuity of the normal component vector of the electrical induction:

$$D_r^i = D_r^{i+1}. \tag{4.73}$$

If the ith and $(i + 1)$th layers are conductors, we have no additional boundary conditions (apply boundary conditions (4.60)).

If the ith layers are piezoelectric and $(i + 1)$th are conducting layers, we get the next boundary conditions (1.178):

$$\Phi^i = 0. \tag{4.74}$$

Hereafter the superscript "i" will be omitted.

The resolving vector-function of mixed type is

$$R = \left\{\Phi, T_1, T_\beta, T_\alpha, u_\alpha, u_\gamma, u_\beta, D_r\right\}. \tag{4.75}$$

The kind of the resolving system depends on the direction of preliminary polarization of the piezoceramic. In the case of radial polarization, the system takes the form:

$$\begin{aligned} \frac{d\mathbf{R}}{dx} &= \mathbf{C} \cdot \mathbf{R}, \\ \mathbf{D}_1 \cdot \mathbf{R}\,(-1) &= \|0, 0, 0, 0, 0\|^T, \\ \mathbf{D}_2 \cdot \mathbf{R}\,(+1) &= \|0, 0, 0, 0, 0\|^T, \end{aligned} \tag{4.76}$$

where **C**, $\mathbf{D}_1$ and $\mathbf{D}_2$ are given by

$$
\left\|\begin{matrix}
0 & \frac{e_{33}}{\delta} & 0 & 0 & 0 & 0 & -l\delta_2\tilde{\varepsilon} & -\frac{c_{33}}{\delta} \\
0 & 2\tilde{\varepsilon}\left(\delta_1 - 1\right) & -l\tilde{\varepsilon} & 0 & 0 & -2\delta_3\tilde{\varepsilon}^2 - \Omega^2 & -l\delta_3\tilde{\varepsilon}^2 & 2\delta_2\tilde{\varepsilon} \\
0 & \delta_1\tilde{\varepsilon} & -3\tilde{\varepsilon} & 0 & 0 & -\delta_3\tilde{\varepsilon}^2 & -\left(l\delta_4 + 2c_{66}\right)\tilde{\varepsilon}^2 - \Omega^2 & 0 \\
0 & 0 & 0 & -3\tilde{\varepsilon} & -c_{66}\left(2 - l\right)\tilde{\varepsilon}^2 - \Omega^2 & 0 & 0 & 0 \\
0 & 0 & 0 & \frac{1}{c_{55}} & \tilde{\varepsilon} & 0 & 0 & 0 \\
0 & \frac{\varepsilon_{33}}{\delta} & 0 & 0 & 0 & -2\delta_1\tilde{\varepsilon} & -l\delta_1\tilde{\varepsilon} & \frac{e_{33}}{\delta} \\
\frac{e_{15}\tilde{\varepsilon}}{c_{55}} & 0 & \frac{1}{\tilde{c}_{55}} & 0 & 0 & \tilde{x} & \tilde{x} & 0 \\
-\frac{le_{15}\tilde{\varepsilon}^2}{c_{55}} & 0 & 0 & 0 & 0 & -l\delta_5\tilde{\varepsilon}^2 & 0 & -2\tilde{\varepsilon}
\end{matrix}\right\|
$$

$$
\mathbf{D}_1 = \mathbf{D}_2 = \left\|\begin{matrix}
1 & 0 & 0 & 0 & 0 & 0 & 0 & 0 \\
0 & 1 & 0 & 0 & 0 & 0 & 0 & 0 \\
0 & 0 & 1 & 0 & 0 & 0 & 0 & 0 \\
0 & 0 & 0 & 1 & 0 & 0 & 0 & 0
\end{matrix}\right\|.
$$

The solution of the boundary-value problem for the systems of ordinary differential equations (4.76) based on the method of discrete orthogonalization in combination with the step-by-step search method was carried out by the method of discrete orthogonalization in combination with the step-by-step search method was carried out (see Chap. 1).

4.4.3 *Analysis of the Calculation Results of the Free Frequencies of Nonaxisymmetric Vibrations of an Inhomogeneous Sphere Made of Metal and Piezoceramic Materials*

Note that there is no wave parameter in the edge problem. Natural frequencies do not depend on the waveform parameter in the circumferential direction (in comparison with the natural frequencies of the cylinder vibrations).

The values of the natural frequencies of an inhomogeneous sphere depend on the geometric parameter ε and on the Legendre polynomial index n.

Let us solve the following problem. The sphere has three layers. The face layers have thickness $h/2$ each, while the thickness of the middle layer is h. The face layers are made of steel with the following characteristics:

$$
E = 21 \cdot 10^{10}\ \mathrm{N/m^2},\ \nu = 0.28,\ \rho_S = 7.85 \cdot 10^3\ \mathrm{kg/m^2}.
$$

The middle layer is made of PZT-4 piezoceramics with the following characteristics:

$$c_{11}^0 = 13.9 \cdot 10^{10}\ \frac{\mathrm{N}}{\mathrm{m}^2}, c_{12}^0 = 7.43 \cdot 10^{10}\ \frac{\mathrm{N}}{\mathrm{m}^2}, c_{13}^0 = 7.78 \cdot 10^{10}\ \frac{\mathrm{N}}{\mathrm{m}^2},$$
$$c_{33}^0 = 11.5 \cdot 10^{10}\ \frac{\mathrm{N}}{\mathrm{m}^2}, c_{55}^0 = 2.56 \cdot 10^{10}\ \frac{\mathrm{N}}{\mathrm{m}^2}, e_{13}^0 = -5.2\ \frac{\mathrm{C}}{\mathrm{m}^2}, e_{15}^0 = 12.7\ \frac{\mathrm{C}}{\mathrm{m}^2},$$
$$e_{33}^0 = 15.1\ \frac{\mathrm{C}}{\mathrm{m}^2}, \varepsilon_{11}^0/\varepsilon_0 = 730,\ \varepsilon_{33}^0/\varepsilon_0 = 635.$$

We have the problem of the free vibrations of the plane (flat, planar) layer at $\varepsilon = 0$. In this case, the system of differential equations that describes the marked vibrations will consist of three independent homogeneous differential equations. The two systems of the equations correspond to axisymmetric vibrations and have been studied previously. Here we will consider the shear vibrations in the axis direction 0φ:

$$\tilde{c}_{55} u_\varphi'' + \Omega^2 u_\varphi = 0, \tag{4.77}$$

with boundary conditions

$$\tilde{c}_{55} u_\varphi' = 0. \tag{4.78}$$

The solution of the problems (4.77), (4.79) for the metal layer reads

$$V(n) = \frac{\pi}{2} n \sqrt{\frac{E}{2(1+\nu)\rho_<}} = 0;\ 1.606;\ 3.211;\ \ldots n = 0, 1, 2 \ldots \tag{4.79}$$

The solution of the problems (4.77), (4.79) for the piezoceramic layer reads

$$V(n) = \frac{\pi}{2} n \sqrt{\frac{c_{55}}{\rho_?}} = 0;\ 0.918;\ 1.835;\ \ldots n = 0, 1, 2 \ldots \tag{4.80}$$

By increasing ε, we can trace where corresponding frequencies go.

For the case of nonaxisymmetric vibrations, the three natural frequencies $U(0)$, $V(0)$, and $W(0)$ go out of zero. In order to determinate which natural frequency will be $U(0)$ or $W(0)$, we use the same expression as for axisymmetric vibrations (as in formula (4.29)). And the natural frequency $V(0)$ is easily specified because it is purely elastic.

In Fig. 4.12, the dependence of the first six natural frequencies of the nonsymmetric vibrations on the geometric parameter ε (the value of the spherical function is $n = 5$) is shown. The frequencies are presented for the free nonaxisymmetric vibration frequencies for laminated and homogeneous piezoceramic spheres. In Fig. 4.12, the solid lines correspond to the frequencies of the vibrations of the inhomogeneous layer piezoceramic sphere and the dashed lines correspond to the free vibration frequency of the homogeneous sphere of the piezoceramic materials of similar geometry.

In Figs. 4.13 and 4.14, the dependence of the first six natural frequencies of the nonaxisymmetric vibrations on the geometric parameter ε (the value of the spherical

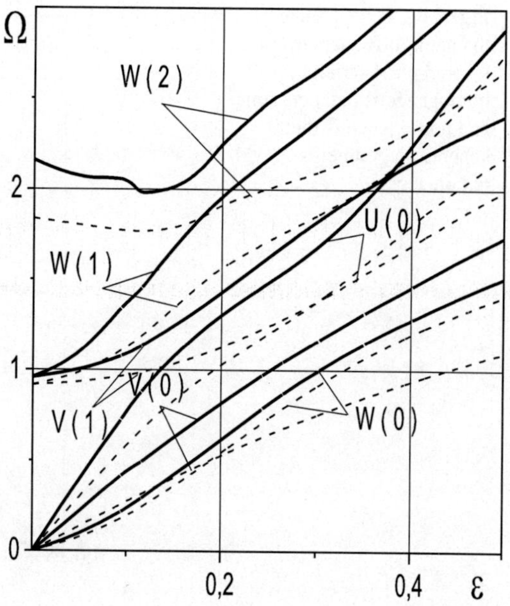

Fig. 4.12 Free nonsymmetrical vibration frequencies for a laminated and homogeneous piezoceramic sphere

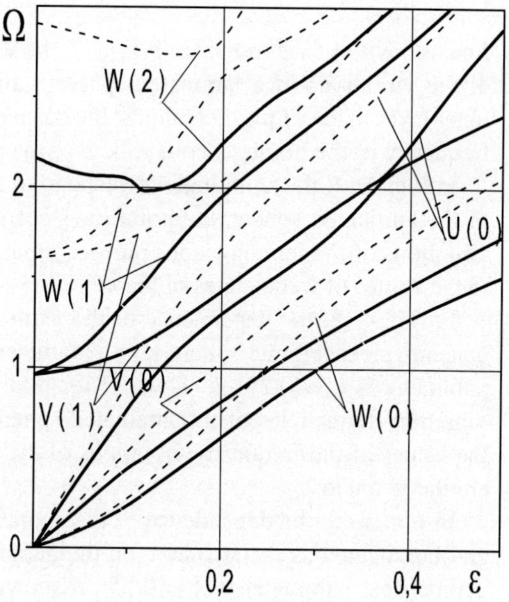

Fig. 4.13 Free nonsymmetrical vibration frequencies for a laminated and homogeneous metal sphere

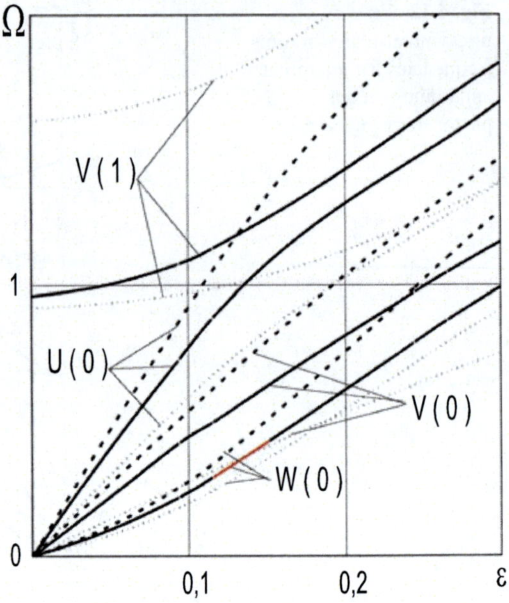

Fig. 4.14 Dependence of the natural frequencies of a layered piezoceramic, homogeneous piezoceramic, and homogeneous metal spheres on geometric parameter

function is $n = 5$) is shown. In Fig. 4.13, the solid lines correspond to the frequencies of the vibrations of a laminated sphere with outer layers made of metal and the inner layer made of piezoceramics; the dashed lines correspond to the free vibration frequency of the homogeneous sphere of the metal materials of similar geometry.

In Fig. 4.14, the solid lines correspond to the values of the vibration frequencies of the laminated sphere, the point lines correspond to the values of the frequency vibrations of the homogeneous piezoceramic sphere, and the dash lines correspond to the values of frequencies of the vibration of the homogeneous metal sphere.

In Fig. 4.15, the dependence of the natural frequencies of a layered and homogeneous piezoceramic sphere on a parameter of spherical function n (geometrical parameter is $\varepsilon = 0.25$) is shown. The solid lines correspond to the values of the vibration frequencies of the laminated sphere, and the dashed lines correspond to the values of the frequency vibrations of the homogeneous piezoceramic sphere of similar geometry.

In Fig. 4.16, the dependence of the natural frequencies of layered piezoceramic and homogeneous metal sphere on the parameter of the spherical functions n (the geometrical parameter is $\varepsilon = 0.25$) is shown. Here, solid lines correspond to the values of the vibration frequencies of the laminated sphere, and the dashed lines correspond to the values of the frequency vibrations of the homogeneous metal sphere of similar geometry.

In Fig. 4.17, the dependence of the natural frequencies of a layered piezoceramic, homogeneous piezoceramic, and homogeneous metal spheres on a parameter of spherical function n (the geometrical parameter is $\varepsilon = 0.25$) is shown. Here, the solid lines correspond to the values of the vibrations of the frequencies of the laminated sphere, the dashed lines correspond to the values of the frequency vibrations

Fig. 4.15 Dependence of the natural frequencies of layered and homogeneous piezoceramic sphere on a parameter of spherical function

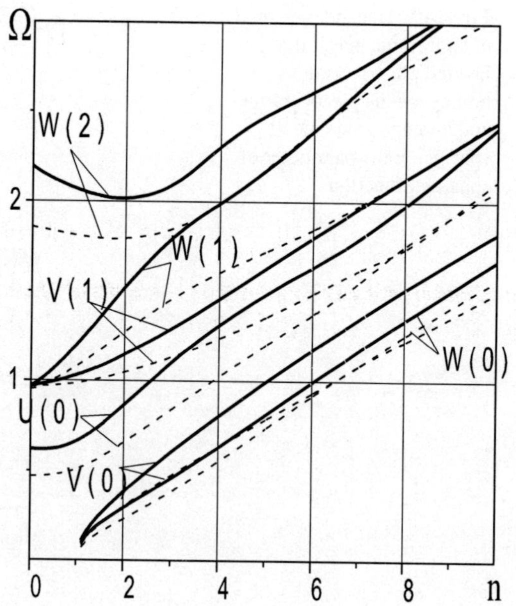

Fig. 4.16 Dependence of first frequencies of free vibrations of a laminated piezoceramic and metallic sphere on the parameter of spherical functions

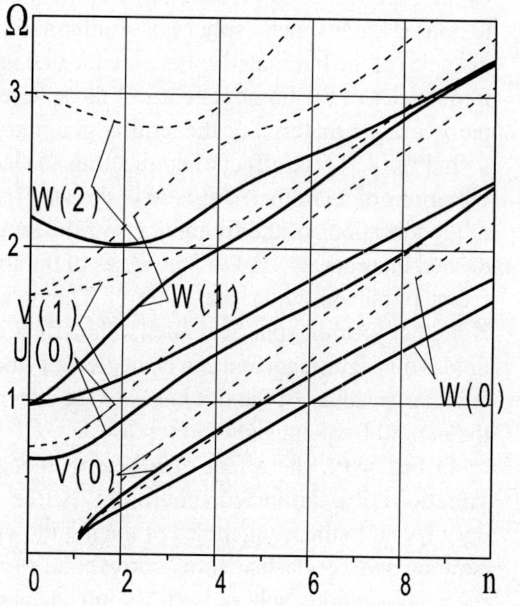

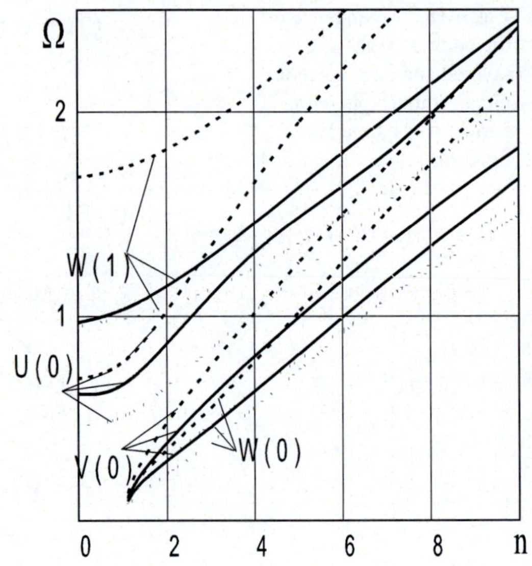

Fig. 4.17 Dependence of natural frequencies of a layered piezoceramic, homogeneous piezoceramic, and homogeneous metal spheres on the parameter of spherical functions

of the homogeneous piezoceramic sphere, and the stroke dashed lines correspond to homogeneous metal sphere of similar geometry. Therefore, the free vibration frequencies of the laminated sphere are located in the interval between the free vibration frequencies of the homogeneous of the sphere made of the metal material and of the piezoceramic material of the similar geometry.

In Fig. 4.18, the effect of connecting an electric field on the frequencies of natural vibrations of a laminated sphere is shown. Here, solid lines correspond to the values of the vibrations of the frequencies of the laminated sphere with the piezoeffect, the dashed lines correspond to the values of the frequencies disregarding the piezoeffect (value of piezomodules are $e_{ij} = 0$). The parameter of the spherical function is n=5. Note that for thin spherical shells, the effect of the bound (coupling) of the electric field is not significant and can be neglected when determining the natural frequencies. With large values of the thickness of the spherical structure, the error in determining the natural frequencies will increase.

In Fig. 4.19, the effect of a heterogeneous structure on frequencies of natural vibrations of a laminated sphere (on its free vibrations) is shown. Here, solid lines correspond to the frequencies of the natural vibrations of the inhomogeneous piezoceramic sphere; dashed lines correspond to the values of the natural vibrations of the homogeneous sphere with the physical–mechanical moduli defined by (4.39). The parameter of the spherical function is n=5. Note that for the first three natural frequencies, their values differ by a very small value. For higher eigenvalues, even in the case of a thin-walled sphere structure, neglecting the heterogeneity factor can lead to significant errors in determining the eigenvalues of the sphere. The first frequency of the natural vibrations in determining neglect of the heterogeneity factor can lead to significant errors.

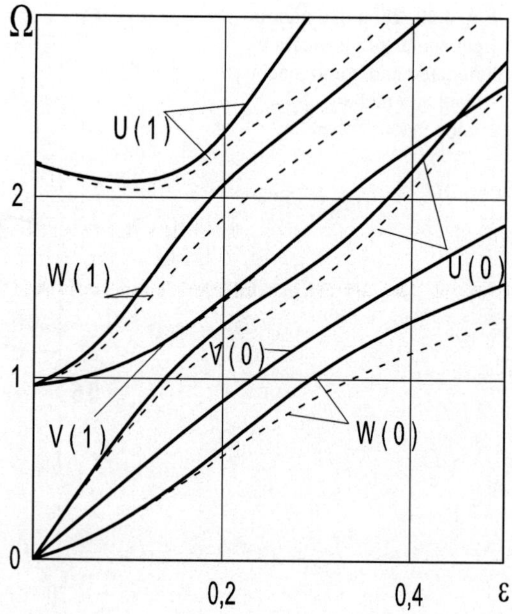

Fig. 4.18 Effect of connected electric field on frequencies of natural vibrations of laminated sphere

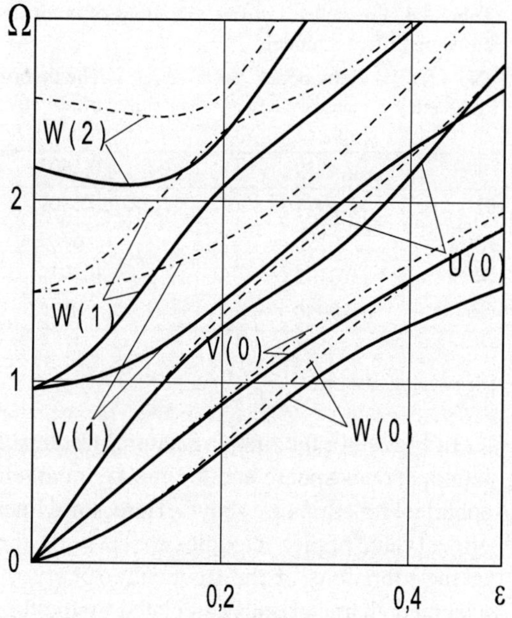

Fig. 4.19 Effect of heterogeneous structure of piezoceramic sphere on its free vibrations

Fig. 4.20 First five natural frequencies of piezoceramic laminated and homogeneous sphere and metal homogeneous sphere

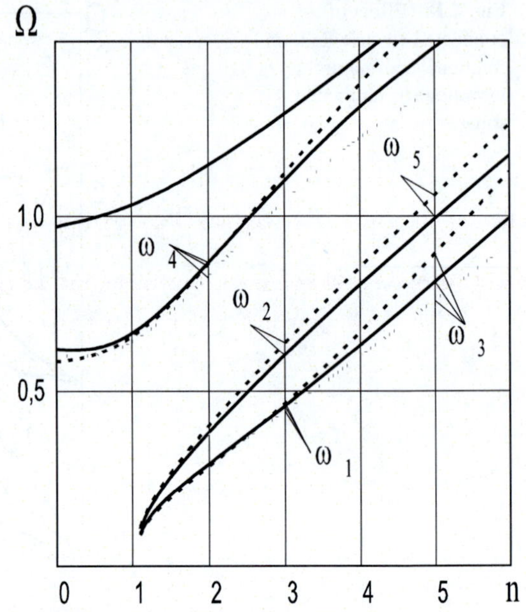

Table 4.4 Frequencies of free vibrations of laminated piezoceramic sphere obtained (calculated) based on different models

No. of frequency	Inhomogeneous sphere	*n*	The theory of effective moduli	Relative error, %	Without piezoeffect	Relative error,
1	0.4559	3	0.4605	1.0	0.4430	2.8
2	0.6074	1	0.6409	5.5	0.6074	0.0
3	0.6796	5	0.6707	1.3	0.6551	3.6
4	0.8133	3	0.8919	9.7	0.7619	6.3
5	0.9950	1	1.0620	6.7	0.9950	0.0

In Fig. 4.20, the first five natural frequencies of the piezoceramic laminated and homogeneous sphere and the metal homogeneous sphere for the odd values of the spherical functions are shown. Here, solid lines correspond to disregarding the piezoeffect (value of piezomodules are $e_{ij} = 0$), the dashed lines correspond to the values of the vibrations of the frequency sphere made of a homogeneous material with averaged characteristics calculated by the theory of effective moduli.

The values of the first five frequencies of the free vibrations of the inhomogeneous piezoceramic sphere are compared with approximate values of the natural frequencies of the sphere made of a homogeneous material with averaged characteristics calculated by the theory of effective moduli and the values of the free vibration frequencies of the sphere disregarding the piezoeffect (the values of piezomoduli are $e_{ij} = 0$) (Table 4.4). Table 4.4 also indicates the index value of the spherical function.

4.4.4 *Analysis of the Calculation Results of the Free Frequencies of the Nonaxisymmetric Vibrations of a Continuously Inhomogeneous Piezoceramic Sphere Made of Piezoceramic Gradient Material*

Let us consider the case of the piezoceramic continuously heterogeneous sphere made of piezoceramic gradient material. The properties of the material are assumed to vary exponentially across the thickness, i.e., $c_{ij} = c_{ij}^{o} e^{sr}, e_{ij} = e_{ij}^{o} e^{sr}, \varepsilon_{ij} = \varepsilon_{ij}^{o} e^{sr}$. The parameter s is determined experimentally and its values are usually in the interval $-2 \leq s \leq 2$. The parameters designated by an index 0 refer to the outside surface of the sphere and have the following values:

$$c_{11}^{0} = 13.9 \cdot 10^{10}\,\frac{\text{N}}{\text{m}^2}, c_{12}^{0} = 7.43 \cdot 10^{10}\,\frac{\text{N}}{\text{m}^2}, c_{13}^{0} = 7.78 \cdot 10^{10}\,\frac{\text{N}}{\text{m}^2},$$
$$c_{33}^{0} = 11.5 \cdot 10^{10}\,\frac{\text{N}}{\text{m}^2}, c_{55}^{0} = 2.56 \cdot 10^{10}\,\frac{\text{N}}{\text{m}^2}, e_{13}^{0} = -5.2\,\frac{\text{C}}{\text{m}^2}, e_{15}^{0} = 12.7\,\frac{\text{C}}{\text{m}^2},$$
$$e_{33}^{0} = 15.1\,\frac{\text{C}}{\text{m}^2}, \varepsilon_{11}^{0}/\varepsilon_0 = 730,\;\; \varepsilon_{33}^{0}/\varepsilon_0 = 635.$$

In Fig. 4.21, the dependence of the first six free frequencies of continuously inhomogeneous piezoceramic sphere made of piezoceramic gradient material and a homogeneous piezoceramic sphere on the geometrical parameter $\varepsilon (n = 5)$ is shown.

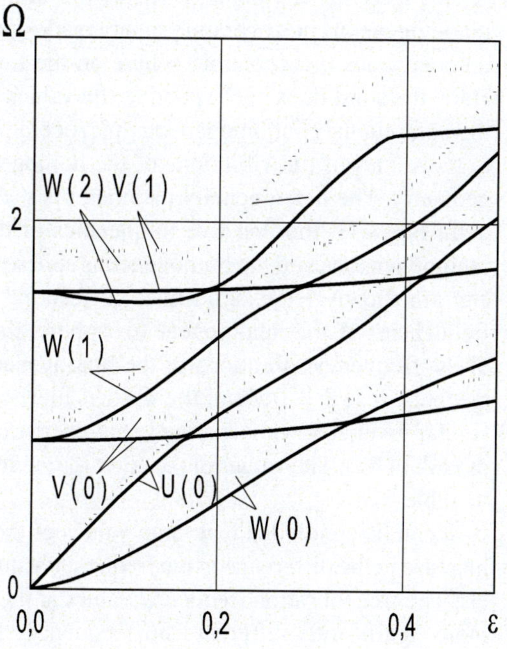

Fig. 4.21 Dependence of free vibration frequencies of a piezoceramic homogeneous and inhomogeneous sphere made of FGPM on geometric parameter

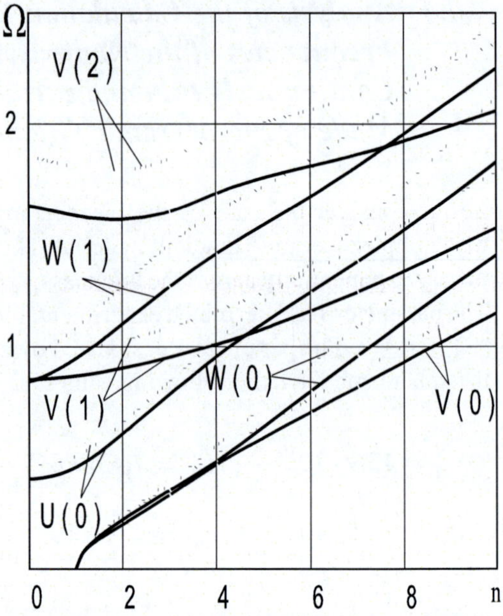

Fig. 4.22 Dependence of free vibration frequencies of piezoceramic homogeneous and inhomogeneous sphere made of FGPM on the Legendre polynomial index

Here, the solid lines correspond to the values of the vibrations of the frequencies of the continuously inhomogeneous piezoceramic sphere, and point lines correspond to the values of the vibrations of the homogeneous piezoceramic sphere of similar geometry. The heterogeneity parameter is $s = 0.5$.

In Fig. 4.22, the dependence of the first six free frequencies of continuously inhomogeneous piezoceramic sphere made of the piezoceramic gradient material and homogeneous piezoceramic sphere on the Legendre polynomial index n is shown. Here, the solid lines correspond to the values of the vibrations of the frequencies of the continuously inhomogeneous piezoceramic sphere, and point lines correspond to the values of the vibrations of the homogeneous piezoceramic sphere of similar geometry. The heterogeneity parameter is $s = 0.5$.

In Fig 4.23, the first five frequencies of the nonaxisymmetric vibrations of the inhomogeneous and the homogeneous sphere for odd values of the Legendre parameter are shown. Here, solid lines correspond to the values of the vibrations of the frequencies of the inhomogeneous sphere, the point lines correspond to the values of the frequency vibrations of the homogeneous piezoceramic sphere disregarding piezoeffect $\left(e_{ij} = 0\right)$, and the dashed lines correspond to homogeneous sphere of similar geometry with averaged characteristics calculated by the theory of effective moduli (4.40). The values of the eigenfrequency presented in Fig. 4.23 are also shown in Table 4.5.

It can be concluded that large values of the Legendre polynomial index show an increase in the difference in the frequency values determined based on different models. The free vibration frequency values at the neglecting the piezoelectric effect are more significantly different from frequency values for the inhomogeneous sphere in

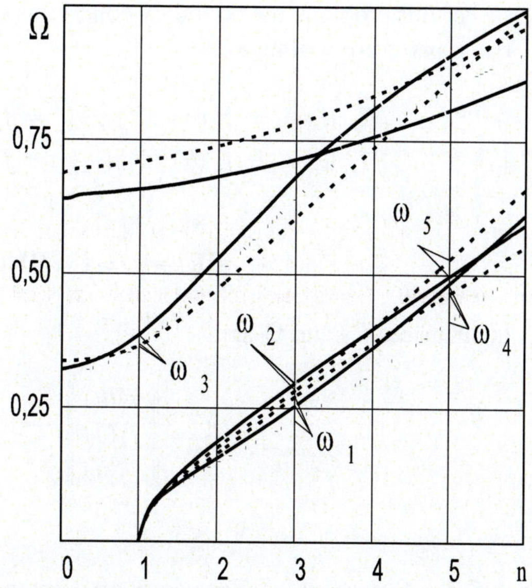

Fig. 4.23 Frequencies of free vibrations of inhomogeneous and homogeneous spheres based on different models

Table 4.5 Free vibration frequencies of piezoceramic homogeneous and inhomogeneous sphere made of FGPM calculation based on different models

No. of frequency	Inhomogeneous sphere	n	The theory of effective moduli	Relative error, %	Without piezoeffect	Relative error, %
1	0.2513	3	0.2728	8.6	0.2260	10.1
2	0.2982	3	0.2805	5.9	0.2982	0.0
3	0.3867	1	0.3664	5.2	0.3662	5.3
4	0.4838	5	0.4650	3.9	0.4126	14.7
5	0.4975	5	0.5291	6.4	0.4975	0.0

comparison with frequencies of the homogenous sphere defined based on calculated by the theory of effective moduli.

4.5 Forced Vibrations of Inhomogeneous Piezoceramic Sphere

In this section, we will investigate the problem of forced vibrations of the second class of a piezoceramic sphere made of radially polarized continuously inhomogeneous material within the framework of three-dimensional electroelasticity theory. In this case, the resolving vector has be form (4.23). A system of differential equations that describes the boundary value-problem (4.27) will be heterogeneous.

For this problem, the boundary conditions on the lateral surfaces of the sphere (at $r = const.$) are written as

$$T_{rr}|_{r=R_-} = f_1(\theta),\ T_{r\theta}|_{r=R_-} = f_2(\theta),\ \Phi|_{r=R_-} = f_3(\theta),$$
$$T_{rr}|_{r=R_+} = f_4(\theta),\ T_{r\theta}|_{r=R_+} = f_5(\theta),\ \Phi|_{r=R_+} = f_6(\theta)$$

or

$$\mathbf{BR}(-1) = \mathbf{C}_1, \mathbf{BR}(+1) = \mathbf{C}_2; \tag{4.81}$$

the matrices $\mathbf{C}_1$ and $\mathbf{C}_2$are:

$$\mathbf{C}_1 = \left\| \begin{matrix} f_1(\theta) \\ f_2(\theta) \\ f_3(\theta) \end{matrix} \right\|, \mathbf{C}_2 = \left\| \begin{matrix} f_4(\theta) \\ f_5(\theta) \\ f_6(\theta) \end{matrix} \right\|. \tag{4.82}$$

4.5.1 Solution of Problem on Forced Vibrations of Inhomogeneous Piezoceramic Sphere for the Case of Nonsplit Electrodes

At present, functionally gradient piezoelectric materials, which combine the advantages of bimorphs and free-interface materials that have different coefficients of thermal expansion, find ever-widening applications. The attempt to allow for continually varying properties of a material leads to the situation where the material moduli are not constants but functions with respect to one coordinate.

Let us consider a hollow sphere made of an FGPM material composed of PZT-4 piezoceramic and steel with $E = 21 \cdot 10^{10}\ \frac{\mathrm{N}}{\mathrm{m}^2}$, and $\nu = 0.28$. The physical and mechanical characteristics of the material vary as follows in thickness:

$$P(r) = (P_m - P_p)V(r) + P_p, \tag{4.83}$$

where $V(r)$ is the volume fraction of the ceramics, which is described as

$$V(r) = \left(\frac{r - R_0}{2h} + \frac{1}{2}\right)^s, \tag{4.84}$$

where s is the heterogeneity parameter. The inner and outer radii of the sphere are equal to 3 and 5 dimensionless values, respectively.

First, we address a homogeneous material. The functions $f_i(\theta)$ have been chosen as follows:

$$f_1(\theta) = f_4(\theta) = 0,\ f_2(\theta) = f_5(\theta) = 0,\ f_3(\theta) = +V_0,\ f_6(\theta) = -V_0. \tag{4.85}$$

In this case, the following electroelastic radial vibrations are excited:

$$\frac{\partial T_1}{\partial r} + \frac{2T_1 - T_2 + T_2 \cot\theta}{r} + \rho\omega^2 u_1 = 0,\ T_1|_{r=R_0 \pm h} = 0. \tag{4.86}$$

The electroelasticity equations are taken into account using the relations:

$$\frac{\partial D_1}{\partial r} + 2\frac{D_1}{r} = 0 \rightarrow \frac{\partial}{\partial r}\left(r^2 D_1\right) = 0 \rightarrow r^2 D_1 = D_0 = \text{const.} \rightarrow D_1 = \frac{D_0}{r^2}. \tag{4.87}$$

The solution of systems (4.86), (4.87) has been known a long time ago and can be expressed in terms of Bessel functions, provided that the components of the strain tensor under radial vibrations are related to the displacement ($u_1 \equiv u$) by

$$S_1 = \frac{\partial u}{\partial r},\ S_2 = S_3 = \frac{u}{r}. \tag{4.88}$$

By considering the solution (4.87) for D_1 and by substituting the expressions for the associated components of the stress tensor into (4.86), we arrive at the equation for harmonic vibrations:

$$\frac{\mathrm{d}^2 u}{\mathrm{d}r^2} + \frac{2}{r}\frac{\mathrm{d}u}{\mathrm{d}r} + \left[\frac{\rho\omega^2}{c_{33}}r^2 - 2\frac{c_{11} + c_{12} - c_{13}}{c_{33}}\right]\frac{u}{r^2} = -\frac{2e_{31}D_0}{c_{33}r^2}. \tag{4.89}$$

After introducing a new searched function

$$w(r) = \frac{1}{\sqrt{kr}}R(kr), \tag{4.90}$$

equation (4.89) reduces to an inhomogeneous Bessel equation

$$\frac{\mathrm{d}^2 w}{\mathrm{d}r^2} + \frac{1}{r}\frac{\mathrm{d}w}{\mathrm{d}r} + \left[k^2 r^2 - \upsilon^2\right]\frac{w}{r^2} = -\frac{2e_{31}D_0}{c_{33}r^2}, \tag{4.91}$$

where $k = \frac{\omega}{c_r} = \omega\sqrt{\frac{\rho}{c_{33}}}$, $\upsilon^2 = 2\frac{c_{11} + c_{12} - c_{13}}{c_{33}}$, c_r is the sound velocity in the direction of a polarization axis. Having solved the homogeneous equation (4.89), we get

$$w = \frac{1}{\sqrt{kr}}Z_\nu\,(kr)\,, \tag{4.92}$$

where $\nu = \sqrt{\upsilon_2^2 + \frac{1}{4}}$, $Z_\nu = C_1 J_\nu + C_2 N_\nu$, C_1, C_2 are the constants which are determined from the boundary conditions:

$$\left.\left(\alpha\frac{\mathrm{d}w}{\mathrm{d}r}+2\beta\frac{w}{r}=\frac{e_{33}D_0}{r^2}\right)\right|_{r=R_0,\pm h}. \tag{4.93}$$

Substituting (4.92) into (4.93), we arrive at the expressions:

$$C_1=\frac{e_{33}D_0}{R_-^2R_+^2}\frac{R_+^2a_{22}-R_-^2a_{12}}{a_{11}a_{22}-a_{12}a_{21}},\ C_2=\frac{e_{33}D_0}{R_-^2R_+^2}\frac{R_+^2a_{11}-R_-^2a_{21}}{a_{11}a_{22}-a_{12}a_{21}}, \tag{4.94}$$

where

$$a_{11}=\left(2\beta-\frac{\alpha k}{2}+\alpha k\nu\right)J_\nu\left(kR_-\right)-\alpha kR_-J_{\nu+1}\left(kR_-\right),$$
$$a_{12}=\left(2\beta-\frac{\alpha k}{2}+\alpha k\nu\right)N_\nu\left(kR_-\right)-\alpha kR_-N_{\nu+1}\left(kR_-\right),$$
$$a_{21}=\left(2\beta-\frac{\alpha k}{2}+\alpha k\nu\right)J_\nu\left(kR_+\right)-\alpha kR_+J_{\nu+1}\left(kR_+\right),$$
$$a_{22}=\left(2\beta-\frac{\alpha k}{2}+\alpha k\nu\right)N_\nu\left(kR_+\right)-\alpha kR_+N_{\nu+1}\left(kR_+\right).$$

If the determinant is equal to zero, the frequencies of free vibrations are defined by

$$a_{11}a_{22}-a_{12}a_{21}=0. \tag{4.95}$$

Since the electrostatic potential is specified on the outside surfaces in terms of electric induction, we obtain with (4.49)

$$\frac{d\Phi}{dr}=\frac{e_{33}}{\varepsilon_{33}}\frac{\mathrm{d}w}{\mathrm{d}r}+\frac{2e_{13}}{\varepsilon_{33}}\frac{w}{r}-\frac{D_0}{\varepsilon_{33}r^2}. \tag{4.96}$$

Considering (4.96), equation (4.91) turns into the following expression:

$$\frac{\mathrm{d}^2w}{\mathrm{d}r^2}+\frac{2}{r}\frac{\mathrm{d}w}{\mathrm{d}r}+\left[k_1^2r^2-\upsilon_1^2\right]\frac{w}{r^2}=-\frac{2e_{31}D_0}{\alpha r^3}, \tag{4.97}$$

where $k_1=\frac{\omega}{k_{33}^2}=\omega\sqrt{\frac{\rho\varepsilon_{33}}{\alpha}}$, $k_{33}^2=c_{33}+\frac{e_{33}^2}{\varepsilon_{33}}$ is the static coefficient of the electromechanical relation, $\upsilon_1^2=\frac{2}{\alpha}\left[\left(c_{11}+c_{12}\right)\varepsilon_{33}+2e_{33}^2-\beta\alpha\right]$. If $\left(e_{ij}=0\right)$, then $k=k_1$, $\upsilon=\upsilon_1$. Having solved the homogeneous equation (4.97), we arrive at the expression:

$$w_0=A_1j_{n+\frac{1}{2}}\left(k_1r\right)+A_2y_{n+\frac{1}{2}}\left(k_1r\right), \tag{4.98}$$

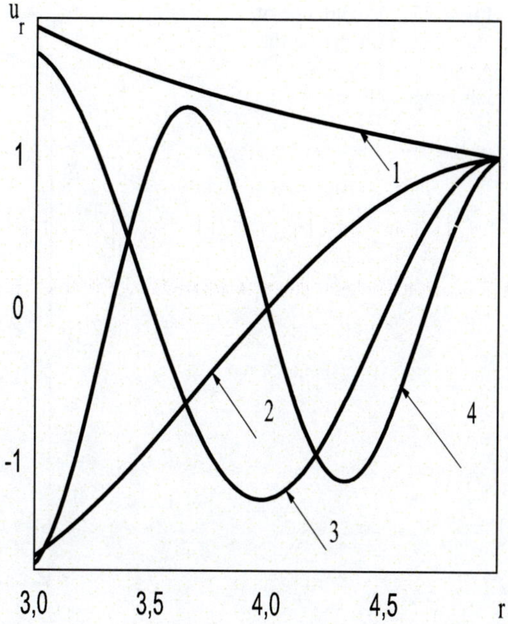

Fig. 4.24 Distribution of radial displacements across the thickness

where $n+\frac{1}{2}=\sqrt{\upsilon_1^2+\frac{1}{4}}$, $j_{n+\frac{1}{2}}(k_1 r)$, $y_{n+\frac{1}{2}}(k_1 r)$ are spherical Bessel functions of the first and second kind, and A_1, A_2 are constants. The partial solution of the inhomogeneous equation (4.97) is found by the Lagrange indefinite multiplier method. However, the closed solution in special functions is difficult for quantitative analysis and is unexplored nowadays. Besides, it is impossible to obtain an analytical solution for a functionally gradient material. The values k, k_1 and υ, υ_1, being variable, are functions of the radial coordinate *r*. For this reason, equations (4.91) and (4.97) cease to present Bessel equations. The natural frequencies and amplitudes of vibrations can be defined with the method proposed above.

The values of the first dimensionless frequencies of sound vibrations with (4.95) are equal to 0.4685, 2.0465, 4.5559, and 6.7269.

Note that the values of frequencies of the sound vibrations in a flat layer occur to be very close to the above values (besides the first frequency). These values follow from the formulas the values of frequencies of the sound vibrations in a flat layer occur to be very close to the above values (besides the first frequency). These values follow from the formulas (4.35) and (4.36).

Figure 4.24 shows the distribution of radial displacements across the thickness. As can be seen, the number of roots of the displacement function increases with the overtone number.

Let us consider vibrations of a hollow sphere with the above geometrical characteristics made of a functionally gradient piezoceramic material. Expression (4.83) represents the general formula for the physical–chemical characteristics of the material, and P_p and P_m are characteristics of the ceramic and the metal, respectively. The

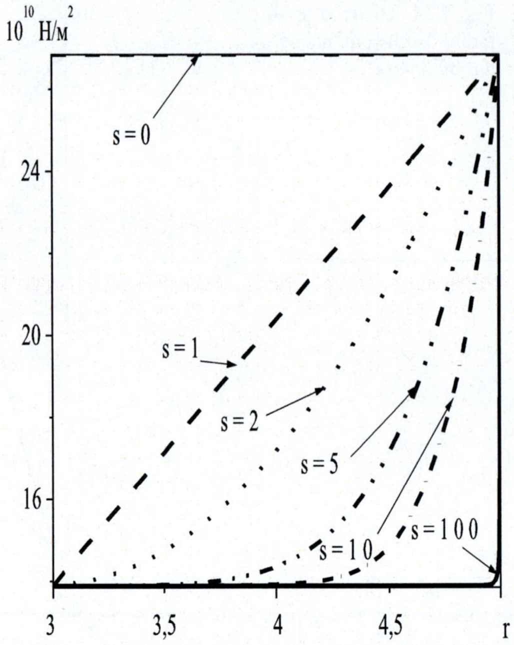

Fig. 4.25 Dependence of elastic modulus c_{11} on the parameter of inhomogeneity s

exponent of the volume fraction of the ceramic appearing in (4.84) can vary within the limits of $0<s<1000$. If $s = 0$, the structure is completely metallic; if $s = \infty$, the structure is completely piezoceramic (see Fig. 4.25).

The Legendre polynomial index for the radial vibrations is equal to zero ($k = 0$). Taking into account that $P_0(\cos\theta) \equiv 1 = \text{const.}$, from (4.25), it follows: $T_6(r, \theta) = u_2(r, \theta) \equiv 0$.

Figures 4.26 and 4.27 show the across-the-thickness distribution of dimensionless amplitude of radial vibrations for the first natural frequency. The solid lines in Fig. 4.26 refer to amplitudes of the radial stresses T_1, and dashed ones refer to the radial displacements u. The amplitudes in the case of the homogeneous material (PZT-4 piezoceramic) are presented by thick lines. Solid lines in Fig. 4.27 represent amplitudes of the electrostatic potential Φ, dashed ones of the component of the electric induction D_1 normal to the surface.

As is seen from the figures, material inhomogeneity influences to a greater extent the electrostatic potential and radial stresses.

The across-the-thickness distribution of dimensionless amplitudes of radial vibrations for the second natural frequency is shown in Figs. 4.28 and 4.29. In this case, the inhomogeneity influences to a great extent the electrostatic potential and radial displacements.

Figures 4.30 and 4.31 show the across-the-thickness distribution of the dimensionless amplitudes of radial vibrations for the third natural frequency. In this case, the

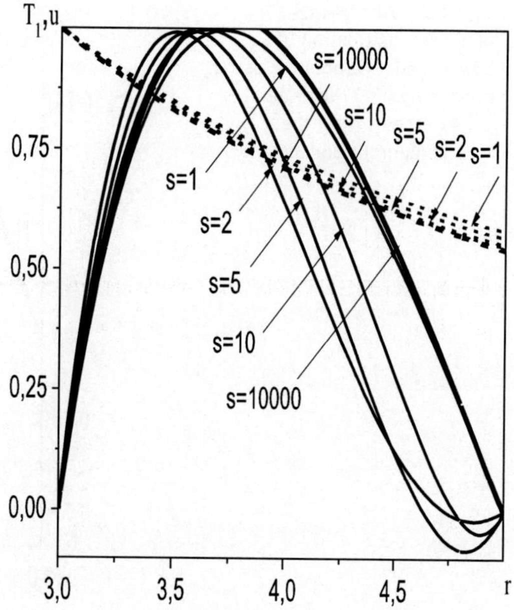

Fig. 4.26 Across-the-thickness distribution of dimensionless amplitude of radial stresses T_1 and radial displacements u for the first natural frequency

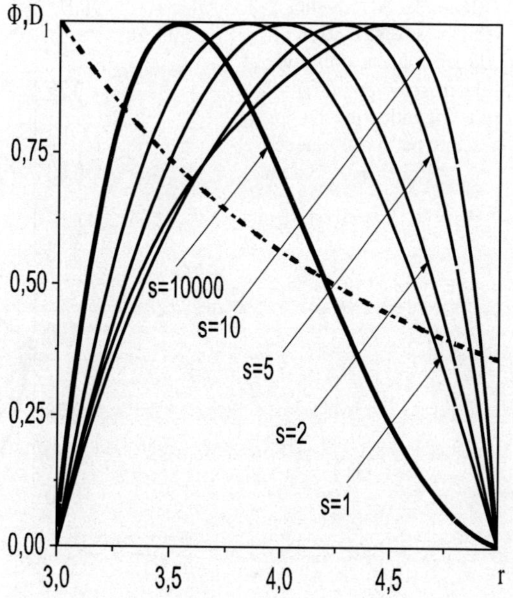

Fig. 4.27 Across-the-thickness distribution of dimensionless amplitude of electrostatic potential Φ and electric induction D_1 for the first natural frequency

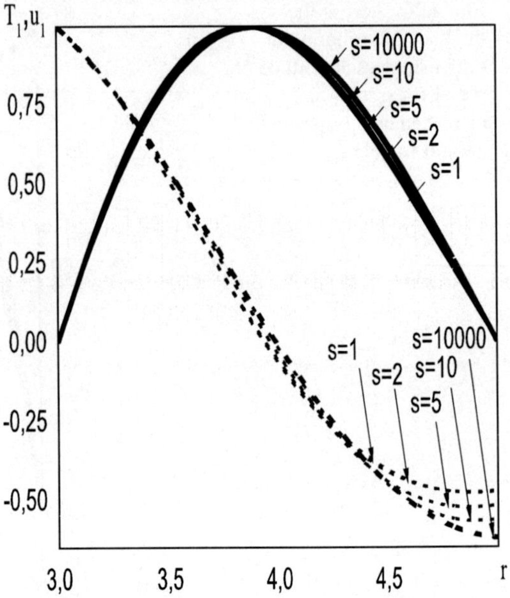

Fig. 4.28 Across-the-thickness distribution of dimensionless amplitude of radial stresses T_1 and radial displacements u for the second natural frequency

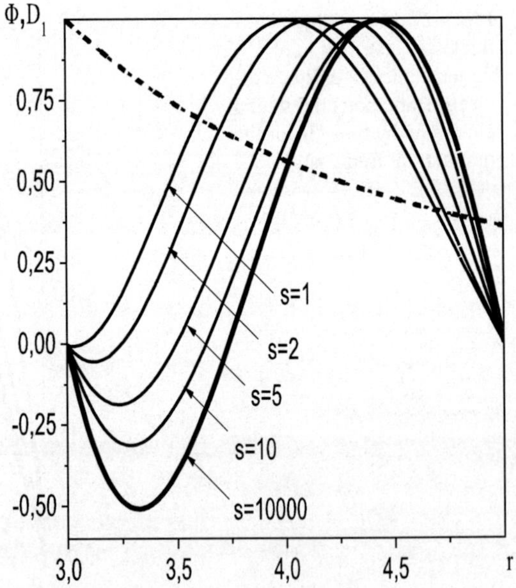

Fig. 4.29 Across-the-thickness distribution of dimensionless amplitude of electrostatic potential Φ and electric induction D_1 for the second natural frequency

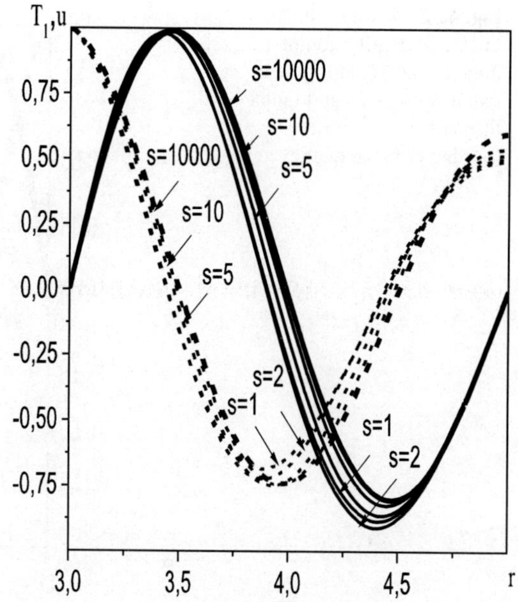

Fig. 4.30 Across-the-thickness distribution of dimensionless amplitude of radial stresses T_1 and radial displacements u for the third natural frequency

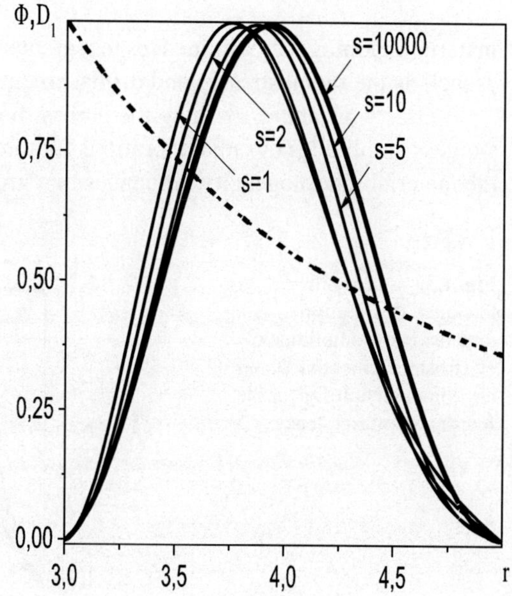

Fig. 4.31 Across-the-thickness distribution of dimensionless amplitude of electrostatic potential Φ and electric induction D_1 for the third natural frequency

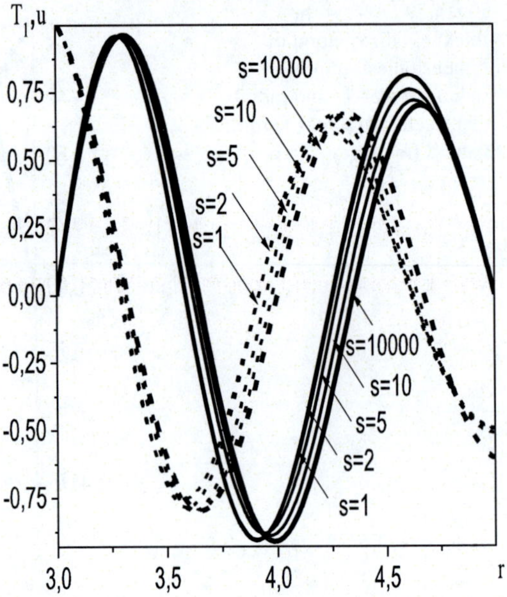

Fig. 4.32 Across-the-thickness distribution of dimensionless amplitude of radial stresses T_1 and radial displacements u for the fourth natural frequency

material inhomogeneity influences to a greater extent both the electrostatic potential as well as the radial stresses and displacements.

Figures 4.32 and 4.33 show the across-the-thickness distribution of the dimensionless amplitudes of radial vibrations for the fourth natural frequency. In this case, the material inhomogeneity influences to a greater extent the electrostatic potential.

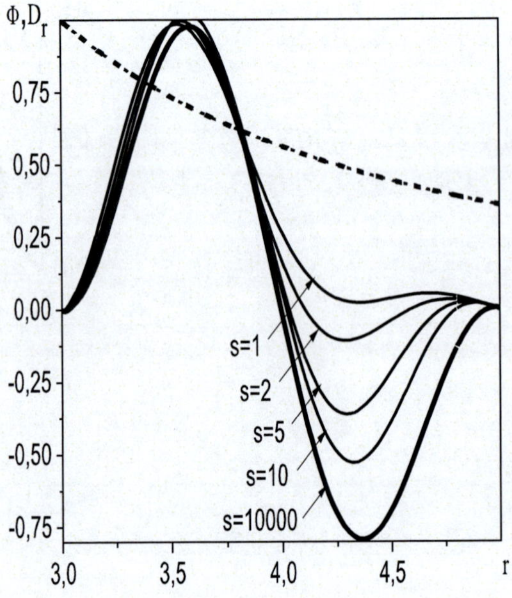

Fig. 4.33 Across-the-thickness distribution of dimensionless amplitude of electrostatic potential Φ and electric induction D_1 for the fourth natural frequency

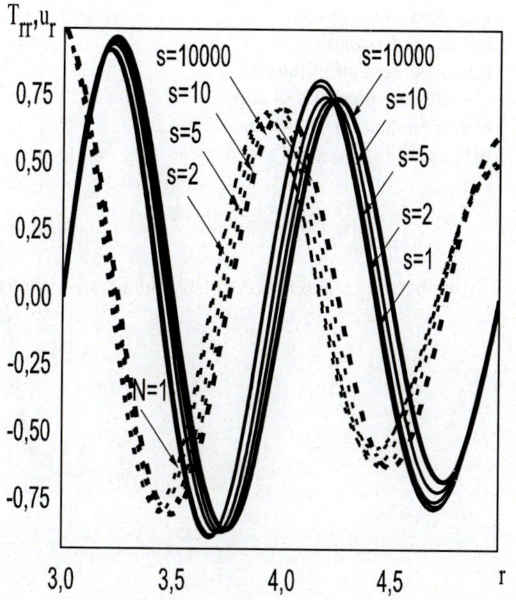

Fig. 4.34 Across-the-thickness distribution of dimensionless amplitude of radial stresses T_1 and radial displacements u for the fifth natural frequency

Figures 4.34 and 4.35 show the across-the-thickness distribution of the dimensionless amplitudes of radial vibrations for the fifth natural frequency. In this case, the material inhomogeneity influences to a greater extent the electrostatic potential. Note that the number of zeros of amplitude functions increases with the number of the overtone.

4.5.2 *Solution of Problem on Forced Vibrations of Non-Native Piezoceramic Sphere for the Case of Split Electrodes*

Let us consider a sphere with a surface covered by electrodes which are split along the line $\theta = {}^{\pi}/_{2}$. The electrodes are supplied with a harmonic potential $V_0 e^{i\omega t}$ with a phase shift between upper and lower half-spheres $\theta = {}^{\pi}/_{2}$. In this case, the functions Φ^+ and Φ^- take the following values:

$$\text{at } 0 < \theta < \frac{\pi}{2}\ \Phi^+ = -V_0 \sum_{j=0}^{\infty} \alpha_j P_j(\cos\theta),\ \Phi^- = +V_0 \sum_{j=0}^{\infty} \alpha_j P_j(\cos\theta),$$

$$\text{at } 0 < \theta < \frac{\pi}{2}\ \Phi^+ = +V_0 \sum_{j=0}^{\infty} \alpha_j P_j(\cos\theta),\ \Phi^- = -V_0 \sum_{j=0}^{\infty} \alpha_j P_j(\cos\theta),$$

$$\text{at } \theta = \frac{\pi}{2}\ \Phi^+ = \Phi^- = 0, \tag{4.99}$$

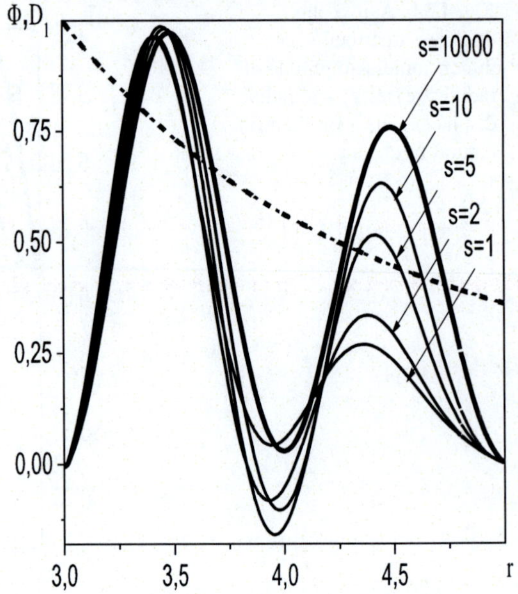

Fig. 4.35 Across-the-thickness distribution of dimensionless amplitude of electrostatic potential Φ and electric induction D_1 for the fifth natural frequency

where α_j are the expansion coefficients of the step-wise Heaviside function in Legendre's polynomials:

$$\alpha_j = \frac{2j+1}{2} \int_{-1}^{1} H(y)P(y)dy. \tag{4.100}$$

Here $y = \cos\theta$, $H(y) = \begin{cases} -1, & -1 < y < 0 \\ 1, & 0 < y < 1 \\ 0, & y = 0 \end{cases}$

As the analysis of formula (4.100) shows, the coefficients with different values of the index of Legendre's polynomial are equal to zero. Thus, the summing in series (4.99) is carried out over the odd values of the index of Legendre's polynomial.

As is known from numerical analysis, in expanding a discontinuous function into Fourier series (Legendre's polynomial series may be considered as a generalized Fourier series), the so-called "Gibbs oscillations" appear at break points. Let us demonstrate this situation on the example with the first nine series terms being kept (see Fig. 4.38).

Here, the red line shows values of the function Φ with the first nine terms of the series (4.60) being kept. The situation can be improved using either the Fejér method (blue line) or Lanczos' multipliers (green line). In our opinion, employment of the Lanczos multipliers is preferable because the magnitude of Φ in this case can be found more quickly.

In the case of a homogeneous sphere, the solution obtained has been compared with the results defined by expanding the solution of the components of a governing

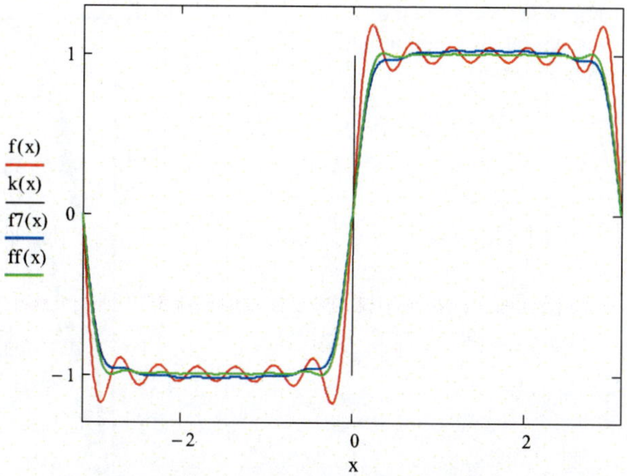

Fig. 4.36 Across-the-thickness distribution of dimensionless amplitude of radial displacements u_1 and meridional displacements u_2 for the second natural frequency

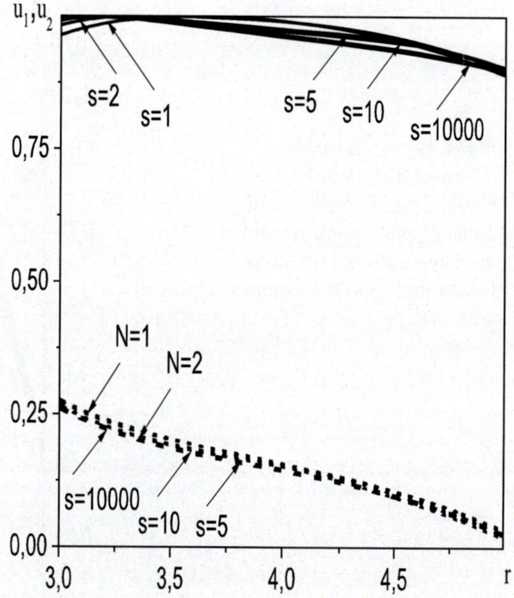

Fig. 4.37 Across-the-thickness distribution of dimensionless amplitude of normal stresses T_1 and tangential stresses T_6 for the second natural frequency

vector in a power series in the radial coordinate [24]. The frequencies are given in Table 4.6.

Let us consider vibrations of a hollow sphere made of a functionally gradient piezoceramic material. Figures 4.39, 4.40 and 4.41 show how the dimensionless amplitudes of axisymmetric vibrations for the first natural frequency ($n = 3$) are distributed across the thickness. The solid lines in Fig. 4.39 refer to the amplitudes of the radial displacements u_1, while the dashed lines refer to the meridional displacements u_2. The thick lines refer to the amplitudes in the case of a sphere made of homogeneous material PZT-4. The solid lines in Fig. 4.40 refer to the amplitudes of the normal stress T_1 while the dashed lines refer to the tangential stress T_6. The solid

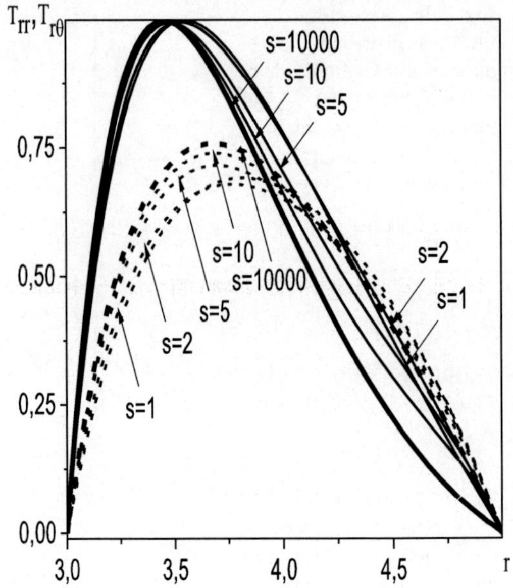

Fig. 4.38 Expanding a discontinuous function into Fourier series

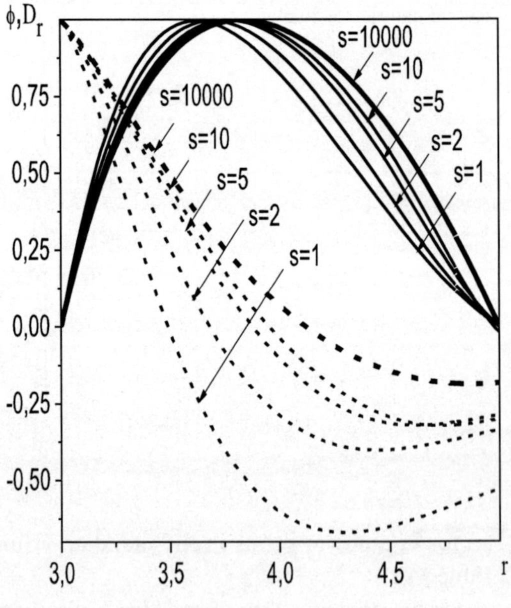

Fig. 4.39 Across-the-thickness distribution of dimensionless amplitude of radial displacements u_1 and meridional displacements u_2 for the first natural frequency

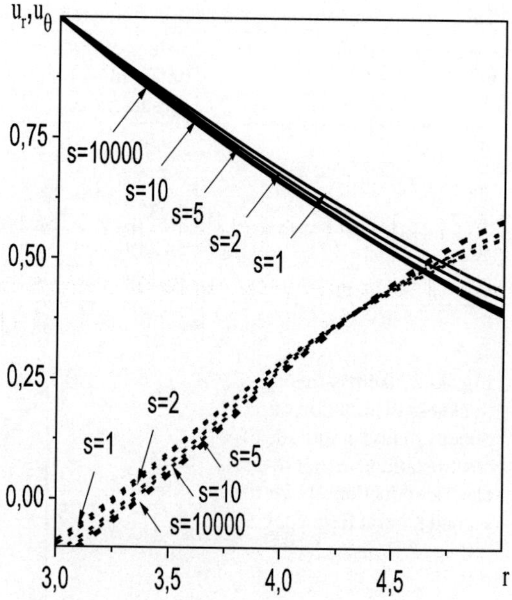

Fig. 4.40 Across-the-thickness distribution of dimensionless amplitude of normal stresses T_1 and tangential stresses T_6 for the first natural frequency

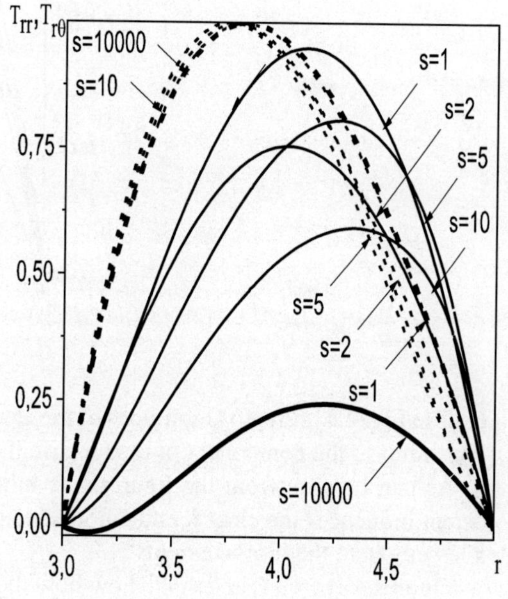

Fig. 4.41 Across-the-thickness distribution of dimensionless amplitude of electrostatic potential Φ and electric induction D_1 for the first natural frequency

Table 4.6 Comparison of the natural frequencies of homogeneous sphere vibration

Number of frequency	Values of the natural frequency obtained by power series in present paper	Values of the natural frequency obtained by power series in the radial coordinate [24]
1	0.4768	0.4770
2	0.7200	0.720
3	0.9325	0.932
4	1.2249	1.225
5	1.4418	1.442

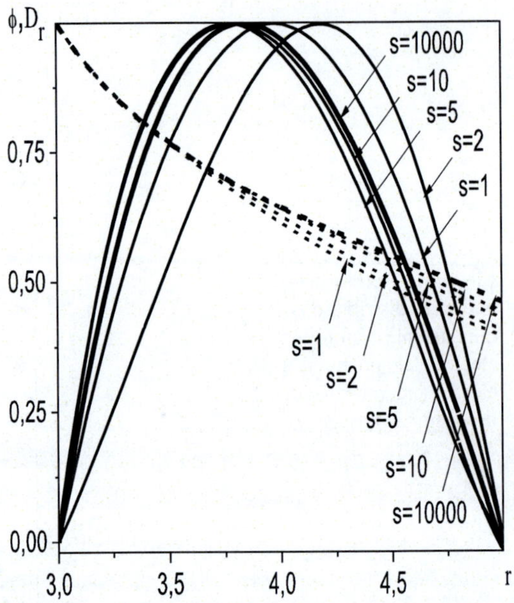

Fig. 4.42 Across-the-thickness distribution of dimensionless amplitude of electrostatic potential Φ and electric induction D_1 for the second natural frequency

lines in Fig. 4.41 refer to amplitudes of the electrostatic potential Φ, while the dashed lines refer to the component of the electric induction D_1 normal to the surface.

As can be seen from the figures, the inhomogeneity of a material to a greater extent influences the electric induction, electrostatic potential, and stresses while to a lesser extent the displacements.

Figures 4.36, 4.37, and 4.42 show how the dimensionless amplitudes of axisymmetric vibrations for the second natural frequency ($n = 1$) are distributed across the thickness. It is shown that the inhomogeneity of a material influences the radial stresses and electrostatic potential to a greater extent.

Figures 4.43, 4.44, and 4.45 show how the dimensionless amplitudes of axisymmetric vibrations for the third natural frequency ($n = 5$) are distributed across the thickness. Evidently, the inhomogeneity of a material influences the electric induction, electrostatic potential, and the radial stresses to a greater extent.

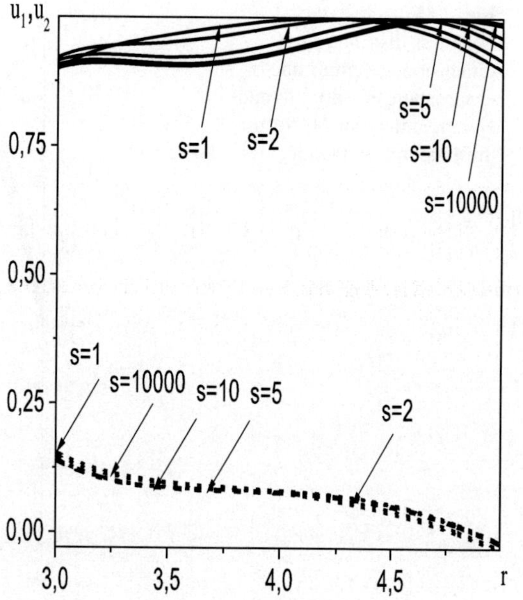

Fig. 4.43 Across-the-thickness distribution of dimensionless amplitude of radial displacements u_1 and meridional displacements u_2 for the third natural frequency

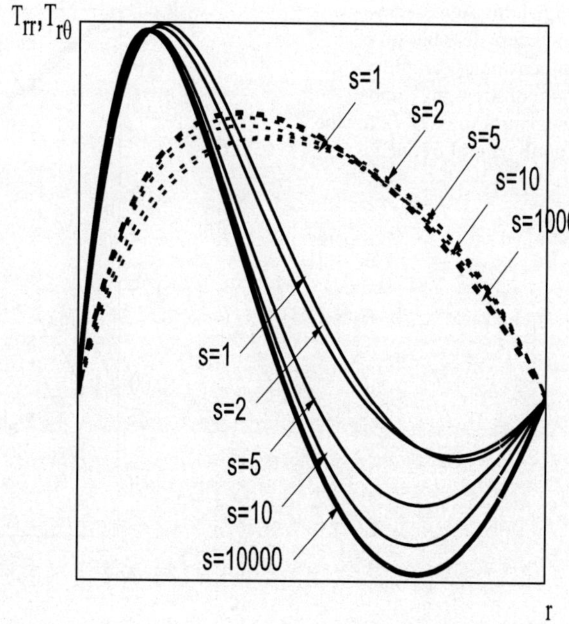

Fig. 4.44 Across-the-thickness distribution of dimensionless amplitude of normal stresses T_1 and tangential stresses T_6 for the third natural frequency

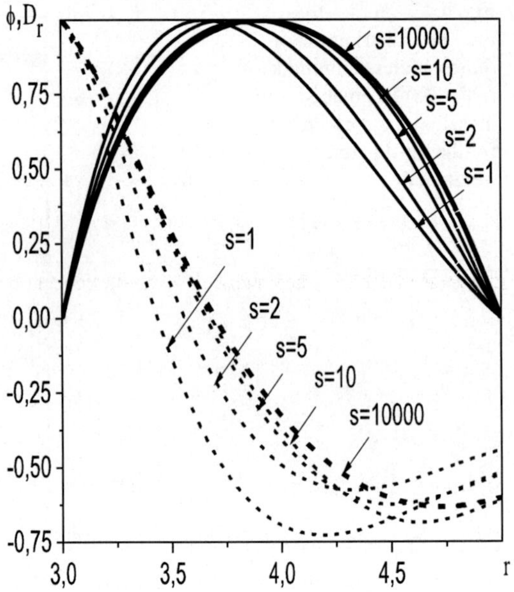

Fig. 4.45 Across-the-thickness distribution of dimensionless amplitude of electrostatic potential Φ and electric induction D_1 for the third natural frequency

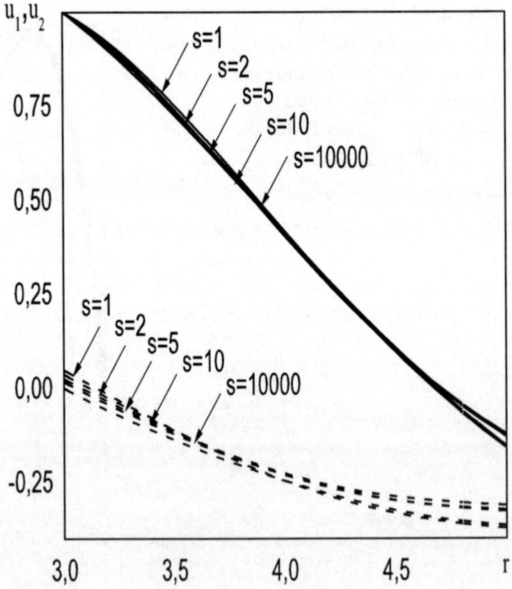

Fig. 4.46 Across-the-thickness distribution of dimensionless amplitude of normal stresses T_1 and tangential stresses T_6 for the fourth natural frequency

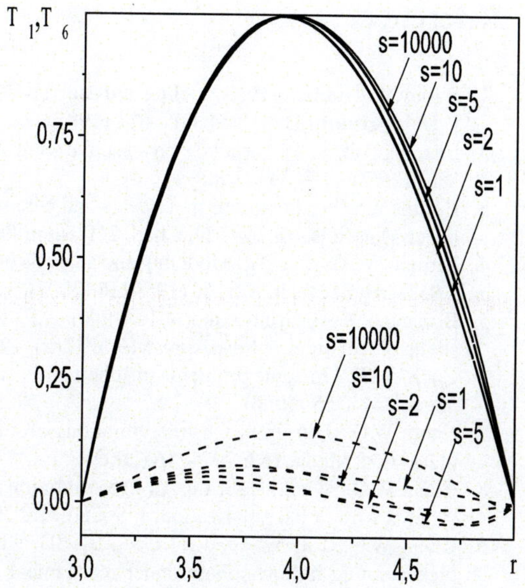

Fig. 4.47 Across-the-thickness distribution of dimensionless amplitude of electrostatic potential Φ and electric induction D_1 for the fourth natural frequency

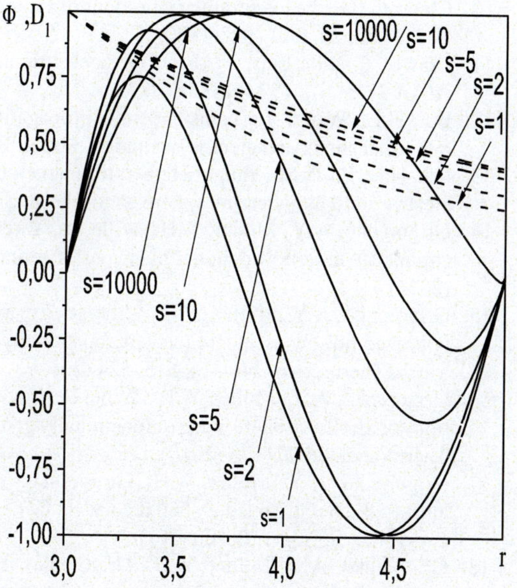

Fig. 4.48 Across-the-thickness distribution of dimensionless amplitude of normal stresses T_1 and tangential stresses T_6 for the fourth natural frequency

Figures 4.46, 4.47, and 4.48 show how the dimensionless amplitudes of axisymmetric vibrations for the fourth natural frequency ($n = 3$) are distributed across the thickness. Obviously, the inhomogeneity of the material influences the electric induction, electrostatic potential, and radial stresses to a greater extent.

References

1. Bailey, T., Hubbard, E.: Modeling and analysis of functionally graded materials and structures. J. Guid. Control Dyn. **8**(5), 605–611 (1985)
2. Baz, A., Poh, S.: Performance on an active control systemwith piezoelectric actuators. J. Sound Vib. **126**(2), 327–343 (1988)
3. Berlincourt, D.: Piezoelectric crystals and ceramics. In: Mattiat, O.E. (ed.) Ultrasonic Transducer Materials, pp. 62–124. Chap. 2. Plenum Press, New York (1971)
4. Birman, V., Byrd, L.W.: Modeling and analysis of functionally graded materials and structures. ASME Appl. Mech. Rev. **60**, 195–216 (2007)
5. Boriseiko, V.A., Grinchenko, V.T., Ulitko, A.F.: Relations of electroelasticity for piezoceramic shells of rrevolution. Sov. Appl. Mech. **12**(2), 126–131 (1971)
6. Chen, W.T.: On some problems in spherically isotropic elastic materials. Trans. ASME Ser. E **33**(3), 347–355 (1966)
7. Chen, W.Q., Din, H.J.: On free vibrations of a functionally graded piezoelectric rectangular plate. Acta Mech. **153**, 207–216 (2002)
8. Chen, W.Q., Wang, L.Z., Lu, Y.: Free vibrations of functionally graded piezoceramic hollow spheres with radial polarization. J. Sound Vib. **251**(1), 103–114 (2002)
9. Chen, W.Q., Lu, Y., Ye, J.R., Ca, J.B., 3D, : Electroelastic fields in a functionally graded piezoceramic hollow sphere under mechanical and electric loading. Arch. Appl. Mech. **72**, 39–51 (2002)
10. Chree, C.: On the longitudinal vibrations of aeolotropic bars with one axis of material symmetry. Q. J. Math. **24**, 340–358 (1890)
11. Crawly, E.F., de Luis, J.: Use of piezoelectric actuators as elements of intelligent structures. AIAA J. **25**(10), 1373–1385 (1987)
12. Dai, H.L., Wang, X.: Transient wave propagation in piezoelectric hollow spheres subjected tothermal shock and electric excitation. Struct. Eng. Mech. **19**(4), 441–457 (2005)
13. Dai, H.L., Fu, Y.M., Yang, J.H.: Electromagnetoelastic behaviors of functionally graded piezoelectric solid cylinder and sphere. Acta Mech. Sinica **23**, 55–63 (2007)
14. Grigorenko, A.Y., Müller, W.H., Wille, R., Yaremchenko, S.N.: Numerical solution of stress-strain state in hollow cylinder by means of spline approximation. J. Math. Sci. **1802**, 135–145 (2012)
15. Grigorenko, A.Y., Müller, W.H., Wille, R., Loza, I.A.: Nonaxisymmetric vibrations of radially polarized hollow cylinders madeof functionally gradient piezoelectric materials. Continuum Mech. Thermodyn. **24**(4–6), 515–524 (2012)
16. Grigorenko, A.Y., Müller, W.H., Wille, R., Loza, I.A.: Nonaxisymmetric electroelastic vibrations of a hollow sphere made of functionally gradient piezoelectric material. Continuum Mech. Thermodyn. 26:771–78 (2014)
17. Grigorenko, A.Y., Müller, W.H., Grigorenko, Y.M., Vlaikov, G.G.: Recent developments in anisotropic heterogeneous shell theory. In: General Theory and Applications of Classical Theory, vol. 1. Springer, Berlin (2016)
18. Grigorenko, A.Y., Müller, W.H., Grigorenko, Y.M., Vlaikov, G.G.: Recent developments in anisotropic heterogeneous shell theory. In: Applications of Refined and Three-dimensional Theory, vol. IIA, p. 42. Springer, Berlin (2016)
19. Grigorenko, A.Y., Müller, W.H., Grigorenko, Y.M., Vlaikov, G.G.: Recent developments in anisotropic heterogeneousshell theory. In: Applications of Refined and Three-dimensional Theory, vol. IIB. Springer, Berlin (2016)
20. Jaerisch, P.: Über die elastischen Schwingungen einer isotropen Kugel. J. Math. (Crelle) **88**, 131–145 (1880)
21. Lamb, H.: On the vibration of an elastic sphere. Proc. Lond. Mat. Soc. **13**, 189–212 (1882)
22. Love, A.E.: A Treatise on the Mathematical Theory of Elasticity. Cambridge University Press, reprint in USA (1952)
23. Loza, I.A., Shul'ga, N.A.: Axisymmetric Vibrations of a hollow piezoceramic sphere with radial polarization. Int. Appl. Mech. **20**(2), 113–117 (1984)

24. Loza, I.A., Shul'ga, N.A.: Forced axisymmetric vibrations of a hollow piezoceramic sphere with an electrical method ofexcitation. Int. App. Mech. **26**(6), 818–821 (1990)
25. Morse, F.M., Feshbach, G.: Methods of Theoretical Physics. McCraw-Hill, NY (1953)
26. Poisson, S.D.: Memoire sur l'equilibre et le movement des corps elastique. L'Académie Royale des Sciences VIII: 357–570 (1829)
27. Sato, Y., Usami, T.: Basic study on the oscillation on a homogeneous elastic sphere. Geophys. Mag. Tokyo **31**, 15–62 (1962)
28. Shul'ga, N.A.: Electroelastic oscillations of a piezoceramic sphere with radial polarization. Int. Appl. Mech. **22**(5), 97–500 (1986)
29. Shul'ga, N.A., Grigorenko, A.Y., Efimova, T.L.: Free non-axisymmetric oscillations of a thick-walled nonhomogeneous transversally isotropic hollow sphere. Int. Appl. Mech. **24**(5), 439–444 (1988)

Conclusion

The dynamic processes in cylindrical and spherical piezoceramic inhomogeneous bodies based on 3D electroelasticity theory are studied in this monograph. Different cases of heterogeneous structures of the corresponding piezoceramic bodies (piecewise continuous or continually continuous structures) were considered. The investigation of the problem defined above is connected with large computational difficulties. The discretely continual analytical numerically approach to solve the wide class of problems of the spatial electroelasticity theory for stationary dynamical processes in piecewise inhomogeneous (laminated) and continually inhomogeneous piezoceramic cylindrical and spherical bodies is proposed. This approach was used efficiently in solving elasticity problems. The principal relations of the refined Maxwell electromagnetic theory of piezoceramic bodies demonstrating a piezoelectric effect are presented. The closed system of electroelasticity equations are stated and applied to cylindrical and spherical coordinate systems. The boundary conditions for mechanical and electrical components of the coupled field on the surfaces of the piezoceramic body and on the interface of layers in the case of the piecewise inhomogeneous material are presented. The axisymmetric and nonaxisymmetric problem of natural and forced vibrations of hollow inhomogeneous piesoceramic cylinders, a sphere, and electroelasticity wave propagation in the cylinder with preliminary polarization is solved. Two types of inhomogeneous materials are considered, i.e., when the properties of the material are piecewise constant (layered structures with metal and dielectric layers) and when they vary continuously (functionally gradient piezoelectric materials–FGPM). New electromechanical effects of the behavior of the spectral characteristics associated with the heterogeneous structure of the studied bodies and the connected electric fields are determined. In the case of free vibrations of the piezoceramic inhomogeneous cylinders of finite length, the value of the first natural frequency from the coupled electric field is insignificant. High eigenfrequency values in the case of thin or long piezoceramic cylinders are observed. For each case of preliminary polarization of the piezoceramic material of the cylinder,

A. Ya. Grigorenko et al., *Selected Problems in the Elastodynamics of Piezoceramic Bodies*, Advanced Structured Materials 154,
https://doi.org/10.1007/978-3-030-74199-0

i.e., the axial, radial, and circular, the effects of the inhomogeneous structure and the coupled electric field on the distribution of dynamic characteristics of the cylinder were analyzed. In the case of a piezoceramic cylinder made of continuously inhomogeneous material (FGPM) and physical and the mechanical properties changing according to an exponential law, the presence of a heterogeneous structure has a smaller effect on the values of the free vibrations frequencies of the cylinder than the presence of a coupled electric field. The analysis of the dispersion propagation curves of the electroelastic waves in piezoceramic inhomogeneous cylinder in both layered and continuously inhomogeneous is performed. In the case of a laminated cylindrical structure made of piezoceramic and metallic materials, the preliminary polarization of the ceramics is of great importance in the distribution of the dynamic characteristics of the cylinders. A significant difference in the behavior of dispersion curves for the case of continuously inhomogeneous cylinder material is observed. The axisymmetric problems of the propagation force acoustoelectric waves in the layered hollow cylinder with piezoceramic and metallic layers under mechanical excitation are considered. The influence of the heterogeneity on the kinematic characteristics of propagating waves are studied. We have established the following. The first branch in the short-wave range in a cylinder made of a homogeneous material tends to a Rayleigh-type wave propagating along the outside surface of the cylinder. The second wave tends to a Rayleigh-type wave propagating along the inside surface of the cylinder. For the higher branches, there are substantial differences in the distribution of displacement amplitudes. If the material is homogeneous, with the decrease in the wavelength, the displacement amplitudes distribute either symmetrically or asymmetrically about the cylinder mid-surface. If the material is inhomogeneous, with the decreasing wavelength, the dispersion branches approach each other pairwise and the distribution of displacement across the thickness becomes pairwise similar. The displacements concentrate in the softer piezoceramic layer with the lower curvature, whereas the stiffer steel layer remains undeformed. The axisymmetric problems of the propagation force acoustoelectric waves in the continually inhomogeneous piezoceramic cylinder under electrical excitation is considered. We have established that the inhomogeneity of the cylinder material causes strong distinctions in the pattern of distribution of the displacement amplitudes of running waves. These distinctions are especially strong in the case of the second branch of the dispersion relations. Thus, in the case of a homogeneous material, the first branch in the short-wave range transforms into a Rayleigh-type surface wave propagating over the outside surface of the cylinder. The second branch transforms into a Rayleigh-type surface wave, which propagates over the inside surface of the cylinder. In the case of an inhomogeneous material, only the first branch transforms into a Rayleigh-type surface wave propagating over the surface with minimum moduli values. The second branch does not transform into a surface wave. In the case of the higher branches, the symmetry in the distribution of displacement amplitudes across the thickness and in their shifting to the domain with minimum values of the moduli is broken. In the case of a homogeneous material, the higher branches in the short-wave range transform into the waves propagating without dispersion. At the same time, the displacement amplitudes are distributed either symmetrically or asymmetrically with respect to the mid-surface when the number of half-waves per unit increases with the wave number. In the case of an inhomogeneous material, the symmetry in distribution of the displacement amplitudes is broken and motions of the cylinder particles shift to the lower values of the moduli. The axisymmetric and nonaxisymmetric problems of natural and forced vibrations of a hollow piezoceramic sphere made of the piecewise inhomogeneous and the continually inhomogeneous materials based on 3D electroelasticity theory are considered. In the case of the free vibrations of the piezoceramic hollow sphere, the influence of the heterogeneous structure and the coupled electric field on the behavior of natural frequencies was studied. At the same time, such laws have been identified: (a) the frequencies of the vibrations of the piezoceramic hollow sphere do not depend on the number of half-waves in the circumferential direction, unlike the natural frequencies of the cylinder; (b) the frequencies of the free vibrations of the piezoceramic hollow sphere significantly depend on the geometric parameters of the sphere and the polynomial index of the spherical function; (c) the heterogeneous structure and the presence of a coupled electric field have a significant effect on the change in the frequencies of free vibrations of the piezoceramic cylinder. The analysis of the influence of inhomogeneous structure of the piezoceramic sphere on the distribution of the electric and the elastic characteristics at the different methods of the electric excitation with split and non-cut electrodes based on the three-dimensional theory of electric elasticity is carried out. It is shown that the influence of the inhomogeneous structure of piezoceramic material on vibrations at several first natural frequencies for specified cases of electric excitation is enhanced. A comparative analysis of the effect of the electric method of excitation of the piezoceramic sphere vibrations on the behavior of the electrical characteristics—electrical induction and electrostatic potential and mechanical characteristics—, the displacement, and the stresses was also carried out. Attention is paid to the validation of the reliability of the results obtained by numerical calculations.